BICENTENNIAL

1807

✦WILEY✦

2007

THE WILEY BICENTENNIAL—KNOWLEDGE FOR GENERATIONS

$\mathcal{E}$ach generation has its unique needs and aspirations. When Charles Wiley first opened his small printing shop in lower Manhattan in 1807, it was a generation of boundless potential searching for an identity. And we were there, helping to define a new American literary tradition. Over half a century later, in the midst of the Second Industrial Revolution, it was a generation focused on building the future. Once again, we were there, supplying the critical scientific, technical, and engineering knowledge that helped frame the world. Throughout the 20th Century, and into the new millennium, nations began to reach out beyond their own borders and a new international community was born. Wiley was there, expanding its operations around the world to enable a global exchange of ideas, opinions, and know-how.

For 200 years, Wiley has been an integral part of each generation's journey, enabling the flow of information and understanding necessary to meet their needs and fulfill their aspirations. Today, bold new technologies are changing the way we live and learn. Wiley will be there, providing you the must-have knowledge you need to imagine new worlds, new possibilities, and new opportunities.

Generations come and go, but you can always count on Wiley to provide you the knowledge you need, when and where you need it!

WILLIAM J. PESCE
PRESIDENT AND CHIEF EXECUTIVE OFFICER

PETER BOOTH WILEY
CHAIRMAN OF THE BOARD

MINERALS AND ROCKS
Third Edition

Exercises in Crystal and Mineral Chemistry, Crystallography, X-ray Powder Diffraction, Mineral and Rock Identification, and Ore Mineralogy

Cornelis Klein

University of New Mexico

JOHN WILEY & SONS, INC.

Cover design by Marc A. Klein

Supplementary Material

A *Solutions Manual* with completely worked answers to most of the assignments in *Minerals and Rocks: Exercises* is available to instructors who have adopted *Minerals and Rocks: Exercises* from John Wiley & Sons. Please visit the book's website at www.wiley.com/college/klein or contact your local Wiley Sales Representative for further information. You can find your local Wiley Sales Representative's name and contact information at www.wiley.com/college/rep.

To order books or for customer service, please call 1-800-CALL-WILEY (225-5945).

ISBN 978-0-471-77277-4

Printed in the United States of America

10 9 8 7 6 5

Printed and bound by Bind-Rite Graphics, Inc.

Laboratory exercises and take-home assignments are an integral part of a course or course sequence in mineralogy-petrology and through them a student's understanding of subject matter is greatly enhanced. Indeed, students learn best in a hands-on, inquiry-based, problem-solving mode.

When the exercises contain all or most of the material needed to complete an assignment, the student is relieved of many noninstructive tasks. In this book the material in each exercise begins with a concise statement of goals, followed by a section on "background information" that should suffice for the successful completion of the assignment. A listing of necessary materials is provided, and in many of the exercises the essential materials are part of the assignment. Each assignment is outlined in easy-to-follow steps, and the parts of the exercise that should be checked or graded by the instructor are printed on one-sided tear-out sheets. Each exercise gives references to several of the most commonly used textbooks, as well as additional listings from other publications. Specific references are also made to locations on the CD-ROM that accompanies the *Manual of Mineral Science*, 2008, 23rd edition by Klein and Dutrow. The sequence of subject coverage in this book is very similar to that of the *Manual of Mineral Science*.

TO THE INSTRUCTOR

This collection of exercises is meant for the student enrolled in an introductory mineralogy course, or a beginning-level mineralogy-introductory petrology sequence, with the goal of increasing the student's understanding of subject material presented in lectures.

The thirty-five exercises cover aspects of mineralogy, crystallography, and hand specimen petrology. Many of them are so self-contained that they can be performed outside the classroom, or laboratory, and as such can be assigned as homework. Several of the exercises, however, are materials-dependent and are most successful when performed in a laboratory setting where morphological crystal models, crystal structure models, and hand specimens can be made available.

Exercises 24 through 33, and exercises 35, are specimen-based, be they mineral specimens (as in 24 through 30) or rock and ore specimens (as in 31, 32, 33, and 35).

Some of the exercises may not fully occupy the student for a standard two- to three-hour laboratory period. For example, exercises 1 and 2 (on physical properties of minerals) can be studied in the same laboratory period. Exercise 26, on the identification of layer silicates, can also be combined, in the same laboratory period, with another assignment on mineral identification.

In a semester period approximately eleven to twelve weeks are available for classroom and laboratory sessions. If twelve of the thirty-six exercises are selected for laboratory assignments, another twelve (or more) can be assigned as homework. In a two-semester mineralogy-petrology sequence all the exercises can be accommodated. In courses based on the quarter system, in which coverage of the material is more intensive than in a semester, the student will again benefit from exposure to the maximum number of exercises.

With thirty-five exercises covering a wide range of topics, there should be enough variety for an instructor to select assignments that are most consistent with his or her approach to the subject and with the time available.

The basic ideas in the majority of exercise and assignment materials in this book have been continually tested and developed over many years of teaching undergraduate mineralogy and introductory petrology. However, their self-contained, self-explanatory, and self-instructional format was specifically designed for this book.

TO THE STUDENT

The subject matter in this book of exercises is divided into three sections: mineralogy, crystallography, and hand specimen petrology.

The overall purpose of this book is to help you, as a student, to have as complete an educational experience in mineralogy-petrology as is possible in a prescribed course or sequence of courses. The use of worked examples, the background information, and the specific instructions that accompany each exercise should make your learning process efficient. The format is specifically designed to relieve you of much noninstructive drudgery, such as locating references in the library, gathering specific study materials, and performing an experiment, as in the X-ray laboratory. My hope is that this book will provide you with a positive challenge that will round out your experience in a mineralogy and introductory petrology course sequence. Good luck in your work!

ACKNOWLEDGEMENTS

I thank three colleagues, La Verne M. Friberg of the University of Akron, and two anonymous reviewers, for their constructive comments on the prior edition of this book. I trust that incorporation of many of their suggestions has improved the usefulness of this new, third edition.

I thank my son, Marc A. Klein of New York City, for the design that is used on the cover of this manual.

Albuquerque, New Mexico   CORNELIS KLEIN

CONTENTS

Physical Properties I: Habit, Twinning, Cleavage, Fracture, Luster, Color, and Streak

PURPOSE OF EXERCISE

The assessment of physical properties of minerals in hand specimens is the basis for their identification. This exercise introduces you to seven important physical properties (seven more are introduced in exercise 2) that you will learn to recognize and use by careful observation of such properties in collections of minerals that are especially selected (by your instructor). Your study must combine the careful observation of the hand specimen with the descriptions that follow in this exercise. Evaluations of these and other physical properties (see exercise 2) form the basis for mineral identification assignments in exercises 24 through 30.

FURTHER READING

Klein, C. and Dutrow, B., 2008, *Manual of Mineral Science*, 23rd ed., Wiley, Hoboken, New Jersey, pp. 2–4 and 19–30.

Nesse, W. D., 2000, *Introduction to Mineralogy*, Oxford University Press, New York, pp. 97–113.

Perkins, D., 2002, *Mineralogy*, 2nd ed., Prentice Hall, Upper Saddle River, New Jersey, pp. 45–61.

Wenk, H. R. and Bulakh, A., 2004, *Minerals: Their Constitution and Origin*, Cambridge University Press, New York, New York, pp. 84–99, and pp. 266–271.

Hurlbut, C. S., Jr. and Sharp, W. E., 1998, *Dana's Minerals and How to Study Them*, 4th ed., Wiley, Hoboken, New Jersey, pp. 63–86.

Background Information: If this exercise is assigned to you at the very beginning of your course in Mineralogy, you will not yet have been introduced to crystallographic concepts about crystal form, and the shorthand notation of their faces, or cleavage directions by Miller or Bravais-Miller index. The Miller index notation consists of three digits enclosed in brackets such as (100), (010), or {100} and {010}. The Bravais-Miller index is used for crystals with a specialized symmetry content (referred to as hexagonal) and consists of four digits, such as in {10$\bar{1}$1}. If you have not yet covered these crystallographic subjects, you should at least refer to Fig. 10.1 and Table 10.1 for a quick introduction to the geometry of common crystal forms and the names applied to such forms. Miller indices are part of exercise 11.

Habit

Our first impression of a mineral, or crystal, is of its *habit*, or *crystal habit*. A crystal habit is defined as the common and characteristic form, or combination of forms, assumed by a mineral, including its general shape and irregularities of growth. Even when a specimen lacks a regular external form, its regular internal (atomic) structure may be reflected in the smooth, plane fracture surfaces known as cleavage. In exercises 9 and 10 you will see highly idealized drawings of "perfect" crystal forms. Examples of such highly symmetrical crystals are generally seen only in the exhibit cases of museums, or in locked cabinets in the laboratory. Such crystals are uncommon and tend to be very expensive, because of their rarity. If real crystals (as opposed to wooden blocks) are available for study in the laboratory, many of them will probably be somewhat *malformed* (see Fig. 1.1) and as such their true symmetry is not apparent. On such malformed crystals, the size of the equivalent faces may vary, and therefore the shape of the crystal as a whole appears distorted. However, the interfacial angles between the faces of a malformed crystal and those of an equivalent perfect crystal remain the same. Three common terms that express the quality of the development of external crystal forms are

euhedral–(from the Greek roots *eu*, meaning good, and *hedra*, meaning plane)–describing a mineral that is completely bounded by crystal faces and whose growth during crystallization was not restrained or interfered with by adjacent crystals or mineral grains.

subhedral–(from the Latin root *sub*, meaning less than)–describing a crystal or mineral grain that is partly bounded by crystal faces and partly by surfaces formed against coexisting grains.

anhedral–(from the Greek root *an*, meaning without)–for minerals that lack crystal faces and that may show rounded or irregular surfaces produced by the crowding of adjacent minerals during crystallization.

All the crystal drawings in exercises 9 and 10 are therefore of *euhedral* crystals, whereas the minerals that are intergrown, for example, in a granite, will tend to be *subhedral* and *anhedral*.

If mineral specimens display well-developed crystal forms, the form names (see Fig. 10.1 and Table 10.1) are used to describe their outward appearance. Examples are

prismatic–for a crystal with one dimension markedly longer than the other two.

rhombohedral–with the external form of that of a rhombohedron, for example, {10$\bar{1}$1}.

cubic–with the external form of a cube, {100}.

octahedral–with the external form of an octahedron, {111}.

pinacoidal–with the pronounced development of one or more two-sided forms, the pinacoid.

However, most mineral specimens will tend to be *aggregates* of many smaller grains, ranging in form from euhedral through subhedral to anhedral. These crystalline aggregates are traditionally defined by descriptive terms, such as the following (see also Fig. 1.2).

massive–applied to a mineral specimen totally lacking crystal faces.

cleavable–applied to a specimen exhibiting one or several well-developed cleavage directions.

granular–made up of mineral grains that are of approximately equal size. The term is mainly applied to minerals whose grains range in size from about 2 to 10 mm. If the individual grains are larger, the aggregate is described as *coarse-granular*; if smaller, it is *fine-granular*.

compact–applied to a specimen so fine-grained that the state of aggregation is not obvious to the eye.

lamellar–made up of layers like the leaves in a book.

foliated–made up of thin leaves or plates that can be separated from each other, as in graphite.

micaceous–applied to a mineral whose separation into thin plates occurs with great ease, as in mica.

columnar–with a crystal habit made up of a subparallel arrangement of columnar individuals, as in some occurrences of calcite.

bladed–with individual crystals (or grains) that are flattened blades or flattened elongate crystals.

fibrous–having a tendency to crystallize in needlelike grains or fibers, as in some amphiboles and in asbestos. In asbestos the fibers are *separable*, that is, they are easy to pull apart.

acicular (from the Latin root *acicula*, meaning needle)–describing a mineral with a needlelike habit.

radiating (or *radiated*)–describing a mineral in which acicular crystals radiate from a central point.

dendritic (from the Greek root *dendron*, meaning tree)–applied to a mineral exhibiting a branching pattern.

banded–describing a mineral aggregate in which a single mineral species may show thin and roughly parallel banding (as in banded malachite) or in which two or more minerals form a finely banded intergrowth (as in quartz and hematite bands in banded iron-formation).

concentric–with bands or layers arranged in parallel positions about one or more centers (as in malachite).

mammillary (from the Latin word *mamma* meaning breast)–with an external form made up of rather large, rounded prominences. Commonly shown by massive hematite or goethite.

botryoidal (from the Greek root *botrys* meaning bunch or cluster of grapes)–having the form of a bunch of grapes. The rounded prominences are generally smaller than those described as mammillary. Botryoidal forms are common in smithsonite, chalcedony, and prehnite.

globular–having a surface made of little spheres or globules (as commonly in prehnite).

reniform (from the Latin *renis*, meaning kidney)–with a rounded, kidney-shaped outer surface as in some massive hematite specimens.

colloform (from the Greek root *collo*, meaning cementing or welding)–Since there is often no clear distinction between the four previous descriptive terms (mammillary, botryoidal, globular, and reniform), the term colloform includes them all.

stalactitic (from the Greek *stalaktos* meaning dripping)–made up of small stalactites that are conical or cylindrical in form as is common on the ceilings of caves.

concretionary–clustering about a center, as in calcium carbonate concretions in clay. Some concretions are roughly spherical, whereas others assume a great variety of shapes.

geode–a rock cavity lined with mineral matter but not wholly filled. Geodes may be banded as in agate, through successive depositions of material, and the inner surface is commonly covered with projecting crystals.

oolitic (from the Greek *oön*, meaning egg)–made up of oolites, which are small, round, or ovate (meaning egg-shaped) accretionary bodies, resembling the roe of fish. This texture is common in some iron-rich specimens, made of hematite, known as oolitic iron ore.

pisolitic (from the Greek *pisos*, meaning pea; therefore, pea-sized)–having a texture similar to that of an oolitic aggregate but somewhat coarser in grain size. Bauxite, the major source of aluminum ore, is commonly pisolitic.

Twinning and Striations

Twinning is very common in crystals, and the size of the twinned units can range from an almost atomic scale (with twin lamellae or twin domains on the order of 10s to 100s

FIGURE 1.1 (*a*) Octahedron and malformed octahedron. (*b*) Dodecahedron and malformed dodecahedrons. (*c*) Cube and octahedron and a malformed combination of cube and octahedron. (From C. S. Hurlbut, Jr. and Sharp, W. E., 1998, *Dana's Minerals and How to Study Them,* 4th edition, Wiley, New York, pp. 49, 50.)

(*a*)

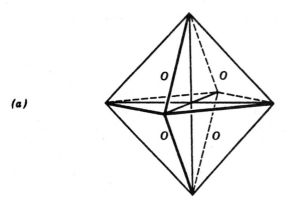 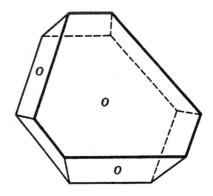

Octahedron and malformed octahedron

(*b*)

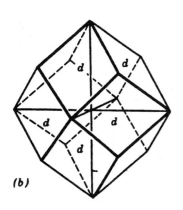 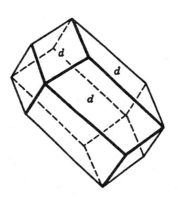 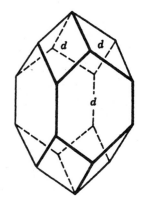

Dodecahedron and malformed dodecahedrons

(*c*)

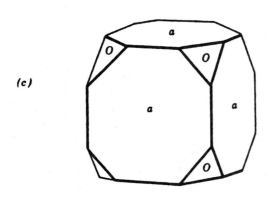 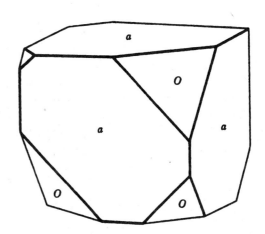

Cube and octahedron and a malformed
combination of cube and octahedron

FIGURE 1.2 Some common mineral habits and occurrences. (Several of these illustrations are modified after J. Sinkankas, 1964, *Mineralogy,* Van Nostrand Reinhold, New York, p. 94.)

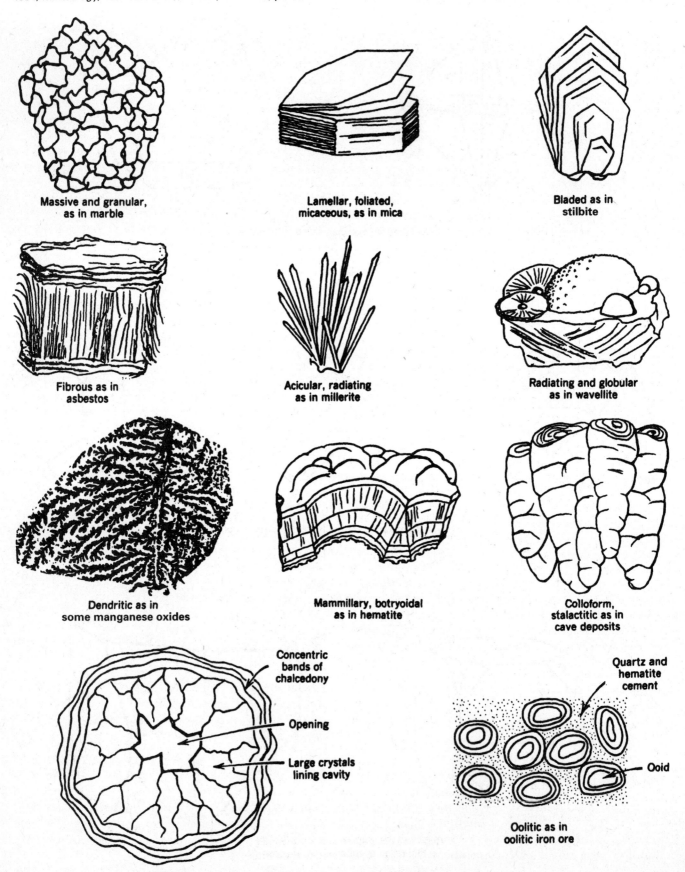

Massive and granular, as in marble

Lamellar, foliated, micaceous, as in mica

Bladed as in stilbite

Fibrous as in asbestos

Acicular, radiating as in millerite

Radiating and globular as in wavellite

Dendritic as in some manganese oxides

Mammillary, botryoidal as in hematite

Colloform, stalactitic as in cave deposits

Concentric bands of chalcedony

Opening

Large crystals lining cavity

Geode

Quartz and hematite cement

Ooid

Oolitic as in oolitic iron ore

of angstroms in size) to such a large scale that the individuals are easily seen by the naked eye. Generally the twins that are easiest to recognize in hand specimen are *contact* twins or *penetration twins*. Although the twinned relationship in such symmetric intergrowths is easily recognized, the twin law that underlies the twinned relationship may not be so obvious. Examples of common contact and penetration twins are shown in Fig. 1.3.

A twin relationship that is more subtle in its appearance is *polysynthetic twinning*. In a polysynthetic twin the successive composition planes of the twin are parallel to each other. When a large number of individuals in a polysynthetic twin are closely spaced, crystal faces or cleavage surfaces cutting across the composition planes show striations owing to the reversed positions of adjacent individuals. A highly diagnostic polysynthetic twin is albite twinning in the plagioclase feldspar series. The twin plane is (010), and the individual twin lamellae that can be seen by the naked eye are commonly quite thin, ranging from 0.1 to several millimeters in thickness. This twin is evidenced by parallel lines or striations seen on cleavage directions that cut across the {010} pinacoid. The striations resulting from this polysynthetic twinning are shown in Figs. 1.4*a* and *b*. Polysynthetic twinning in a magnetite crystal is shown in Fig. 1.4*c*.

By no means are all striations, as seen on crystal faces, the result of polysynthetic twinning, however. Figures 1.4*d* and *e* show striations that result from the intergrowths of two forms. Pyrite cubes (Fig. 1.4*d*) typically show striations that are the result of successive combinations of other faces or of another form (pyritohedral) in narrow lines on the cube face. The magnetite crystal in Fig. 1.4*e* shows striations on dodecahedral faces caused by the stepwise growth of octahedral faces.

Cleavage, Parting, and Fracture

Cleavage, parting, and fracture are the responses of a mineral to hammering or crushing. *Cleavage* is the natural and easy fracture that yields more or less smooth, plane surfaces in some crystallographic directions. *Parting* is commonly less obvious than well-developed cleavage and is a breaking parallel to rational crystallographic planes. It is restricted to specimens that are subjected to pressures or have been twinned. *Fracture* is the breaking of a mineral in response to a hammerblow along an irregular surface. Cleavage, parting, and fracture all mark directions in crystal structures where the forces binding the atoms (or ions) are relatively weak (see Fig. 1.5). *Cleavage* may be strongly developed in some minerals, or it may be fairly obscure in some other minerals. The range of cleavage development is expressed as *perfect cleavage*, as, for example, the cubic cleavage of galena; *good cleavage*, as, for example, the prismatic cleavage in pyroxenes; and *poor cleavage*, as in beryl or apatite. A mineral such as quartz shows no cleavage. When we determine cleavage in a mineral, it is important to note (1) the number of cleavage directions in a single mineral specimen, (2) the angles

between these directions, and (3) the quality and ease of cleavage. Because cleavage planes are parallel to possible rational crystal faces, the planes are commonly identified by Miller indices. For example, cubic cleavage is noted as {100}, octahedral cleavage as {111}, dodecahedral cleavage as {011}, rhombohedral cleavage as {10$\bar{1}$1}, prismatic cleavage as {110}, and basal cleavage as {001}. (See Fig. 1.5 for illustrations of these cleavage directions.)

Directions of *parting* can similarly be expressed by Miller index notation. Basal parting in pyroxene is parallel to {100}, and rhombohedral parting in corundum is parallel to {10$\bar{1}$1}.

Because *fracture* surfaces are not parallel to any specific crystallographic directions, they can be described only in qualitative terms such as

conchoidal – showing smooth, curved fracture surfaces resembling the interior surface of a shell. Conchoidal fracture is diagnostic of amorphous materials such as glass (see Fig. 1.5*i*), but it is also shown by quartz.

fibrous and *splintering*.

hackly – showing fractures with sharp edges.

uneven or *irregular*.

Luster

The term *luster* refers to the general appearance of a mineral surface in reflected light. The two distinct types of luster are *metallic* and *nonmetallic*, but there is no sharp division between them. Although the difference between these types of luster is not easy to describe, the eye discerns it easily and, after some experience, seldom makes a mistake. *Metallic* is the luster of a metallic surface such as chrome, steel, copper, and gold. These materials are quite opaque to light; no light passes through even at very thin edges. Galena, pyrite, and chalcopyrite are common minerals with metallic luster. *Nonmetallic* luster is generally shown by light-colored minerals that transmit light, if not through thick portions at least through their edges. The following terms are used to describe further the luster of nonmetallic minerals.

vitreous – with the luster of a piece of broken glass. This is commonly seen in quartz and many nonmetallic minerals.

resinous or waxy – with the luster of a piece of resin. This is common in samples of sphalerite.

pearly – with the luster of mother of pearl. An iridescent pearllike luster. This is characteristic of mineral surfaces that are parallel to well-developed cleavage planes. The cleavage surface of talc and the basal plane cleavage of apophyllite show pearly luster.

greasy – appears as though covered with a thin layer of oil. This luster results from light scattered by a microscopically rough surface. Some milky quartz and nepheline specimens may show this.

FIGURE 1.3 (a) Octahedron with possible twin plane b–b (11$\bar{1}$). This is one of four octahedral directions in the form {111}. (b) Octahedral twinning {111} as shown by spinel. (c) Right-and left-handed quartz crystals twinned along (11$\bar{2}$2), the Japanese twin law. (d) Two interpenetrating cubes of fluorite twinned on [111] as the twin axis. (e) Two pyritohedral crystals (of pyrite) forming an iron cross, with twin axis [001]. (f) Orthoclase exhibiting the Carlsbad twin law in which two interpenetrating crystals are twinned by a 180° rotation about the c axis, [001] direction. The schematic cross section, parallel to (010), reveals the presence of the 2-fold twin axis along [001]. (From *Manual of Mineral Science*, 23rd ed., Fig. 10.14, p. 229.)

Contact Twins

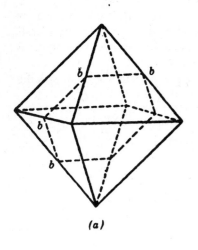

(a)

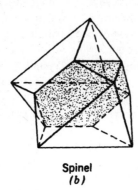

Spinel
(b)

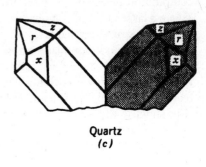

Quartz
(c)

Penetration Twins

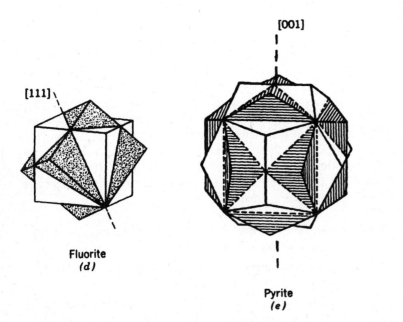

Fluorite
(d)

Pyrite
(e)

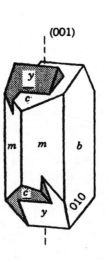

Orthoclase
(f)

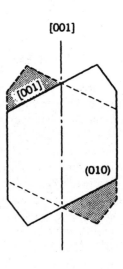

FIGURE 1.4 Polysynthetic twinning and striations. (*a*) Albite polysynthetically twinned on (010). (*b*) The appearance of albite twinning as striations or parallel groovings across a cleavage or crystal face that crosses {010}. (*c*) Octahedral crystal of magnetite with twinning lamellae appearing as striations on an octahedral face. (*d*) Striations on a cube of pyrite. (*e*) Striations on dodecahedral faces of magnetite caused by the presence of octahedral faces (*o*). (Parts *c, d* and *e* from C. S. Hurlbut, Jr., and Sharp, W. E., 1998, *Dana's* 1949, *Minerals and How to Study Them*, 4th ed., Wiley, New York, p. 53.)

POLYSYNTHETIC TWINS

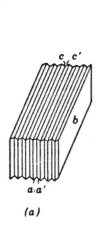

(*a*)

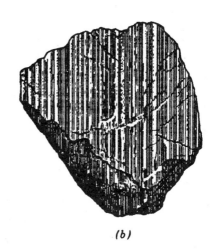

(*b*)

STRIATIONS

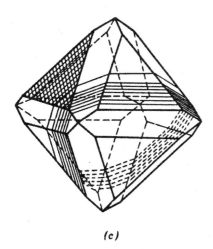

(*c*)

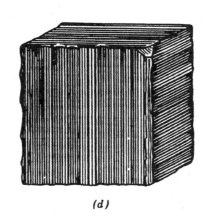

(*d*)

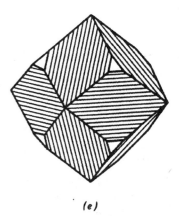

(*e*)

silky, or *silklike* – describing the luster of a skein of silk or a piece of satin. This is characteristic of some minerals in fibrous aggregates. Examples are fibrous gypsum, known as satin spar, and chrysotile asbestos.

adamantine (from the Greek word *adamas*, meaning diamond) – with the luster of the diamond. This is the brilliant luster shown by some minerals that also have a high refractive index and as such refract light strongly as does diamond. Examples are the carbonate of lead, cerussite, and the sulfate of lead, anglesite.

Color

The variation in the color of minerals is very great, and the terms used in describing the color are so familiar that they explain themselves.

Because color varies not only from one mineral to another, but also within the same mineral (or mineral group), the mineralogy student must learn in which minerals it is a constant property and can therefore be relied upon as a distinguishing criterion. Most minerals with a metallic luster vary little in color, but most nonmetallic minerals vary widely in color. Although the color of a freshly broken surface of metallic minerals is often highly diagnostic, these same minerals may become tarnished with time. Such a tarnish may dull some minerals such as galena, which has a bright, bluish lead-gray color on a fresh surface, but may become dull upon long exposure to air. Bornite, which on a freshly broken surface has a brownish-bronze color, may be so highly tarnished on an older surface that it shows variegated purples and blues; hence, it is called *peacock ore*. In other words, in the identification of minerals with a metallic luster, it is important to have a freshly broken surface to which a more tarnished surface can be compared.

Of all the minerals with nonmetallic luster, a few have such a *constant color* that their coloration can be used as a truly diagnostic property. Examples are malachite, which is green; azurite, which is blue; rhodonite, which is red; and turquoise, which gives its name to the turquoise color, a greenish blue to blue-green. Most nonmetallic minerals have a relatively narrow *range in colors*, although some show an unusually large range. Members of the plagioclase feldspar series range from almost pure white in albite, through light gray to darker gray toward the anorthite end-member. Most common garnets show various shades of red to red-brown to brown. Members of the monoclinic pyroxene group range from almost white in pure diopside, to light green in diopside with a little iron in substitution for magnesium in the structure, through dark green in hedenbergite, to almost black in many augites. Members of the orthopyroxene series (enstatite to ferrosilite) range from light beige to darker brown. On the other hand, tourmaline may show many colors (red, blue, green, brown, and black) as well as distinct color zonation from colorless through pink to green within a single crystal. Similarly, gem minerals such as corundum, beryl, quartz, and numerous others occur in many colors;

the gemstones cut from them are given varietal names. In short, in most nonmetallic minerals color is a helpful property, but not commonly a truly diagnostic (and therefore unique) property.

Streak

It is commonly very useful, especially in metallic minerals, to test the color of the fine powder or the color of the streak. To determine the color of the streak of a mineral, you will use a piece of unglazed white porcelain called a *streak plate*. Minerals with a hardness higher than that of the streak plate will not powder by rubbing them against it. The hardness of a streak plate is about 7 (see exercise 2). This test will show that black hematite with a silvery luster has a red streak. Most minerals with a nonmetallic luster will have a whitish streak, even though the minerals themselves are colored.

Other Properties Depending On Light

Minerals are commonly described in terms of the amount of light they can transmit. Such properties are grouped under the term *diaphaneity*, meaning the light-transmitting qualities of a mineral, from the Greek word *diaphanes*, meaning transparent. Examples follow.

transparent – describing a mineral that is capable of transmitting light, and through which an object may be seen. Quartz and calcite are commonly transparent. Most gem materials are highly transparent and commonly priced on the basis of the quality of their transparency.

translucent – said of a mineral that is capable of transmitting light diffusely, but is not transparent. Although a translucent mineral allows light to be transmitted, it will not show a sharp outline of an object seen through it. Some varieties of gypsum are commonly translucent.

opaque – describing a mineral that is impervious to visible light, even on the outer edges of the mineral. Most metallic minerals are opaque.

As a mineral is turned in a light source, the light may be reflected (or diffracted) in such a way that a series of unusual light patterns are produced. The following properties may be observed.

Play of colors – the result of the production of a series of colors as the angle of the incident light is changed. Precious (gem) opals may show such striking color displays.

Opalescence – the pearly reflection of a range of colors from the interior of a mineral. It may be well developed in common opal and in varieties of Na-rich plagioclase known as *moonstone*.

Iridescence (also referred to as *schiller*, or *labradorescence*) – a show of rainbow colors in shifting patterns, the result of light scattered by extremely thin and closely spaced (microscopic) cleavage planes, twin or exsolution lamellae within the mineral. The plagioclase feldspar, *labradorite*,

FIGURE 1.5 Cleavage: (*a*) cubic, (*b*) octahedral, (*c*) dodecahedral, (*d*) rhombohedral, (*e*) prismatic and pinacoidal, (*f*) pinacoidal (basal). (From *Manual of Mineral Science,* 23rd ed., Fig. 2.12, p. 29.) (*g*) Basal parting in pyroxene, and (*h*) rhombohedral parting in corundum with twinning striations, which are potential parting planes. (From *Manual of Mineral Science,* 23rd ed., Fig. 2.13, p. 29), (*i*) conchoidal fracture in obsidian (volcanic glass). (From *Manual of Mineral Science,* 23rd ed., Fig. 2.14, p. 30.)

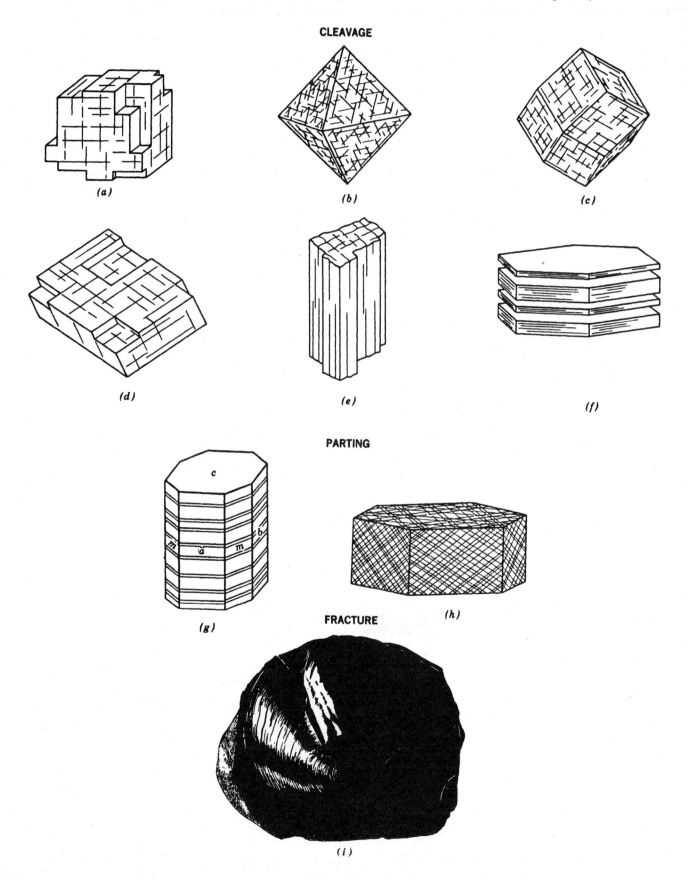

CLEAVAGE

(a)   *(b)*   *(c)*

(d)   *(e)*   *(f)*

PARTING

(g)   *(h)*

FRACTURE

(i)

may show colors ranging from blue to green to yellow with the changing angle of light.

Chatoyancy – a movable wavy or silky sheen concentrated in a narrow band of light that changes position as the mineral is turned. It results from the reflection of light from minute parallel fibers, cavities, or needlelike inclusions within the mineral. The effect is best seen on cabochon-cut gemstones, which may display a thin band of light at right angles to the length of inclusions or fibers inside the cut stone. Examples are *cat's eye*, a gem variety of chrysoberyl, and *tiger's eye*, fibrous crocidolite (blue or oxidized to brown) replaced by quartz.

Asterism – the optical phenomenon of a rayed or star-shaped figure of light displayed by some minerals when viewed in reflected light. It is generally the result of minute, needlelike inclusions arranged in three crystallographic directions at 120° to each other. This is beautifully shown by cabochon-cut gemstones such as star sapphires and star rubies.

MATERIALS

Various collections of mineral specimens that display a range of habits, aspects of twinning, types of cleavage, metallic and nonmetallic luster, and a range of colors. Such specialized collections, if not already available in the laboratory, can be bought from mineral supply houses such as Ward's Natural Science Establishment, Inc., Rochester, New York. For the determination of streak an unglazed white porcelain plate is needed. A hand lens, with about 10x magnification, is very helpful for the detailed observation of various physical properties. A binocular microscope with a good light source is also very useful. Note that some of the physical properties introduced in this exercise form the basis for determinative tables in several textbooks (e.g., Chapter 22 in *Manual of Mineral Science*, 23rd ed.; Appendix B in *Introduction to Mineralogy*, by Nesse; Appendix B in *Mineralogy*, by Perkins; and Appendix 1 in *Minerals: Their Constitution and Origin*, by Wenk and Bulakh).

ASSIGNMENT

1. *Habit.* If a specific mineral collection with a range of mineral habits is available, observe and carefully handle each of the specimens and familiarize yourself with the various descriptive terms. Read through the listing of such terms, as just given, and at the same time look over the illustrations in Fig. 1.2. If a specific habit collection is not available, look through any mineral exhibits that are available in the laboratory, in the hallways of your building, or in museum collections, and locate in these collections the various habits as described earlier.

2. *Crystal form.* If you have already completed some assignments in crystallography, you will have little difficulty in recognizing the forms displayed by well-developed crystals. Good crystal collections are usually housed in locked cabinets, and you will probably have little chance to handle good-sized, well-developed crystals. If, however, you have been asked to do the present assignment without prior knowledge of various aspects of crystallography, you will find it helpful to consult at least Fig. 10.1 and Table 10.1 for a brief introduction to the most common geometrical shapes and form names of crystals. You can further acquaint yourself with crystal form descriptions by consulting any of the references quoted in exercise 10.

3. *Twinning.* The most important type of twinning that you should learn to recognize is that displayed by members of the plagioclase feldspar series. This mineral group commonly shows well-developed polysynthetic twinning, as illustrated in Fig. 1.4*b*. The presence of this twinning, commonly seen as closely spaced striations on a cleavage face, is highly diagnostic of the plagioclase feldspars. Several specimens ranging from albite to anorthite should be handled and observed. A hand lens or binocular microscope will be very helpful in recognizing this type of twinning.

4. *Cleavage and fracture.* A collection showing minerals with a range of cleavage types and perfection is very helpful. The collection should include minerals such as galena, calcite, fluorite, halite, tremolite (or some other amphibole), diopside (or some other pyroxene), muscovite (or some other mica), microcline, and quartz. For each of the mineral specimens in such a collection, note the number of cleavage directions and the quality of the cleavage (as described, under background information). If more than one cleavage direction is present, estimate the angles between them; this is especially diagnostic in amphiboles, pyroxenes, and feldspars. *Conchoidal fracture* is generally very well illustrated by volcanic glass specimens.

5. *Luster.* A collection of minerals illustrating various types of luster is very instructive. In such a collection observe the obvious distinction between *metallic* and *nonmetallic* luster, and among the nonmetallic minerals familiarize yourself with the various descriptive terms given under background information.

6. *Color, streak, and some other properties dependent on light.* Here again a collection specifically directed to various colors and color ranges is helpful for introductory study. The concept of *streak* is best introduced by scratching black, metallic hematite on an unglazed porcelain plate and observing the color of the streak. *Iridescence* is commonly well developed in specimens of the plagioclase feldspar known as labradorite. Properties such as chatoyance, and asterism are best seen in exhibits of some gem materials.

7. Figure 2.3 in exercise 2 is a handy note sheet for recording all of the observations you have made with regard to the physical properties of a single mineral specimen.

More Physical Properties and a Chemical One II: Hardness, Tenacity, Specific Gravity, Magnetism, Fluorescence, Radioactivity, and Solubility in Hydrochloric Acid

PURPOSE OF EXERCISE

To gain familiarity with some additional diagnostic properties used in hand specimen identification of minerals and rocks. Evaluations of such properties are the basis for mineral identification assignments in exercises 24 through 30.

FURTHER READING

Klein, C. and Dutrow, B., 2008, *Manual of Mineral Science*, 23rd ed., Wiley, Hoboken, New Jersey, pp. 30–36.

Nesse, W. D., 2000, *Introduction to Mineralogy*, Oxford University Press, New York, pp. 97–113.

Perkins, D., 2002, *Mineralogy*, 2nd ed., Prentice Hall, Upper Saddle River, New Jersey, pp. 45–61.

Wenk, H. R. and Bulakh, A., 2004, *Minerals: Their Constitution and Origin*, Cambridge University Press, New York, New York, pp. 84–99, and pp. 266–271.

Hurlbut, C. S., Jr. and Sharp, W. E., 1998, *Dana's Minerals and How to Study Them*, 4th ed., Wiley, Hoboken, New Jersey, pp. 63–86.

Background Information: In exercise 1 you were introduced to several physical properties that are relatively easy to observe in hand specimens without specific testing tools (except for the streak plate needed to obtain a streak). In this exercise you will be introduced to additional physical properties all of which require various kinds of testing tools or apparatus.

Hardness (H)

Hardness is the resistance of a mineral to scratching. It is a property by which minerals may be described, relative to a standard scale of ten minerals known at the *Mohs scale*. The ten minerals making up the *Mohs scale of hardness* are listed in Table 2.1. The degree of hardness is determined by observing the comparative ease or difficulty with which one mineral is scratched by another, or by a pocket knife or a steel file. Figure 2.1 shows the relationship of the Mohs *relative* hardness scale to *absolute* measurements of hardness.

For measuring the hardness of a mineral, several common objects that can be used for scratching are very helpful, such as a fingernail, a copper coin, a steel pocket knife, a glass plate or window glass, the steel of a file, and a streak plate. The approximate hardness of these materials is listed next to the minerals of the Mohs scale in Table 2.1.

In the practical determination of the relative hardness of a mineral, it is necessary to decide which of the minerals, or other materials (as listed in Table 2.1) it can or cannot scratch. In making hardness tests, keep the following in mind: sometimes when one mineral is softer than another, portions of the softer mineral may leave a mark on the harder material, which may be mistaken for a scratch. Such a mark can be rubbed off, whereas a real scratch is permanent. The surface of some minerals may have been altered to material that is softer than the original mineral. Therefore, a *fresh surface* of the specimen must be used in hardness testing. In addition, the physical and aggregate nature of a mineral may prevent a correct determination of its hardness. For example, if the mineral is finely granular or splintery, it may be broken down and apparently scratched by a mineral softer than itself. It is therefore advisable when making a hardness test to confirm it by reversing the procedure. That is, do not try just to scratch mineral *A* with mineral *B*, but also try to scratch sample *B* with sample *A*.

Because there is a general link between hardness and chemical composition, the following generalizations can be made.

1. Most hydrous minerals are relatively soft ($H < 5$).

2. Halides, carbonates, sulfates, and phosphates are also relatively soft ($H < 5\frac{1}{2}$).

3. Most sulfides are relatively soft ($H < 5$) with pyrite being an exception ($H < 6$ to $6\frac{1}{2}$).

4. Most anhydrous oxides and silicates are hard ($H > 5\frac{1}{2}$).

Because hardness is a highly diagnostic property in mineral identification, most determinative tables use relative hardness as a sorting parameter (see, for example, Chapter 22 in *Manual of Mineral Science*, 23rd ed.; Appendix B in *Introduction to Mineralogy*, by Nesse; Appendix B in *Mineralogy*, by Perkins; and

FIGURE 2.1 Comparison of the Mohs relative hardness scale to absolute measurement of hardness. (From *Manual of Mineral Science*, 23rd ed., p. 30.)

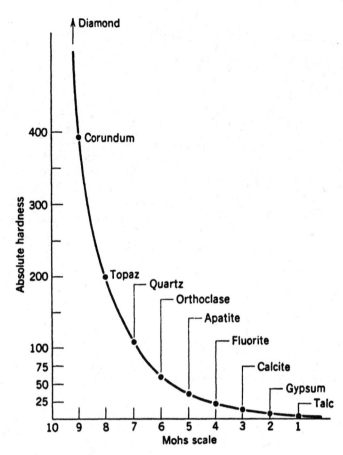

Appendix 1 in *Minerals: Their constitution and Origin*, by Wenk and Bulakh).

Tenacity

Under tenacity are grouped several other mineral properties that depend on the cohesive force between atoms (and ions) in mineral structures. The following terms are used to describe a mineral's tenacity.

malleable–capable of being flattened under the blows of a hammer into thin sheets without breaking or crumbling in fragments. Malleability is conspicuous in gold, silver, and copper. Indeed, most of the native elements show various degrees of malleability.

sectile–capable of being severed by the smooth cut of a knife. Copper, silver, and gold are sectile, whereas chalcocite and gypsum are both imperfectly sectile.

ductile–capable of being drawn into the form of a wire. Gold, silver, and copper are ductile.

flexible–bending easily and staying bent after the pressure is removed. Talc is flexible.

brittle–showing little to no resistance to breakage, and as such separating into fragments under the blow of a hammer or with the cut of a knife. Most silicate minerals are brittle.

elastic–capable of being bent or pulled out of shape but, returning to the original form when relieved. Mica is elastic.

Specific Gravity (G)

Specific gravity is a number that expresses the ratio between the weight of a substance and the weight of an equal volume of water at 4°C. Thus a mineral with a specific gravity (**G**) of 2 weighs twice as much as the same volume of water.

The specific gravity of a mineral depends on (1) the atomic weights of all the elements of which it is composed and (2) the manner in which the atoms (and ions) are packed together. In mineral series whose species have essentially identical structures (known as *isostructural*), those with elements of higher atomic weight have higher specific gravities. If two minerals (as in the two polymorphs of carbon, namely graphite and diamond) have the same chemical composition, the differences in specific gravity reflect the difference in internal packing of the

TABLE 2.1   Mohs Hardness Scale and Additional Observations

Mineral	Mohs Hardness	Other Materials	Observations on the Minerals
Talc	1		Very easily scratched by the fingernail; has a greasy feel
Gypsum	2	~2.2 fingernail	Can be scratched by the fingernail
Calcite	3	~3.2 copper penny	Very easily scratched with a knife and just scratched by a copper coin
Fluorite	4		Easily scratched with a knife but not as easily as calcite
Apatite	5	~5.1 pocket knife ~5.5 glass plate	Scratched with a knife with difficulty
Orthoclase	6	~6.5 steel file	Cannot be scratched with a knife, but scratches glass with difficulty
Quartz	7	~7.0 streak plate	Scratches glass easily
Topaz	8		Scratches glass very easily
Corundum	9		Cuts glass
Diamond	10		Used as a glass cutter

atoms or ions (diamond with $\mathbf{G} = 3.51$ has a much more densely packed structure than graphite with $\mathbf{G} = 2.23$).

Most people, from everyday experience, have acquired a sense of relative weight even about nonmetallic and metallic minerals. For example, borax ($\mathbf{G} = 1.7$) seems light for a nonmetallic mineral, whereas barite ($\mathbf{G} = 4.5$) appears heavy. This means that people have developed an idea of an *average specific gravity* for a nonmetallic mineral, or a sense of what a nonmetallic mineral of a given size should weigh. The average specific gravity is considered to be somewhere between 2.65 and 2.75, which is reflected by the range of specific gravities of quartz ($\mathbf{G} = 2.65$), feldspar ($\mathbf{G} = 2.60$ to 2.75), and calcite ($\mathbf{G} = 2.72$). People have the same sense about metallic minerals: graphite ($\mathbf{G} = 2.23$) seems light, whereas silver ($\mathbf{G} = 10.5$) appears heavy. The average specific gravity for metallic minerals is about 5.0, that of pyrite. Thus, with some practice, a person can, by merely lifting specimens, distinguish minerals that have comparatively small differences in specific gravity.

A simple spring balance allows us to determine specific gravity with ease and with much greater accuracy than is possible by assessing the average heft of a mineral by hand. Such a balance, known as the *Jolly balance* (see Fig. 2.2), provides numerical values for a small mineral specimen (or fragment) in air as well as in water. When the mineral is weighed immersed in water, it is buoyed up and weighs less than it does in air; this weight loss is equal to the weight of water it displaces. Hence, if we find first the weight of a mineral fragment on a pan of the balance in air, and subsequently its weight while immersed in water (it being suspended on a pan by a thin wire thread), and subtract the two weights, the difference is the weight of the equal volume of water. For example, the weight of a small quartz fragment is 4.265 grams in air; in water it is 1.609 grams. The loss of weight, or weight of an equal volume of water

exactly equal to it, is therefore 2.656 grams; hence the specific gravity is

$$\frac{4.265}{4.265 - 2.656} = \frac{4.265}{1.609} = 2.65$$

which is indeed that of quartz.

In other words, the specific gravity of a mineral ($\mathbf{G}$) can be expressed as follows,

$$\mathbf{G} = \frac{W_a}{W_a - W_w}$$

in which W_a is the weight in air and W_w is the weight in water.

Because specific gravity is merely a ratio, it is not necessary to determine the absolute weight of a specimen but merely values proportional to the weights in air and in water. The *Jolly balance* allows us to do this directly by measuring the stretch of a spiral spring. In using the balance, first place a fragment on the upper scale pan and note the elongation of the spring (this gives the weight in air, W_a). The fragment is subsequently transferred to the lower pan and immersed in water. The elongation of the spring is now propositional to the weight of the fragment in water (W_w). More specific instructions regarding the use of the Jolly balance are given in the assignment section.

For the accurate determination of the specific gravity of a mineral fragment, one must use a *homogeneous* and *pure* specimen. The specimen must also be compact, without cracks or cavities within which bubbles or films of air may be trapped. For routine mineralogical work, the specimen should have a volume of about one cubic centimeter. If these conditions are not met, a specific gravity determination with the Jolly balance will have little meaning.

FIGURE 2.2 Jolly balance. (From *Manual of Mineral Science,* 23rd ed., p. 33.)

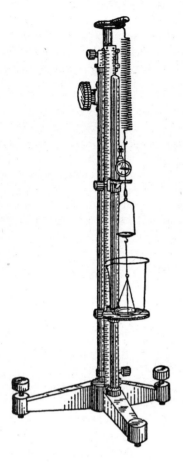

Magnetism

The two common minerals, magnetite and pyrrhotite, are highly magnetic. That is, both are easily attracted to a small hand magnet. Both are opaque minerals that may occur as minor constituents in a wide range of mineral associations and rock types. Even if present in small quantities, or in small-sized grains, the removal of one or several grains from a specimen with a needle or a pocket knife allows for the testing of magnetism of individual grains. Magnetite is very strongly attracted to a magnet, pyrrhotite less so.

Fluorescence

Some minerals when exposed to ultraviolet light will emit visible light during irradiation; this is known as fluorescence. Some minerals fluoresce only in shortwave ultraviolet, whereas others may fluoresce only in longwave ultraviolet, and still others fluoresce under either. The color and intensity of the emitted light vary considerably with the wavelengths of ultraviolet light. Fluorescence is an unpredictable property, because some specimens of a mineral show it, whereas other apparently similar specimens, even those from the same locality, do not. Several minerals that may show fluorescence are fluorite,

sheelite, calcite, scapolite, willemite, and autunite. Specimens of willemite and calcite from the Franklin district, New Jersey, may show brilliant fluorescence colors.

Radioactivity

Minerals containing uranium and thorium will continually undergo *decay reactions* in which radioactive isotopes of U and Th form various daughter elements and also release energy in the form of alpha and beta particles and gamma radiation. The radiation produced can be measured in the laboratory or in the field using a Geiger counter or a scintillation counter. A radiation counter, therefore, is helpful in the identification of U- and Th-containing minerals. Examples are uraninite, pitchblende, thorianite, and autunite.

Solubility in HCl

The positive identification of carbonate minerals is much aided by the fact that the carbon–oxygen bond of the (CO_3) group in carbonates becomes unstable and breaks down in the presence of hydrogen ions available in acids. This is expressed by the reaction $2H^+ + CO_3 \rightarrow H_2O + CO_2$, which is the basis for the "fizz" test with dilute hydrochloric acid. Calcite, aragonite, witherite, and strontianite as well as Cu-carbonates show bubbling or effervescence (fizz) when a drop of dilute HCl is placed on the mineral. The fizz is the result of the release of CO_2. Other carbonates such as dolomite, rhodochrosite, magnesite, and siderite will show effervescence only in hot HCl.

MATERIALS

The minerals of Mohs' scale for testing hardness are essential to this exercise. Such a specialized collection, if not already available, can be bought from mineral supply houses, such as Ward's Natural Science Establishment, Inc., Rochester, New York. The students should be provided with glass plates for hardness testing, and they should have a pocket knife for this exercise as well as for any subsequent assignments on mineral identification. A collection of small, clean, homogeneous, and single mineral specimens for specific gravity determination with the Jolly balance is highly instructive in teaching the diagnostic aspects of specific gravity and the technique itself. Small plastic bottles with eye droppers, filled with dilute HCl (concentrated HCl diluted with distilled water in a proportion of 1 to 10) are needed for carbonate fizz tests, and horseshoe magnets are essential in the evaluation of magnetic properties. An ultraviolet light source (preferably with shortwave and longwave uv radiation) and selected fluorescent mineral specimens will quickly illustrate the usefulness of this diagnostic technique. Similarly, a Geiger counter or a scintillation counter in conjunction with some radioactive mineral specimens will illustrate the powerful diagnostic aspects of radioactivity and the need for safety precautions. A hand lens, with about 10x magnification, is very helpful for the detailed observation of the various physical properties. A binocular microscope with a good light is also very helpful. Note that some of the physical properties introduced in this exercise form the basis for determinative tables in various textbooks (e.g., Chapter 22 in *Manual of Mineral Science*, 23rd ed.; Appendix B in *Introduction to Mineralogy*, by Nesse; Appendix B in *Mineralogy*, by Perkins; and Appendix 1 in *Minerals: Their Constitution and Origin*, by Wenk and Bulakh).

ASSIGNMENT

1. *Hardness.* Determine the relative hardness of a set of unknown minerals, and arrange them into groups of soft (**H** = 1 to 3), medium (**H** = 4 to 6), and hard (**H** = 7 to 9). Do this by using any of the simple tools available, such as a fingernail, copper penny, pocket knife, glass plate, and streak plate. Subsequently you may wish to arrange the same minerals in an even better-defined hardness sequence by comparing the hardness of each of the unknowns with the values of a specific hardness collection based on the Mohs scale. Generally such kits have nine minerals but lack diamond.

One small scratch is really all that is needed for making a hardness test. Additional scratching or even gouging of softer specimens is no more diagnostic than the one small scratch; one scratch leaves a much nicer specimen for others to work with. Because hardness is a property of a single mineral grain, it is best to try to scratch a single grain instead of an aggregate of grains. When a fine-grained aggregate of quartz grains, as in a sandstone, is scratched with a pocket knife, the relatively loosely cemented grains may disaggregate, giving an erroneous impression of relatively low hardness.

2. *Specific Gravity.* As an introduction to the evaluation of the relative "heft" of minerals, weigh several (metallic and nonmetallic) mineral specimens by hand, while at the same time estimating the relative sizes of the specimens. This will help you develop a general sense of relative specific gravity.

A much more accurate method of determining specific gravity is by the Jolly balance. The data obtained by this technique can be used for quite unambiguous identification of unknown materials; refer to Table 2.2. You should have available a few unknown, small, single-mineral specimens for use on the balance. For measuring with the Jolly balance you

TABLE 2.2 Minerals Arranged According to Increasing Specific Gravity*

G	Name	G	Name	G	Name
1.6	Carnallite	2.23	**Graphite**	2.54–2.57	**Microcline**
1.7	**Borax**	2.25	**Natrolite**	2.57	**Othoclase**
1.95	**Kernite**	2.26	Tridymite	***2.6–2.79***	
1.96	**Ulexite**	2.27	**Analcime**	2.55–2.65	**Nepheline**
1.99	Sylvite	2.29	Nitratite	2.6–2.63	**Kaolinite**
2.0–2.19		2.30	Cristobalite	2.62	**Albite**
2.0–2.55	**Bauxite**	2.30	**Sodalite**	2.60–2.66	Cordierite
2.0–2.4	**Chrysocolla**	2.32	**Gypsum**	2.65	**Oligoclase**
2.05–2.09	**Sulfur**	2.33	Wavellite	2.65	**Quartz**
2.05–2.15	**Chabazite**	2.3–2.4	**Apophyllite**	2.69	**Andesine**
1.9–2.2	**Opal**	2.39	**Brucite**	2.6–2.8	Alunite
2.09–2.14	Niter	***2.4–2.59***		2.6–2.8	Turquoise
2.1–2.2	**Stilbite**	2.0–2.55	**Bauxite**	2.71	**Labradorite**
2.16	**Halite**	2.2–2.65	**Serpentine**	2.65–2.74	**Scapolite**
2.18–2.20	**Heulandite**	2.42	**Colemanite**	2.65–2.8	**Beryl**
2.2–2.39		2.42	Petalite	2.72	**Calcite**
2.0–2.4	**Chrysocolla**	2.4–2.45	**Lazurite**	2.6–3.3	**Chlorite**
2.2–2.65	**Serpentine**	2.45–2.50	**Leucite**	2.62–2.76	**Plagioclase**
		2.2–2.8	**Garnierite**	2.6–2.9	**Collophane**

* The names printed in boldface type are those of the most common minerals.
SOURCE: From *Manual of Mineral Science*, 23rd ed., pp. 635 and 636.

TABLE 2.2 (continued)

G	Name	G	Name	G	Name
2.74	**Bytownite**	3.2–3.4	**Augite**	3.5–4.3	**Garnet**
2.7–2.8	Pectolite	3.25–3.37	Clinozoisite	3.9	Antlerite
2.7–2.8	**Talc**	3.26–3.36	Dumortierite	3.9–4.03	**Malachite**
2.76	**Anorthite**	3.27–3.35	Axinite	3.95–3.97	**Celestite**
		3.27–4.37	**Olivine**		
2.8–2.99		3.2–3.5	**Enstatite**	**4.0–4.19**	
2.6–2.9	**Collophane**			3.9–4.1	**Sphalerite**
2.8–2.9	Pyrophyllite	**3.4–3.59**		4.02	**Corundum**
2.8–2.9	**Wollastonite**	3.27–4.27	Olivine	3.9–4.2	**Willemite**
2.85	**Dolomite**	3.3–3.5	**Jadeite**		
2.86	Phlogopite	3.35–3.45	**Diaspore**	**4.2–4.39**	
2.76–2.88	**Muscovite**	3.35–3.45	**Epidote**	4.1–4.3	**Chalcopyrite**
2.8–2.95	**Prehnite**	3.35–3.45	**Vesuvianite**	3.7–4.7	**Romanechite**
2.8–3.0	**Datolite**	3.4–3.5	**Hemimorphite**	4.18–4.25	**Rutile**
2.8–3.0	**Lepidolite**	3.45	Arfvedsonite	4.3	**Manganite**
2.89–2.98	**Anhydrite**	3.40–3.55	Acmite	4.3	**Witherite**
2.9–3.0	Boracite	3.4–3.55	**Titanite**	4.37	**Goethite**
2.95	**Aragonite**	3.48	**Realgar**	4.35–4.40	**Smithsonite**
2.95	Erythrite	3.42–3.56	Triphylite		
2.8–3.2	**Biotite**	3.49	**Orpiment**	**4.4–4.59**	
2.95–3.0	**Cryolite**	3.4–3.6	**Topaz**	4.43–4.45	**Enargite**
2.97–3.00	Phenacite	3.5	**Diamond**	4.5	**Barite**
		3.45–3.60	**Rhodochrosite**	4.55	Gahnite
3.0–3.19		3.5–4.3	**Garnet**	4.52–4.62	**Stibnite**
2.97–3.02	Danburite				
2.85–3.2	Anthophyllite	**3.6–3.79**		**4.6–4.79**	
3.0–3.1	**Amblygonite**	3.27–4.37	Olivine	3.7–4.7	**Romanechite**
3.0–3.1	**Lazulite**	3.5–4.2	Allanite	4.6	**Chromite**
3.0–3.2	**Magnesite**	3.5–4.3	**Garnet**	4.58–4.65	**Pyrrhotite**
3.0–3.1	Margarite	3.6–4.0	**Spinel**	4.7	**Ilmenite**
3.0–3.25	**Tourmaline**	3.56–3.66	**Kyanite**	4.75	**Pyrolusite**
3.0–3.3	**Tremolite**	3.58–3.70	**Rhodonite**	4.6–4.76	Covellite
3.09	Lawsonite	3.65–3.75	**Staurolite**	4.62–4.73	**Molybdenite**
3.1–3.2	Autunite	3.7	**Strontianite**	4.68	**Zircon**
3.1–3.2	Chondrodite	3.65–3.8	**Chrysoberyl**		
3.15–3.20	**Apatite**	3.75–3.77	Atacamite	**4.8–4.99**	
3.15–3.20	**Spodumene**	3.77	**Azurite**	4.6–5.0	Pentlandite
3.16–3.20	**Andalusite**			4.6–5.1	**Tetrahedrite-Tennantite**
3.18	**Fluorite**	**3.8–3.99**		4.89	**Marcasite**
		3.7–4.7	**Romanechite**		
3.2–3.39		3.6–4.0	**Spinel**	**5.0–5.19**	
3.1–3.3	Scorodite	3.6–4.0	Limonite	5.02	**Pyrite**
3.2	**Hornblende**	3.83–3.88	**Siderite**	4.8–5.3	**Hematite**
3.23	**Sillimanite**	3.5–4.2	Allanite	5.06–5.08	**Bornite**
3.2–3.3	**Diopside**			5.15	**Franklinite**

TABLE 2.2 (continued)

G	Name	G	Name	G	Name
5.0–5.3	Monazite	**6.0–6.49**		**7.0–7.49**	
5.18	**Magnetite**	5.9–6.1	Crocoite	7.0–7.5	**Wolframite**
		5.9–6.1	**Scheelite**	7.3	**Acanthite**
5.2–5.39		6.0	**Cuprite**	**7.5–7.99**	
5.4–5.59		6.07	**Arsenopyrite**	7.4–7.6	**Galena**
5.5	Millerite	6.0–6.2	Polybasite	7.3–7.9	Iron
5.5±	**Chlorargyrite**	6.2–6.4	**Anglesite**	7.78	**Nickeline**
5.55	**Proustite**	5.3–7.3	Ferrocolumbite	**>8.0**	
5.6–5.79		6.33	**Cobaltite**	8.0–8.2	Sylvanite
5.5–5.8	**Chalcocite**	**6.5–6.99**		8.10	**Cinnabar**
5.68	**Zincite**	6.5	**Skutterudite**	8.9	**Copper**
5.7	Arsenic	6.55	**Cerussite**	9.0–9.7	Uraninite
5.5–6.0	Jamesonite	6.78	Bismuthinite	9.35	Calaverite
5.3–7.3	Ferrocolumbite	6.5–7.1	Pyromorphite	9.8	Bismuth
5.8–5.99		6.8	**Wulfenite**	10.5	**Silver**
5.8–5.9	Bournonite	6.7–7.1	Vanadinite	15.0–19.3	**Gold**
5.85	**Pyrargyrite**	6.8–7.1	**Cassiterite**	14–19	**Platinum**

also need a glass beaker filled with water and two weighing pans, one in the water, the other above it in air (and dry). Specific instructions for the use of the Jolly balance follow.

a. With the two pans in the arrangement described, bring all scales to 0. You do this by loosening the locking screws at the left and bottom right and by turning the large knurled screw until you arrive at 0.

b. When all scales are at 0, and one pan is dry, the other in water, align the horizontal marking on the mirror very carefully with a reference point at the bottom of the spring. Now everything is at a reference position.

c. Insert the unknown in the upper pan, being careful that your pan does not hit the water under the weight; if it gets wet, dry it. Use the large knurled knob to raise the pan until the reference point is again in line with the horizontal marking on the mirror. The number you now read on the left-hand scale is W_{air} ($=W_a$).

d. Insert the unknown in the lower pan, in water. Lock both locking screws. Use the large knurled knob to adjust the spring enough so that you again align the reference mark below the spring with the horizontal marking on the mirror. The right-hand scale now gives you a difference in weight directly, that is, $W_{air} - W_{water}$.

e. Using $G = W_a / W_a - W_w$ calculate the **G** of your unknown, and using Table 2.2, identify the unknown.

3. *Magnetism.* Using a small horseshoe magnet, evaluate the difference in magnetism of magnetite and pyrrhotite.

4. *Fluorescence.* Irradiate a collection of fluorescent minerals with two different wavelengths of ultraviolet light (shortwave uv and longwave uv) and note the various responses as displayed by differences in color and its intensity.

5. *Radioactivity.* Measure the different intensities of radioactivity (short wavelength radiation in the X-ray region) emitted from various radioactive mineral specimens using a Geiger counter or a scintillation counter.

6. *Carbonate fizz test.* With a drop of dilute HCl on each, test several different carbonate minerals and evaluate their different chemical responses. This is a useful diagnostic technique in distinguishing among various species in the carbonate mineral group. So as not to deface the specimen, test a tiny fragment of the mineral on a glass plate or on a watch glass.

Student Name

FIGURE 2.3   Mineral identification sheet

Crystal Morphology/Symmetry/Twinning/Striations:

Cleavage/Parting/Fracture:

Hardness:

Luster:

Streak:

Color(s):

Tenacity:

Specific gravity (heft):

Other distinguishing features (effervescence in HCl, magnetism, smell, exsolution lamellae, etc.):

Name:

Mineral classification; Formula:

Association:

Please make photocopies of this form if more copies are needed.

Atomic Packing, Radius Ratios, Coordination Numbers and Polyhedra, and Electrostatic Valency

PURPOSE OF EXERCISE

Recognition of various ways in which spheres can be packed in two- and three-dimensional arrays. Equal-sized spheres arrayed in three-dimensions form close-packed hexagonal and cubic structures. Inorganic crystal structures can be analyzed and described in terms of the nearest neighbor atoms about an interstice (a small space between spheres or atoms) or a specific atom; such analysis leads to the concepts of nearest-neighbor coordination, coordination polyhedron, and radius ratio. Once a coordination polyhedron has been selected about a specific atom (or ion), an average bond strength (the electrostatic valency) can be calculated.

FURTHER READING AND CD-ROM INSTRUCTIONS

Klein, C. and Dutrow, B., 2008, *Manual of Mineral Science*, 23rd ed., Wiley, Hoboken, New Jersey, pp. 66–80.

CD-ROM, entitled *Mineralogy Tutorials*, version 3.0, that accompanies *Manual of Mineral Science*, 23rd ed., click on "Module I", and subsequently on the buttons "Coordination of Ions", and "Closest Packing".

Nesse, W. D., 2000, *Introduction to Mineralogy*, Oxford University Press, New York, pp. 57–64.

Perkins, D., 2002, *Mineralogy*, 2nd ed., Prentice Hall, Upper Saddle River, New Jersey, pp. 271–284.

Wenk, H. R. and Bulakh, A., 2004, *Minerals: Their Constitution and Origin*, Cambridge University Press, New York, New York, pp. 84–99, and pp. 12–31.

Bloss, F. D., 1994, 2nd printing with minor revisions, *Crystallography and Crystal Chemistry*, Mineralogical Society of America, Chantilly, Virginia, pp. 201–218 and 221–256.

Background Information: Many inorganic crystalline substances have structures that can be viewed as orderly packings of spheres, where the spheres represent atoms or ions. The simplest structures are built of packings of equally sized spheres; structures are more complex when spheres of different sizes are packed. In using the term *sphere* we assume that it has a fixed size, such as, for example, a plastic ball with a diameter of 1 cm, or a Ping-Pong ball with a diameter of 4 cm. Ions and atoms may be treated as spheres, but not always as rigid spheres with constant diameters. Table 3.8 in *Manual of Mineral Science*, 23rd ed., p. 49 gives effective ionic radii as a function of coordination number. For example, the radius of O^{2-} ranges from 1.36 Å (in 3-fold coordination) to 1.42 Å (in 8-fold coordination). Such variations in radii of specific ions are mainly the result of variations in the local arrangement of ionic neighbors around the oxygen ion in question; these are a function of the *coordination number* of oxygen (as will be discussed more fully later). As such, *real* ions and atoms cannot be treated as having a rigid outer surface with a fixed spherical radius. However, in this exercise we will approximate ions (or atoms) as *rigid spheres of constant diameter* (or *radius*). Even though this approach is, in detail, an oversimplification, it allows for an introduction to, and understanding of, some very basic aspects of the packing of ions (and atoms) in crystal structures.

Ions (or atoms) in crystal structures are bonded to their nearest neighbors by various types of bonds, described as ionic, covalent, metallic, van der Waals, and so on. In this exercise we will concentrate on the geometrical arrangement of nearest neighbors about a central position (which may or may not be occupied by an atom or ion), and we will be less concerned with the bonding mechanisms. The space between an arrangement of spheres is known as an *interstice* (or *interstitial space*). The geometric arrangement of the closest nearby spheres about such an interstitial space (if empty), or about a central (interstitial) sphere (or atom), is known as the *coordination polyhedron*. The number of first (or closest) neighbors is referred to as the *coordination number*. Figure 3.1 shows several such coordination polyhedra for three-dimensional arrays consisting of equal-sized spheres touching their closest neighbors. The central space (in Fig. 3.1 this is an empty interstice) may or may not be occupied (in actual crystal structures) by an appropriately sized spherical atom (or ion). In the references given above, you will find discussions of *radius ratios* in which the radius of an interstitial, spherical atom is commonly referred to as R_X (the radius of a cation) and the radii of the touching spheres (which define the coordination polyhedron) are referred to as R_Z (radius of an anion). Values of R_X/R_Z are useful predictors of the number of nearest neighbors and the geometric shape of the coordination polyhedron (see Fig. 3.2). Methods for the calculation of limiting radius ratios (that is, radius ratios

FIGURE 3.1 Examples of the close packing of equal-sized spheres in three dimensional arrays. In each of the three drawings the interstitial space (the interstice between the spheres) is empty. The number of spheres that touch one another about the interstice is known as the coordination number (C.N.). The shape that is outlined by connecting the centers of the closest spheres is known as the coordination polyhedron.

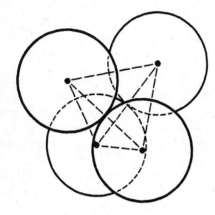

(a) Four equal-sized spheres with each sphere touching the other three outline a regular tetrahedron (so shown by connecting the centers of the spheres). The coordination about the interstice is four closest neighbors; therefore the C.N. is 4.

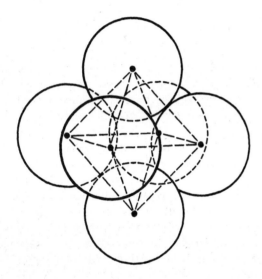

(b) Six equal-sized spheres with each sphere touching four other spheres outline a regular octahedron (so shown by connecting the centers of the spheres). The coordination about the interstice is six closest neighbors; therefore the C.N. for the interstitial space is 6.

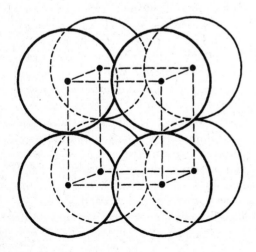

(c) Eight equal-sized spheres with each sphere touching three other spheres outline a cube (so shown by connecting the centers of the spheres). The coordination about the interstitial space is eight closest neighbors; therefore the C.N. of the interstice is 8.

FIGURE 3.2 Examples of various radius ratios. R_X is the radius of the smaller sphere (or atom, or ion) that can be fitted into the interstice between larger spheres with radius R_Z. The five diagrams with large spheres represent plan views; the smaller diagrams represent vertical views. To help you understand more fully what is illustrated here, use four Ping-Pong balls and several small plastic spheres to make identical arrangements.

Triangular coordination
C.N. = 3

Triangular coordination
C.N. = 3

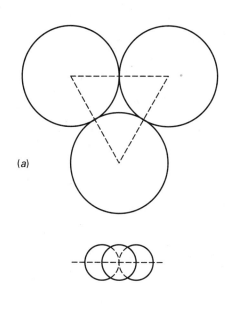

(a)

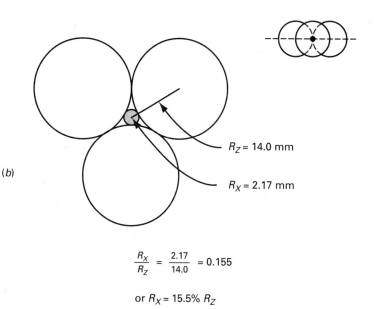

(b)

$R_Z = 14.0$ mm

$R_X = 2.17$ mm

$$\frac{R_X}{R_Z} = \frac{2.17}{14.0} = 0.155$$

or $R_X = 15.5\% \ R_Z$

Triangular coordination
C.N. = 3

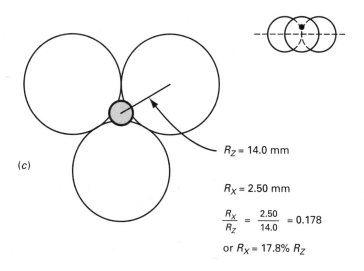

(c)

$R_Z = 14.0$ mm

$R_X = 2.50$ mm

$$\frac{R_X}{R_Z} = \frac{2.50}{14.0} = 0.178$$

or $R_X = 17.8\% \ R_Z$

(a) Triangular coordination (in a two-dimensional array) about an empty interstitial space, **C. N. = 3**.

(b) Here the interstitial space is filled exactly with the largest possible small sphere. The center of this small sphere lies in the plane of the centers of the three larger spheres. R_X is the radius of the small sphere, R_Z the radius of the larger spheres. Their ratio, $R_X : R_Z = 0.155$, means that thee radius of the small sphere, as compared to the radius of the large sphere, must be 15.5% that of the large sphere. If it is larger than this, it will not fit between the three large spheres in the plane defined by the centers of the large spheres (see the small vertical view).

(c) Here the radius of the small sphere is too large to fit between the three touching large spheres in the plane defined by the centers of the large spheres. The small sphere now has to be moved up (vertically) to be accomodated (see small vertical view).

FIGURE 3.2 *(continued)*

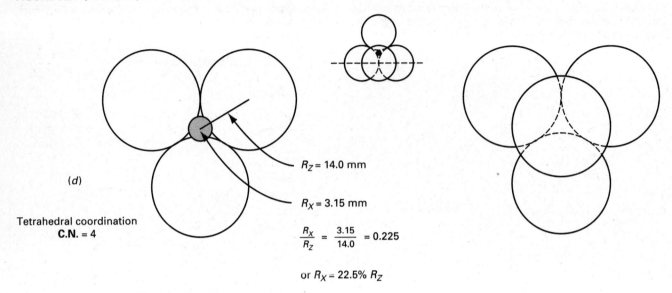

(d)

Tetrahedral coordination
C.N. = 4

$R_Z = 14.0$ mm

$R_X = 3.15$ mm

$\dfrac{R_X}{R_Z} = \dfrac{3.15}{14.0} = 0.225$

or $R_X = 22.5\%\ R_Z$

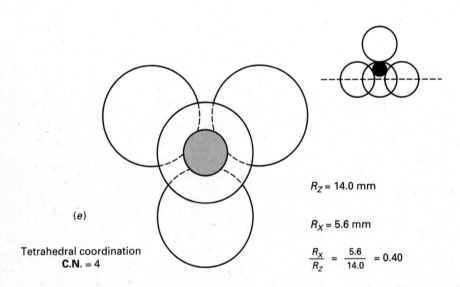

(e)

Tetrahedral coordination
C. N. = 4

$R_Z = 14.0$ mm

$R_X = 5.6$ mm

$\dfrac{R_X}{R_Z} = \dfrac{5.6}{14.0} = 0.40$

(d) Here the smaller sphere is larger than in parts *b* and *c* and has been drawn so that $R_X : R_Z = 0.225$. This ratio of the radius of the small sphere to that of the large spheres allows not only the small sphere to touch all four surrounding large spheres but the four large spheres to touch one another. This is tetrahedral co-ordination (**C. N.** = 4). We can continue to increase the size of the small sphere until the radius ratio is 0.414. Between 0.225 and 0.414 there is tetrahedral coordination, but a ratio such as 0.400

implies that the large spheres no longer touch one another; all four touch only the central smaller sphere (see part *e*).
(e) There is tetrahedral coordination, with four large spheres each touching the smaller central sphere. The radius ratio for this arrangement is 0.400. If the central sphere becomes so large, relative to the surrounding spheres, that the ratio is larger than 0.414, the coordination becomes octahedral.

TABLE 3.1 Atomic Packing Schemes

Minimum Radius Ratio[a] $R_z : R_z$	Coordination Number C.N.	Packing Geometry		
<0.155	2	Linear		
0.155	3	Corners of an equilateral triangle (triangular coordination)		
0.225	4	Corners of a tetrahedron (tetrahedral coordination)		
0.414	6	Corners of an octahedron (octahedral coordination)		
0.732	8	Corners of a cube (cubic coordination)		
1.0	12	Corners of a cuboctahedron (close packing)		

[a]Geometric derivations of these limiting ratio values are given in *Manual of Mineral Science*, 23rd ed., p. 72.

representing a change from one type of coordination to another) can be found in several of the references given. Here, it will suffice to list such radius ratios and the various packing geometries (see Table 3.1).

Until now we have concerned ourselves with various sizes of neutral spheres and their packing arrangements. If we wish to extend this approach to real ions, we must also consider the various charges the ions (or in this discussion the spheres) carry. For example, the $(AsO_4)^{3-}$ group in arsenates occurs as a tetrahedron because of the radius ratio of $R_{arsenic} : R_{oxygen}$ is 0.25 (ionic radius of As^{5+} = 0.34 Å; ionic radius of O^{2-} = 1.36Å; taken from, e.g., Table 3.8 in *Manual of Mineral Science*, 23rd ed., p. 49). But as is clearly implied by the charge on the arsenic (5+) and on the oxygen (2−) ions, the spheres in such packing schemes are not neutral. Indeed, it is appropriate to ask how the charge of the cation (generally the smaller ion in a coordination polyhedron) is distributed over its closest ionic neighbors. The attraction between oppositely charged ions is the main bonding force in ionically bonded structures. In the tetrahedrally coordinated $(AsO_4)^{3-}$ group, the As^{5+} is surrounded by four closest-neighbor O^{2-} ions, and the question "What on average is the bond strength (or the electrostatic valency) of one of the four As−O bonds?" can be answered by a simple calculation. *Electrostatic valency* (abbreviated as e.v.) is defined as the ion's valence charge divided by its coordination number. In the $(AsO_4)^{3-}$ group the e.v. of one of the As−O bonds is 5 divided by 4 (5 being the +5 charge on the As^+ cation and 4 being the **C.N.**, the number of closest neighbors to the As) resulting in an e.v. of $1\frac{1}{4}$. If the average bond strength reaching an anion (in a coordination polyhedron) is exactly half the charge carried by the anion, numerous linking patterns (or coordination polyhedra) can develop (see exercise 4).

MATERIALS

Figures 3.3 through 3.6 and Table 3.2 form the basis for this assignment. You will also need a ruler and pencil. Access in the laboratory to models of coordination polyhedra and closest-packed structures will help you to recognize the various geometrical arrangements. If such models are not available, various coordination polyhedra can be constructed very simply by gluing together Ping-Pong balls, or tennis balls, and various-sized small plastic balls.

ASSIGNMENTS

The assignments are based on the illustrations in Figs. 3.3 through 3.6, and Table 3.2.

1. Two-dimensional packing of equal-sized spheres, Fig. 3.3. In this figure equal-sized (identical) spheres are arranged in various planar patterns (arrays). For each of the arrays (*a* through *e*) you are asked to describe the geometry of the packing. Do this by connecting the centers of neighboring spheres with lines (in doing this you are outlining unit cell[*] choices of two-dimensional plane groups); always work on spheres that are located in the inner part of the drawing, because if you select spheres at the edge you lose the continuity of the pattern (or array). Descriptive geometric terms for your answers include *oblique*, *square*, and *hexagonal*. For each of the drawings you are also asked to give the coordination number (**C.N.**) by counting the number of neighbors that touch any chosen sphere. You are furthermore asked to outline the geometric shape of the coordination pattern by connecting the centers of the spheres that touch some central sphere. Finally, you are asked to sketch the shape between touching spheres, that is, the shape of the interstice.

2. Three-dimensional packing of equal-sized spheres, Fig. 3.4. In the first three parts of this figure (*a* through *c*) you are asked questions about unit cells and their shapes; about coordination numbers for specifically identified spheres or interstices; and about cross sections through the three-dimensional models along specific planes. In Fig. 3.4*d* you are introduced to a stacking sequence described as *ABABAB* . . . where the spheres in the second layer lie on top of the *B* voids of the first layer; here you are also asked about coordination numbers. In Fig. 3.4*e* you are asked to identify the stacking sequence with the appropriate symbolic letter sequence and to locate the two different types of coordination arrangements throughout this two-layer sequence. In Fig. 3.4*f* you are introduced to a three-layer stacking sequence *ABCABC* . . . in which the first-layer spheres are represented by *A*, the second-layer spheres lie on the top of the *B* voids, and the third-layer spheres lie on top of the C voids.

3. Two-dimensional packing of two types of equal-sized spheres, Fig. 3.5. The two types of spheres (labeled *X* and *Y*) are packed in various arrays (*a* through *d*). In the drawings the two types are distinguished by differences in shading. If these were to represent atoms in real crystals (of, for example, alloys), *X* might represent Cu atoms and *Y* Zn atoms. In the four drawings you are asked to outline unit cells and to determine the composition of your chosen unit cell; concentrate on spheres in the internal part of the drawing, not at the edge. After you have outlined the unit cell, count how many spheres there are of *X* and of *Y*. Clearly you must sum the various fractions that each of the spheres contributes to the unit cell as well. For example, four quartered segments (of a sphere, or atom in a unit cell) add up to one whole sphere. You express your composition in terms of *XY* with subscripts to either *X* or *Y*, or to both. If the composition consists of equal amounts of *X* and *Y*, the final notation is X_1Y_1, or just *XY*. A similar composition for a Cu−Zn alloy would be CuZn. In Fig. 3.5 outline the unit cells in the drawings and write the resultant compositions to the right of the drawings.

---

[*] A unit cell is the smallest unit of repeat in a pattern; it will yield the entire pattern when translated repeatedly without rotation.

4. Calculation of average bond strength (electrostatic valency; e.v.) in various coordination polyhedra, Fig. 3.6. For each of the five polyhedra calculate the e.v. by dividing the cationic charge by the number of nearest neighbors (**C.N.**). Which of the five are mesodesmic bonds? Label these with the term "mesodesmic."

5. Calculation of radius ratios for various given compounds, and deriving these from the predicted coordination number about a specific cation, Table 3.2. Consult Table 3.1 for limiting values of radius ratios.

EXERCISE 3

Student Name

FIGURE 3.3 Assignment on two-dimensional packing of equally-sized spheres.

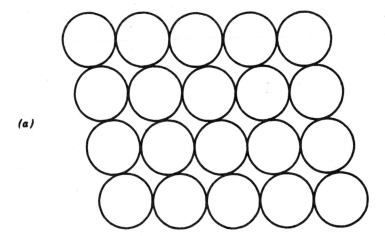

(a)

1. Geometry of array _ _ _ _ _ _ _ _

2. C.N. _ _ _ _ _ _ _

3. In the array outline the shape of coordinating spheres

4. Sketch the shape of the interstice

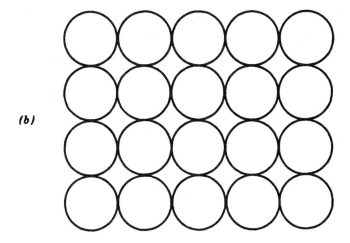

(b)

1. Geometry of array _ _ _ _ _ _ _ _

2. C.N. _ _ _ _ _ _ _

3. In the array outline the shape of coordinating spheres

4. Sketch the shape of the interstice

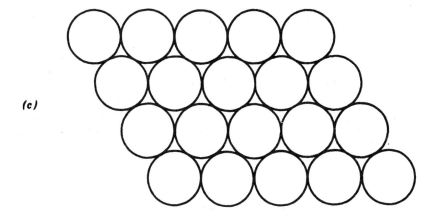

(c)

1. Geometry of array _ _ _ _ _ _ _ _

2. C.N. _ _ _ _ _ _ _

3. In the array outline the shape of coordinating spheres

4. Sketch the shape of the interstice

FIGURE 3.3 *(continued)*

Student Name

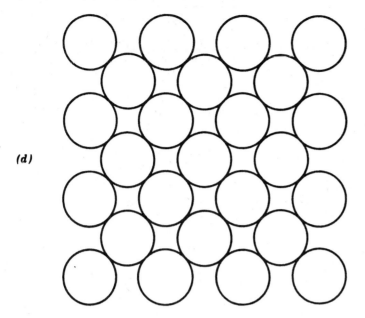

(d)

1. Geometry of array _____

2. C.N. _____

3. In the array outline the shape of coordinating spheres

4. Sketch the shape of the interstice

5. How does this array relate to that in fig. 3.3*b*?

   _____.

(e)

1. Geometry of array _____

2. C.N. _____

3. In the array outline the shape of coordinating spheres

4. Sketch the shape of the interstice

5. How does this array relate to that in fig. 3.3*a*?

   _____

EXERCISE 3

Student Name

FIGURE 3.4 Assignment on three-dimensional packing of equally-sized spheres.

Plan view	Three-dimentional stacking

(a)

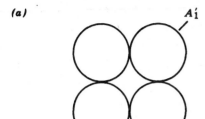
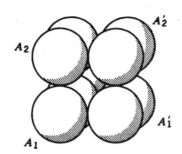

1. Outline unit cell in model and give name of its shape

2. C.N. of interstice of model _ _ _ _ _

3. Sketch of packing along the plane $A_1 A_1' A_2 A_2'$

(b)

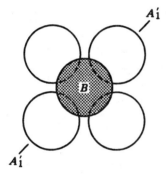
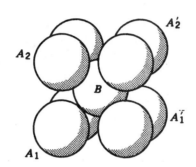

1. Outline unit cell in model and give name of its shape

2. C.N. of central (shaded) sphere (B)
   _ _ _ _ _ _ _ _

3. Sketch of packing along the plane $A_1 A_1' A_2 A_2'$.

4. What differentiates this packing from that shown in part *a*?

   _ _ _ _ _ _ _ _

(c)

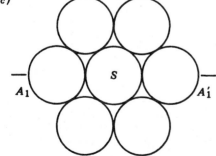
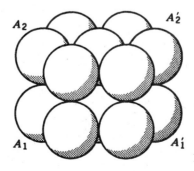

1. Outline smallest unit cell choice in plan view and give name of its shape

2. C.N. of sphere S in the two–layer sequence

   _ _ _ _ _ _ _ _

3. Sketch of packing along $A_1 A_2 A_1' A_2'$

4. After adding a third layer below the plane $A_1 A_1'$, what becomes of the C.N. of sphere S?

   _ _ _ _ _ _ _ _

FIGURE 3.4 *(continued)*

(d)

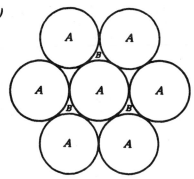

One layer of spheres;
B is the location of voids

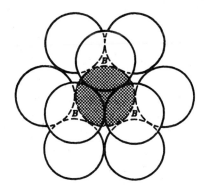

Two-layer stacking *AB*

C.N. of shaded sphere is

– – – – – – –

In three-layer stacking *ABA* the
C.N. of any sphere is

– – – – – – –

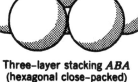

Three-layer stacking *ABA*
(hexagonal close-packed)

(e)

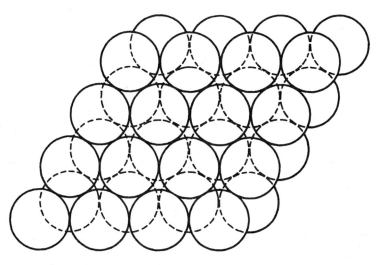

1. Give stacking
   sequence of this
   two-layer stack
   (using *A* and *B* notation)

   – – – – – – –

2. This stacking has two
   different types of interstices
   with different coordination
   numbers. Locate these two
   types throughout the drawing and
   give their C.N.:

   – – – – – – –

FIGURE 3.4 *(continued)*

(f)

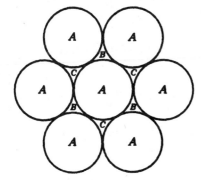

One layer of spheres with
all void spaces identified

△ **B voids**

▽. **C voids**

Instead of stacking a sequence of *AB AB AB*. . . .
(as in Fig. 3.4*d*), we can stack a somewhat
different sequence *ABC ABC*. . . . .

What is the C.N. of any sphere in such
a sequence?

— — — — — — —

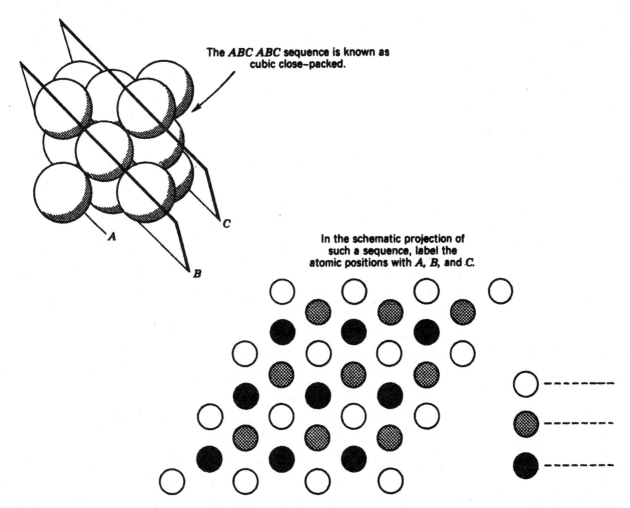

The *ABC ABC* sequence is known as
cubic close–packed.

In the schematic projection of
such a sequence, label the
atomic positions with *A, B,* and *C.*

FIGURE 3.5 Assignment on the two-dimensional packing of two types of equal-sized spheres.

(a)

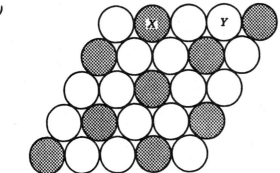

In each of these four drawings outline an appropriate unit cell* and determine the composition of this unit cell (e.g., XY_2, etc.).

*A unit cell is the smallest unit of repeat in a pattern; it will yield the entire pattern when trannslated repeatedly without rotation

-- -- -- -- -

(b)

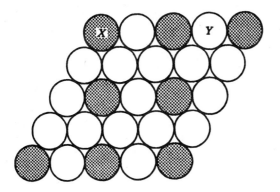

-- -- -- -- -

(c)

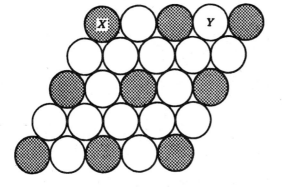

-- -- -- -- -

(d)

-- -- -- -- -

Single layers of equally-sized spheres (= atoms)

FIGURE 3.6 Assignment on the calculation of average bond strengths.

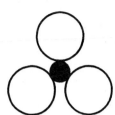

Average bond strength, or e.v. =

- - - - - - -

Triangular: CO_3 with C^{4+} and O^{2-}

e.v. = - - - - - - -

Triangular: BO_3 with B^{3+} and O^{2-}

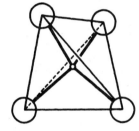

e.v. = - - - - - - -

Tetrahedral: SO_4 with S^{6+} and O^{2-}

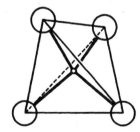

e.v. = - - - - - - -

Tetrahedral: PO_4 with P^{5+} and O^{2-}

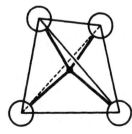

e.v. = - - - - - - -

Tetrahedral: SiO_4 with Si^{4+} and O^{2-}

TABLE 3.2 Assignment on the Calculation of the Radius Ratios for Various Compounds.

Compound	Radius Ratio $R_X : R_Z$ (Radii Given Below)	Predicted Coordination Number [a] (C.N.)
ZnS		
SiO_2		
Al_2O_3		
Fe_2O_3		
FeO		
CaO		
MgO		
TiO_2		
NaCl		
CaF_2		
CsCl		
CuCu (metal)		

$Zn^{2+}[6] = 0.74$ Å; $S^{2-}[4] = 1.84$ Å; $O^{2-}[3] = 1.36$Å; $Si^{4+}[4] = 0.26$ Å; $Al^{3+}[4] = 0.39$Å; $Fe^{3+}[6] = 0.65$Å; $Fe^{2+}[6] = 0.78$Å; $Ca^{2+}[6] = 1.00$Å; $Mg^{2+}[6] = 0.72$Å; $Ti^{4+}[6] = 0.61$Å; $Na^+[6] = 1.02$Å; $Cl^-[6] = 1.81$Å; $F^-[6] = 1.33$Å; $Cs^+[6] = 1.67$Å; Cu(atomic) = 1.28Å; ionic and atomic radii taken from Tables 3.7 and 3.8, respectively, in *Manual of Mineral Science*, 23rd ed., p. 47 and p. 49. Numbers inside square bracket are coordination numbers.

[a] Predicted coordination numbers may not equal the coordination number in the actual crystal structure. This is in large part because in real structures the atoms and ions do not behave as perfect spheres. Indeed ions and atoms, especially large ones, are easily *polarized*, which means that their shape has become somewhat ellipsoidal, instead of being truly spherical.

On the Linking of Anionic Groups: Triangles in Borates and Tetrahedra in Silicates

PURPOSE OF EXERCISE

Understanding the linkages that may develop between $(BO_3)^{3-}$ triangles in borates and between $(SiO_4)^{4-}$ tetrahedra in silicates. The various linkage schemes of such co-ordination polyhedra are known as *polymerization*. Each of these combinations of polyhedra has a unique chemical composition.

FURTHER READING

Klein, C. and Dutrow, B., 2008, *Manual of Mineral Science*, **23rd ed., Wiley, Hoboken, New Jersey, p. 88 and 89.**

Nesse, W. D., 2000, *Introduction to Mineralogy*, Oxford University Press, New York, pp. 62–64.

Perkins, D., 2002, *Mineralogy*, 2nd ed., Prentice Hall, Upper Saddle River, New Jersey, pp. 284–286.

Wenk, H. R. and Bulakh, A., 2004, *Minerals: Their Constitution and Origin*, Cambridge University Press, New York, New York, pp. 425–428.

Bloss, F. D., 1994, 2nd printing with minor revisions, *Crystallography and Crystal Chemistry*, Mineralogical Society of America, Chantilly, Virginia, pp. 258–271.

Background Information: In exercise 3 you were introduced to the concept of the electrostatic valency (abbreviated as e.v.) of a bond, as being the ion's valence charge divided by its coordination number (see Fig. 3.6). If the average bond strength reaching an anion in a coordination polyhedron (such a bond strength is expressed as e.v.) is exactly half the charge carried by that anion, numerous linking patterns (of coordination polyhedra) can develop.

For example, in a tetrahedral coordination polyhedron of $(SiO_4)^{4-}$ the average bond strength (e.v.) of one of the four Si–O bonds is 4 divided by the coordination number, **C.N.** = 4; this results in an e.v. of 1 per bond. However, an oxygen ion has a negative charge of 2–; as such exactly half of the charge of the oxygen ion is neutralized by one of the Si–O bonds. This means that the other half of the charge on the O^{2-} can be neutralized by linking to another $(SiO_4)^{4-}$ tetrahedron. As such the oxygen in question becomes a bridging ion between two tetrahedra with its 2– charge completely and exactly neutralized by two Si–O bonds from two different $(SiO_4)^{4-}$ tetrahedra. Such a bridging mechanism is fundamental to the linking of coordination polyhedra (in this case tetrahedra) and is the basis for what

is commonly referred to as *polymerization*. Such linking patterns are restricted to coordination polyhedra where the e.v. of the cation–anion bond is exactly half that of the charge of the anion; such bonds are known as *mesodesmic*, from the Greek words *mesos*, meaning middle, and *desmos*, meaning bond.

Such linkages result in geometric motif combinations made of two or more polyhedra, as well as in infinitely extending polyhedral patterns. In these two-dimensional or three-dimensional patterns of polyhedral linkages, one can outline a unit cell for the pattern, and one can express the chemical composition of the unit cell (as a function of the types and numbers of specific ions making up the polyhedron, that is, the motif).

MATERIALS

Figures 4.1 and 4.2 form the basis for this assignment.

ASSIGNMENT

1. *Linkage of BO₃ triangles as in borates.* In Fig. 4.1*a* the triangular coordination of the BO_3 group is shown; this consists of B^{3+} at the center of the triangle and O^{2-} at the three corners. The overall electrical charge on the BO_3 group is calculated on the basis of the number of cations and anions in the group; one B^{3+} and three O^{2-} results in an overall charge for the BO_3 group of 3– which is noted as a superscript outside the brackets around the group, $(BO_3)^{3-}$. Because the e.v. of each bond that connects B to O is 1, it neutralizes exactly half of the 2– charge on the oxygen. This then is a *mesodesmic* bond and oxygen is the bridging ion between two (BO_3) groups. This leads to various patterns of linking of BO_3 groups and is known as polymerization. In Fig. 4.1 you are asked to derive the chemical composition for the various linked patterns, many of which are found in natural borates. To arrive at the chemical notation, count all the B and O ions that are part of the pattern. Once the subscripts for B and O are known, you can calculate the overall charge for the group, or the subunit of the repeat pattern (that is, the unit cell of the repeat). A unit cell is the smallest unit of repeat in a pattern; it will yield the entire pattern when translated repeatedly without rotation.

2. *Linkage of SiO₄ tetrahedra as in silicates.* In Fig. 4.2 various linkage patterns of SiO_4 tetrahedra are shown.

Figure 4.2a shows the tetrahedral coordination of the SiO_4 group. For each of the schematic patterns you are asked to derive the chemical notation of the whole motif (the linkage unit) or the unit cell of some indefinitely extending linkage. Once you have established the subscripts to Si and O in the chemical formula, you can calculate the overall charge of the group, or unit cell. That is, if the unit has the composition Si_3O_9, the overall charge on this would be $3 \times 4+ = 12+$ (from three Si^{4+}) balanced by $9 \times 2- = 18-$ (from nine O^{2-}), resulting in $(Si_3O_9)^{6-}$. In some instances you are asked to substitute a specific amount of Al for Si in the formula unit and to record the final composition of such Al–Si substitution and the overall charge. For example, beginning with the composition SiO_2, you could imagine substituting one-fourth of Si^{4+} by Al^{3+}. This would lead to the formula $(Si_{0.75}Al_{0.25})O_2$, which if we multiply all subscripts by 4 (to clear the fractions) results in Si_3AlO_8. The overall charge of this Si–Al–O tetrahedral arrangement is $(3 \times 4+) + (1 \times 3+) - (8 \times 2-) = 1-$, resulting in $(Si_3AlO_8)^{1-}$. You probably recognize this as the composition of the tetrahedral framework of orthoclase, $KAlSi_3O_8$.

In Fig. 4.2b through g report the compositions and electrical charges as requested; also outline unit cells in Figs. 4.2d through g.

EXERCISE 4

FIGURE 4.1 Assignment on linkages of BO$_3$ groups in borate structures.

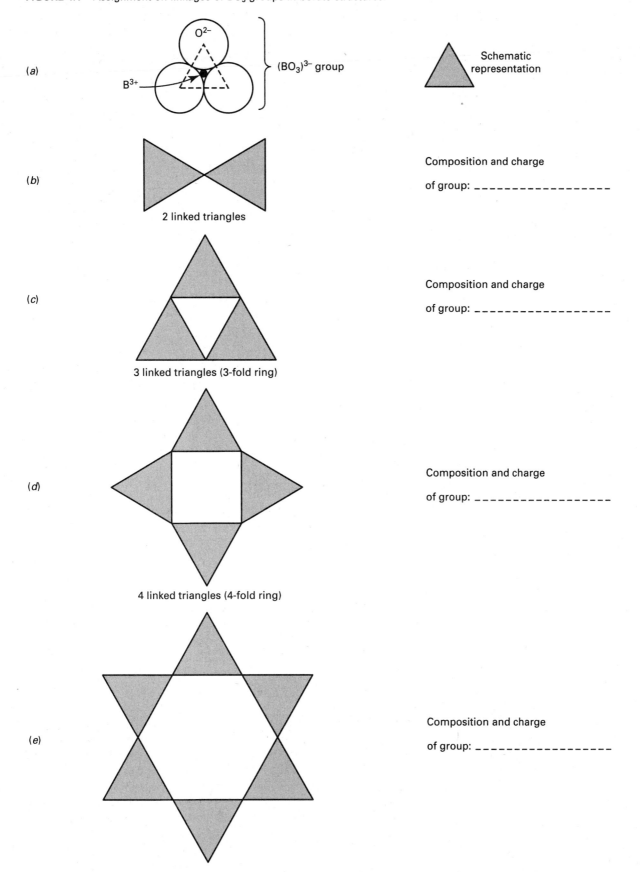

(a) (BO$_3$)$^{3-}$ group

Schematic representation

(b) 2 linked triangles

Composition and charge of group: _____

(c) 3 linked triangles (3-fold ring)

Composition and charge of group: _____

(d) 4 linked triangles (4-fold ring)

Composition and charge of group: _____

(e) 6 linked triangles (6-fold ring)

Composition and charge of group: _____

FIGURE 4.1 *(continued)*

(f)

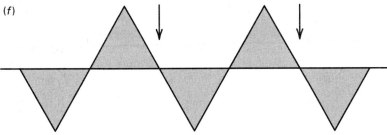

Infinitely extending single chain of triangles

Composition and charge
of the repeat between arrows:

- - - - - - - - - - - - - - - - - - - -

(g)

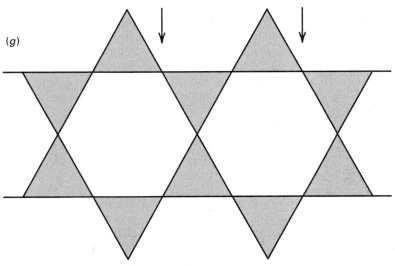

Infinitely extending double chain of triangles

Composition and charge
of the repeat between the arrows:

- - - - - - - - - - - - - - - - - - - -

(h)

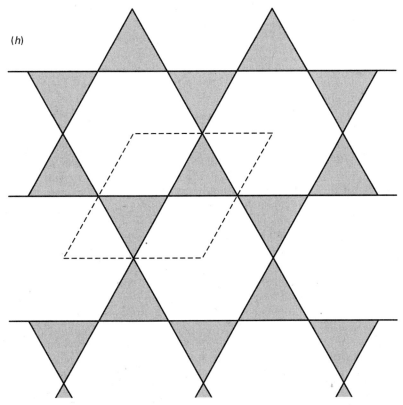

Infinitely extending sheet of triangles

Composition and charge
of the repeat
inside the unit
cell (outlined with
dashed lines)

- - - - - - - - - - - - - - - - - - - -

EXERCISE 4

Student Name

FIGURE 4.2 Assignment on linkages of SiO_4 groups in silicate structures.

(a)

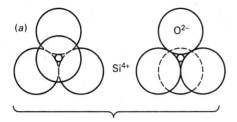

Si^{4+}

O^{2-}

SiO$_4$ tetrahedral group

(b)

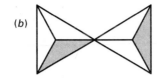

- -
Composition and charge

(c)

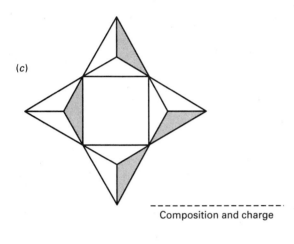

- -
Composition and charge

(e)

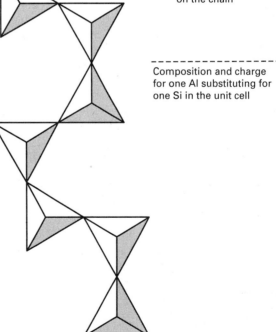

Infinite chain

- -
Composition and charge

Mark the
unit cell choice
on the chain

- -
Composition and charge
for one Al substituting for
one Si in the unit cell

(d)

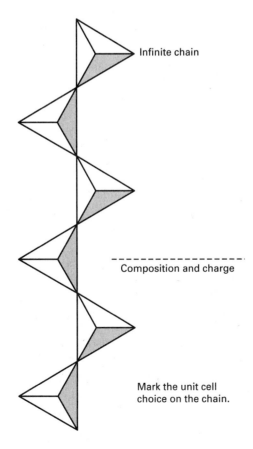

Infinite chain

- -
Composition and charge

Mark the unit cell
choice on the chain.

FIGURE 4.2 *(continued)*

Student Name

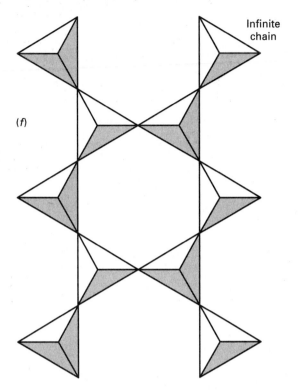

(f)

Infinite chain

_____

Composition and charge

Mark the unit cell choice on the chain

_____

Composition and charge for one Al substituting for one Si in the unit cell

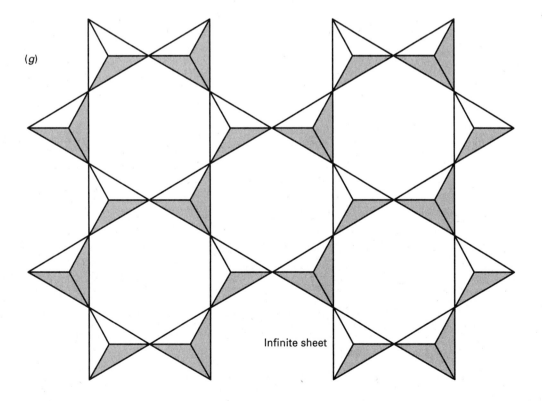

(g)

Infinite sheet

_____

Composition and charge

Mark the unit cell choice on the sheet

_____

Composition and charge for 25% Al substituting for Si per unit cell

Derivation of the Chemical Formula of a Mineral from its Chemical Analysis

PURPOSE OF EXERCISE

Understanding the various steps involved in converting a mineral analysis (generally reported in weight percentages of elements or oxides) into a mineral formula.

FURTHER READING

Klein, C. and Dutrow, B., 2008, *Manual of Mineral Science*, 23rd ed., Wiley, Hoboken, New Jersey, pp. 99–108.

Nesse, W. D., 2000, *Introduction to Mineralogy*, Oxford University Press, New York, pp. 71–72.

Perkins, D., 2002, *Mineralogy*, 2nd ed., Prentice Hall, Upper Saddle River, New Jersey, pp. 21–25.

Wenk, H. R. and Bulakh, A., 2004, *Minerals: Their Constitution and Origin*, Cambridge University Press, New York, New York, pp. 258–263.

Bloss, F. D., 1994, 2nd printing with minor revisions, *Crystallography and Crystal Chemistry*, Mineralogical Society of America, Chantilly, Virginia, pp. 296–299.

Background Information: The chemical compositions of minerals, ranging from simple to complex, are invariably presented in the form of chemical analyses. Such chemical analyses commonly consist of a listing of the relative weight percentages of several elements or oxides. Examples of minerals with simple chemical analyses are

1		2	
Element	Weight %	Element	Weight %
Na	39.34	Si	46.75
Cl	60.66	O	53.25
Total	100.00	Total	100.00

Analysis 1 is for the mineral halite, with the chemical formula NaCl, and analysis 2 is for the mineral quartz, with the chemical formula SiO_2.

Somewhat more complex analyses follow for the minerals chalcopyrite (analysis 3) and alkali feldspar (analysis 4):

3		4	
Element	Weight %	Oxide	Weight %
Cu	34.64	SiO_2	65.67
Fe	30.42	Al_2O_3	20.84
S	34.94	CaO	0.50
		Na_2O	7.59
		K_2O	5.49
Total	100.00	Total	100.09*

* Deviations from 100 percent in the analysis total are commonly the result of experimental error, that is, analytical error.

The chemical formula of chalcopyrite is $CuFeS_2$, and that of the alkali feldspar $(Na,K)AlSi_3O_8$. The reasons for reporting some analyses (such as for oxide and silicate minerals) in oxide form, instead of in elemental form (as was done for analyses 1 through 3), is mainly historical. Irrespective of the form of the original analysis (be it in elemental or oxide form), one is faced with the question of how does such an analysis relate to a standard chemical (mineral) formula?

A mineral formula such as $CuFeS_2$, or $(Na,K)AlSi_3O_8$, records (in chemical shorthand) the ratios of the various elements in the structure. For example, chalcopyrite has a total of two metal cations (Cu^{2+} and Fe^{2+}) and two sulfur anions in its structure; this leads to a metal ions: anions ratio of 1:1. The alkali feldspar, on the other hand, has a ratio of five metal ions to eight oxygens as deduced from the formula $[(Na,K)_1{}^+ + Al^{3+} + Si_3{}^{4+}]$, that is, $1 + 1 + 3 = 5$ cations. Of these, Al and Si substitute for each other in a tetrahedral position in the structure (known as the T site). As such, an alkali feldspar formula could be written as $(Na,K)_1(Al,Si)_4O_8$. The Na and K ions are located in a specific position in the structure, commonly referred to as the M site, a relatively large space outside the Si–Al–O tetrahedral framework.

This brief introduction to specific atomic locations ("sites") in crystal structures is very relevant in the conversion of a weight percent (chemical) analysis to a chemical formula. Indeed, many silicates have structures that are considerably more complicated than those of feldspars; not only may they have a considerable number of sites (the atomic sizes of which are controlled by the packing of oxygen ions about the sites), but they may also show extensive chemical substitution of various elements within the same site. For example, a typical pyroxene, diopside, with formula $CaMgSi_2O_6$, contains tetrahedral sites (known as T sites for the housing of Si^{4+} and Al^{3+}) but also two quite distinct metal (cation) sites, known as M_1 and M_2. The principle difference between the two M sites is the size of interstitial space between various oxygen atoms; the M_2 site is larger than the M_1 site. A projection of the typical monoclinic pyroxene structure is shown in Fig. 5.1.

In the various references listed above you will see worked examples of recalculations from weight percentage analyses to various chemical formulas. This exercise will do much the same, but here we will concentrate on a mineral group, the pyroxenes, in which you must make decisions about the allocation of various cations to the T, M_1, and M_2 sites, in

FIGURE 5.1 The crystal structure of a monoclinic pyroxene projected down [001], onto the (001) plane, the basal pinacoid. Note the locations of the *T*, M_1 and M_2 sites. The shaded areas outline the infinitely extending Si–O chains that face each other across the two M_1 sites. These shaded shapes are referred to in the literature as I-beams. (After M. Cameron and J. J. Papike, 1980, *Reviews in Mineralogy*, vol. 7, *Pyroxenes*, Mineralogical Society of America, Washington, D. C., p. 12).

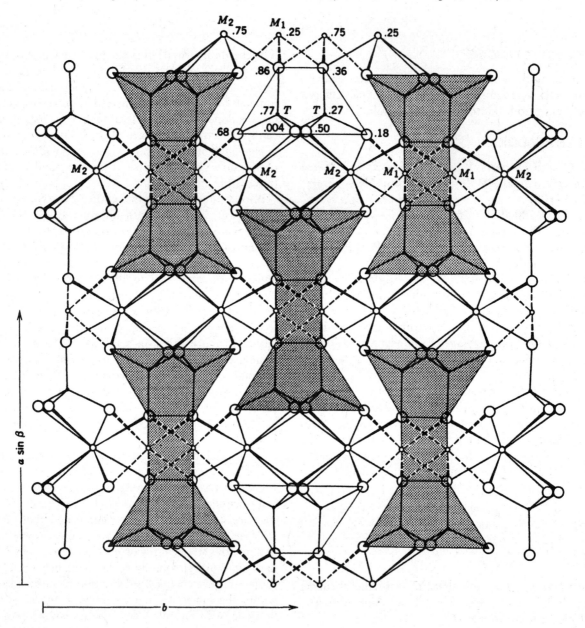

addition to understanding the steps in converting a weight percent analysis to a chemical formula.

First, you might ask what does a weight percent analysis mean? The answer to that is that it lists the relative weight percentages of elements or oxides. To help you understand this answer, we will begin with the formula of chalcopyrite, $CuFeS_2$, and work "backward" to a chemical analysis. From X-ray structural study we know that the unit cell of chalcopyrite contains four such formula units, which means that it has the composition of $Cu_4Fe_4S_8$. Because this second formula contains a common factor of 4, it is normally divided by the factor resulting in $CuFeS_2$. In either case, the ratio of all metal ions to sulfur is constant in both notations (8:8 or 2:2, both of which represent 1:1). The chemical formula states "There is one Cu ion, one Fe ion, and two sulfur ions per formula." We know what each of these ions (or atoms) weighs from its atomic weight. Copper is by far the heaviest with an atomic weight of 63.54; iron is somewhat lighter with an atomic weight of 55.85; and sulfur is by far the lightest with an atomic weight of 32.08. Because the formula states that there is one Cu, one Fe, and two S, we can now calculate the relative weight contributions for these various ions (or atoms).

one Cu weighs	63.55
one Fe weighs	55.85
and two S weigh	64.12 (that is, 2×32.06)
Total weight	183.52

The relative weight of any of the atoms, as referred to the total weight, is derived as follows:

$$\text{Cu contributes } \frac{63.55}{183.52} \times 100\% = 34.62\% \text{ in weight}$$

$$\text{Fe contributes } \frac{55.85}{183.52} \times 100\% = 30.43\% \text{ in weight}$$

$$\text{S contributes } \frac{64.12}{183.52} \times 100\% = 34.94\% \text{ in weight}$$

These calculated weight percentage values compare closely to the numbers listed in analysis 3. This short calculation should give you a better understanding of the earlier statement "A chemical analysis lists the relative weight percentages of elements or oxides."

Let us now inspect analysis 4, that of an alkali feldspar. This is a typical example of the listing of oxide components in a silicate. The order, in going from top to bottom, is fairly standard with the oxides arranged in order of increasing radius of the metal ion (Si^{4+} being the smallest with a radius of 0.26 Å, and K^+ the largest with a radius of 1.51 Å). An inspection of such a listing of oxides and their weight percentages informs the reader of the major elements (Si, Al, Ca, Na, K, and O) that must be part of the chemical formula, but it gives absolutely no direct insight into metal ratios (or subscripts to the elements) in the chemical formula. Various steps in the transformation from weight percentages of the oxides to the final chemical formula lead finally to a quantitative statement of metal ratios (or chemical subscripts). However, in order to complete such a transformation correctly, we must know something about the feldspar structure. The largest ions in feldspar are the oxygens (radius of $O^{2-} = 1.40$ Å) with cations distributed among the various interstitial spaces between the oxygens in close packing. In silicate and oxide structures the packing of oxygen is generally without omissions (that is, without open spaces in the oxygen packing), so that the number of oxygen ions (or atoms) per unit cell of a feldspar is a constant. This number turns out to be 32 oxygens, for four formula units of alkali feldspar, $(Na,K)_4Al_4Si_{12}O_{32}$, per unit cell. Because of the common factor of 4 in this formulation, the formula is normally written as $(Na,K)AlSi_3O_8$, with the number of oxygens, eight, being a *constant in the structure*. In any recalculation sequence, we must know to how many oxygens (as in silicates or oxides) or sulfurs (as in sulfides) the final formula must refer. For feldspar the number is eight oxygens.

We will now go through the various steps that are necessary to convert the chemical analysis of an alkali feldspar to its formula; see Table 5.1. Column 1, in Table 5.1, lists the oxide components; column 2 lists the weight percentages of the oxides. If we wish to know how may "molecules" of oxide this represents for each of the weight percent values, we need to divide the values in column 2 by the appropriate molecular weight (listed in column 3). (The word molecule is put in quotation marks because most inorganic structures do not contain identifiable molecular groupings; they are instead continuous packings or networks of atoms or ions.) The division of the oxide weight percent value (column 2) by the appropriate molecular weight (column 3) leads to a listing of relative numbers of "oxide molecules" (column 4: molecular proportions). Before proceeding further, let us inspect the formulas of the oxide components in column 1. SiO_2 contributes one Si and two O per formula or per molecule of SiO_2; Al_2O_3 contributes two Al and three O per formula or per molecule of Al_2O_3; all the remaining oxide components (CaO, Na_2O, and K_2O) contribute each only one oxygen per formula or per molecule; however, CaO contributes only one Ca, whereas Na_2O and K_2O each contribute two metal ions. In advancing our conversion further, we must ask how many metals and how many oxygens does each of the "oxide molecules," as recorded in column 4, contribute to the overall chemistry of the alkali feldspar? We just determined these numbers by inspection of column 1, and indeed these numbers become multipliers in going from column 4 to columns 5 and 6, respectively. Column 5, labeled "cation proportions," records the number of metals contributed by each oxide compound. Column 6, labeled "number of oxygens," records the number of oxygens contributed by each "molecule." These oxygens will total to some number, 2.9889 in this case. This number is not meaningful in itself, but you must remember that this is the total of all the relative

TABLE 5.1 Example of the Recalculation of an Alkali Feldspar Analysis

1 Oxide Components	2 Weight Percentages	3 Molecular Weights[a]	4 Molecular Proportions	5 Cation Proportions	6 Number of Oxygens	7 Number of Cations per Eight Oxygens	
SiO_2	65.67	60.08	1.0930	1.0930	2.1860	2.925	$\}$ 4.02 ≈ 4.0
Al_2O_3	20.84	101.96[b]	0.2044	0.4088	0.6132	1.094	
CaO	0.50	56.08	0.0089	0.0089	0.0089	0.024	
Na_2O	7.59	61.98	0.1225	0.2450	0.1225	0.655	$\}$ 0.99 ≈ 1.0
K_2O	5.49	94.20	0.0583	0.1166	0.0583	0.312	

Total = 100.09

Total oxygens = 2.9889

Oxygen Factor: $\dfrac{8}{2.9889} = 2.67656$

Final alkali feldspar formula: $(Na,K,Ca)_1 \underbrace{(Al,Si)_4 O_8}$

M site   T site

---

[a] Molecular weights given in Table 3.2, *Manual of Mineral Science*, 23rd ed.

[b] Example of calculation of molecular weight of, e.g., Al_2O_3: atomic weight Al = 26.98; atomic weight O = 16; two Al weigh 53.96; three O weigh 48; totaling 101.96.

oxygens contributed by the various "molecules." Earlier, we noted that structural analysis gave us the number 8 for the total number of oxygens per formula of feldspar. We can use this number 8 (for oxygen) by dividing the total value of 2.9889 into 8, providing us with a multiplier, or oxygen ratio, of 2.67656. We multiply the cation contributions (values in column 5) by this factor to put the metal content of the analysis on the basis of eight oxygens. This gives us, in column 7, the number of metal ions of each kind on the basis of eight oxygens.

In alkali feldspar the number of silicons (in column 7) will be considerably less than 4, generally close to 3.0 with Al substituting for Si in the tetrahedral framework (the T sites of the structure). Indeed Si + Al totals to 4.0, as is to be expected for a framework silicate structure with (Si + Al) : O = 4 : 8. The total for the remaining cations amounts to approximately 1, with Na, K, and Ca substituting for one another in the one metal (M) site of the feldspar structure.

This stepwise conversion should suffice as a preamble to an assignment on the recalculation of pyroxene analyses. However, it may be helpful to inspect the various steps that you must complete in a pyroxene recalculation by studying Table 5.2, which gives an example of a pyroxene analysis. Every column in this table is arrived at in a manner identical to that for the feldspar recalculation, except for the difference in the oxygen factor. The clinopyroxene formula (that is, that of a monoclinic pyroxene) is based on six oxygens per formula. For example, the simplest diopside formula is $CaMgSi_2O_6$. The formula for the clinopyroxene in Table 5.2 is not so simple because various elements substitute for each other in the various sites (M_1, M_2, and T) of the clinopyroxene structure. This is the reason for a somewhat more involved cation allocation scheme in column 7 of Table 5.2 than there is in the equivalent column of Table 5.1. Please refer to Fig. 5.1 for a two-dimensional projection on (001) of the clinopyroxene structure. The smallest cation sites are those of tetrahedral coordination (T) with oxygen; these house Si^{4+} and Al^{3+}. The other two cation sites are marked as M_1 and M_2. The M_1 is the smaller of the two, with the cations in this site housed in regular octahedral coordination with respect to the surrounding oxygens. The coordination of the M_2 site is more variable, ranging from 6-fold (octahedral) to 8-fold (toward cubic configuration, but generally somewhat irregular in geometry) depending on the overall size of the cations occupying this M_2 site. In clinopyroxenes the M_2 site is definitely 8-fold in coordination (as such *not* regular and *not* octahedral) with the larger cations of the structure (Ca^{2+}, Na^+, and K^+) occupying this site. If, on the other hand, as in orthopyroxenes (orthorhombic pyroxenes) smaller cations such as Fe^{2+} and Mg^{2+} are housed in the M_2 site, the coordination polyhedron of M_2 becomes more regularly octahedral. As a result of the size restrictions on the M_1 and M_2 structural sites in clinopyroxenes, the following generalization can be made.

M_1 normally houses: all Al^{3+} that is "left over" after the Si (tetrahedral) position has been filled to a total of 2.00 (see column 7, Table 5.2); all Fe^{3+}; all Fe^{2+}; all Mn^{2+}; and all Mg^{2+} (these ions represent a size range from 0.39 Å for Al^{3+} to 0.83 Å for Mn^{2+}; see table of ionic radii, Table 3.8, *Manual of Mineral Science*, 23rd ed., p. 49.

The total of these "intermediate"-sized cations is 1.037, very close to 1.0 as the structural formula requires (with M_1 completely filled). As noted earlier, the larger cations (those with the Ca^{2+} size of 1.12 Å or larger) are housed in the M_2 site. Indeed, in column 7 of Table 5.2 their total is 0.934, which is close to the ideal of 1.0. If one were interested in making the ionic allocations for M_1 and M_2 in column 7 even closer to the ideal 1 for both, one might allocate all the Mn (0.027 cations) to the M_2 site, it being the largest of the intermediate cations that can commonly be found somewhat preferentially concentrated in the M_2 site. With Mn so reallocated, the final site occupancy totals become $M_1 = 1.037 - 0.027 = 1.010$ and $M_2 = 0.934 + 0.027 = 0.961$; when both these numbers are rounded off, they are very close to 1.0 each. Deviations from "ideality" are generally (but not always) the result of analytical error; if they are not due to analytical error, they may represent cation omissions (if totals are low) or interstitial additions (if totals are high).

The final structural formula for the clinopyroxene known as hedenbergite is given at the bottom of Table 5.2.

Quite independent of the complete formula recalculation schemes that we have presented, a question commonly asked is "What are the relative molecular percentages of the major cations in a mineral?" In a mineral such as a pyroxene, such a question commonly relates to the components FeO, MgO, and CaO (in molecular percentages) or the equivalent values for Fe^{2+}, Ca, and Mg (in atomic percentages). Now that we have recalculated the formula, we can obtain the Fe^{2+} cation contribution of 0.767 (notice that we distinguish between Fe^{2+} and Fe^{3+} contributions); the Mg cation contribution is 0.131; the Ca cation contribution is 0.899. Together these three cations total 1.797. The relative contributions of these three cations, expressed in percentages, can be obtained by dividing each value by the total and multiplying by 100%. The results of such calculations are shown in column 8 of Table 5.2.

Identical results for the relative percentages of the molecular contributions of FeO, MgO, and CaO can be obtained very quickly without completing the whole formula recalculation. We can take from column 4 (Table 5.2) the values for the molecular proportions of FeO (=0.310), for MgO (0.053), and for CaO (=0.363). If we total these three values, we can obtain the relative percentage contributions of these three oxide "molecules" by a calculation analogous to that given in Column 8. This is shown in column 9 of Table 5.2. Indeed, the values obtained are identical to those in column 8; minor variations in the decimal places are due to rounding in the various steps toward column 7. This illustrates that relative molecular percentages can be obtained

TABLE 5.2   Example of the Recalculation of a Clinopyroxene (Hedenbergite) Analysis.

1 Oxide Components	2 Weight Percentages	3 Molecular Weights	4 Molecular Proportions	5 Cation Proportions	6 Number of Oxygens	7 Number of Cations per Six Oxygens	8 Atomic Proportions of Fe^{2+}, Mg and Ca	9 From Column 4
SiO_2	48.00	60.08	0.799	0.799	1.598	1.978		
Al_2O_3	0.63	101.96	0.006	0.012	0.018	0.030		
Fe_2O_3	3.32	159.70	0.021	0.042	0.063	0.104		
FeO	22.85	71.85	0.310	0.310	0.310	0.767		
MnO	0.81	70.94	0.011	0.011	0.011	0.027		
MgO	2.12	40.30	0.053	0.053	0.053	0.131		
CaO	20.35	56.08	0.363	0.363	0.363	0.899		
Na_2O	0.34	61.98	0.005	0.010	0.005	0.025		
K_2O	0.18	94.20	0.002	0.004	0.002	0.010		

Total 98.00[a]

Column 7 bracketing: $\left.{1.978 \atop 0.022}\right\}2.00$; $0.030 \rightarrow {0.022 \atop 0.008}$; $\left.{0.104 \atop 0.767} {0.027 \atop 0.131}\right\}1.037$; $\left.{0.025 \atop 0.010}\right\}0.934$

Column 6: Total oxygens = 2.423

Oxygen factor: $\dfrac{6}{2.423} = 2.476$

Column 8 — Atomic Proportions of Fe^{2+}, Mg and Ca

$Fe^{2+} = 0.767$
Mg $= 0.131$
Ca $= 0.899$
Total $= 1.797$

$Fe^{2+} = \dfrac{0.767}{1.797} \times 100\% = 42.68\%$

Mg $= \dfrac{0.131}{1.797} \times 100\% = 7.29\%$

Ca $= \dfrac{0.899}{1.797} \times 100\% = 50.02\%$

Total 99.99%

Column 9 — From Column 4

Molecular proportions of:

FeO $= 0.310$
MgO $= 0.053$
CaO $= 0.363$
Total $= 0.726$

Relative percentage contributions:

FeO $= \dfrac{0.310}{0.726} \times 100\% = 42.69\%$

MgO $= \dfrac{0.053}{0.726} \times 100\% = 7.30\%$

CaO $= \dfrac{0.363}{0.726} \times 100\% = 50.00\%$

Structural formula: $\underbrace{(Ca,Na,K)_{0.9}}_{M_2}\ \underbrace{(Al,Fe^{3+},Fe^{2+},Mn,Mg)_{1.0}}_{M_1}\ \underbrace{(Si,Al)_{2.0}O_6}_{T}$

[a] Analysis contains 1.72 H_2O, with overall total of 99.72.

with very few conversions; indeed, a complete mineral formula recalculation is unnecessary. Because we already had obtained the pyroxene formula, however, we had access to the numbers in column 7.

MATERIALS

Chemical analyses given for two different pyroxenes, in Tables 5.3 and 5.4, respectively. An electronic calculator is essential, and you must consult a tabulation of atomic weights (e.g., Table 3.2 in *Manual of Mineral Science*, 23rd ed.) so that you can compute the appropriate molecular weights.

ASSIGNMENT

Given the chemical analyses for augite (a monoclinic pyroxene) and hypersthene (a member of the orthorhombic pyroxenes, or orthopyroxenes) in Tables 5.3 and 5.4, respectively, complete the various formula recalculation steps. The column headings are the same as those in the example calculation in Table 5.2.

1. In Table 5.3 the weight percentage analysis shows some TiO_2. In the assignment of cations to structural sites (column 7, and the structural formula), this is normally grouped with Fe^{2+}, Mg, Mn, and Fe^{3+} in the M_1 site. The recalculation scheme must be based on six oxygens in the augite formula. The oxygen factor therefore is 6 divided by the total for the number of oxygens obtained in column 6. The number of cations in column 7 provides you with the data to write a proper augite formula with cations assigned to the T, M_1, and M_2 sites. Write this structural formula below the columns, at the bottom of the page in Table 5.3, and identify which cations you group together in which site. In column 8 you are asked for the atomic proportions of Fe^{2+}, Mg, and Ca. This you obtain from the numbers in column 7. You are asked to do a similar calculation for the molecular proportions of FeO, MgO, and CaO (in column 9) as determined from the molecular proportions in column 4.

2. The analysis in Table 5.4 also shows a small amount of TiO_2. In the assignment of cations allocate this to the M_1 site. The recalculation scheme in this table must be based on three oxygens because the formulas of members of the orthopyroxene series are most generally reported in the form of $MgSiO_3$ to $FeSiO_3$. All the various recalculation steps are analogous to those you performed in Table 5.3 or were shown in the example of Table 5.2. When you reach column 7, be careful in your assignment of cations to the M_1 and M_2 sites. In an orthopyroxene, the M_2 is essentially octahedral and as such is smaller than the M_2 site in a clinopyroxene. The small amounts of Ca, Na, K, and Mn and a large proportion of Fe^{2+} can be assigned to M_2 in orthopyroxene because the M_2 site is still somewhat larger than the M_1 site. Record your final structural formula at the bottom of the table and clearly mark which cations you have grouped in which structural sites. Columns 8 and 9 request that you obtain atomic percentages of Fe, Mg, and Ca, and molecular percentages of FeO, MgO, and CaO, respectively.

Student Name _____

TABLE 5.3 Assignment on the Recalculation of the Chemical Analysis of an Augite.

1 Oxide Components	2 Weight Percentages	3 Molecular Weights	4 Molecular Proportions	5 Cation Proportions	6 Number of Oxygens	7 Number of Cations per Six Oxygens	8 Atomic Proportions of Fe^{2+}, Mg, and Ca	9 Molecular Proportions of FeO, MgO, CaO
SiO_2	48.11							
TiO_2	1.14							
Al_2O_3	7.26							
Fe_2O_3	3.13							
FeO	4.86							
MnO	0.11							
MgO	14.04							
CaO	20.46							
Na_2O	0.66							
K_2O	0.04							
Total	99.81							

Structural formula _____

(Indicate what cations are assigned to which sites in the structure.)

TABLE 5.4 Assignment on the Recalculation of the Chemical Analysis of a member of the orthopyroxene series.

1 Oxide Components	2 Weight Percentages	3 Molecular Weights	4 Molecular Proportions	5 Cation Proportions	6 Number of Oxygens	7 Number of Cations per Three Oxygens	8 Atomic Proportions of Fe^{2+}, Mg, and Ca	9 Molecular Proportions of FeO, MgO, CaO
SiO_2	50.26							
TiO_2	0.16							
Al_2O_3	3.13							
Fe_2O_3	0.65							
FeO	26.54							
MnO	0.76							
MgO	16.36							
CaO	1.76							
Na_2O	0.24							
K_2O	0.13							
Total	99.99							

Structural formula _____

(Indicate what cations are assigned to which sites in the structure.)

Graphical Representation of Mineral Analyses (In Terms of Molecular Percentages of Selected Components) on Triangular Diagrams

PURPOSE OF EXERCISE

The plotting of mineral analyses in terms of some selected (generally major) components on triangular plots. Such plots illustrate the extent of chemical substitution of the selected components in mineral groups that show solid solution.

FURTHER READING AND CD-ROM INSTRUCTIONS

Klein, C. and Dutrow, B., 2008, *Manual of Mineral Science*, 23rd ed., Wiley, Hoboken, New Jersey, pp. 105–108.

CD-ROM, entitled *Mineralogy Tutorials*, version 3.0, that accompanies *Manual of Mineral Science*, 23rd ed., click on "Module I", and subsequently on the button "Graphical Representation".

Nesse, W. D., 2000, *Introduction to Mineralogy*, Oxford University Press, New York, pp. 72–73.

Perkins, D., 2002, *Mineralogy*, 2nd ed., Prentice Hall, Upper Saddle River, New Jersey, pp. 97, 124, 165–166.

Wenk, H. R. and Bulakh, A., 2004, *Minerals: Their Constitution and Origin*, Cambridge University Press, New York, New York, pp. 255–258.

Bloss, F. D., 1994, 2nd printing with minor revisions, *Crystallography and Crystal Chemistry*, Mineralogical Society of America, Chantilly, Virginia, pp. 294–297.

Background Information: As indicated in exercise 5, common minerals such as pyroxenes have mineral analyses that can be quite complex and that show substitution of several elements in a specific site of the structure. In order to illustrate the independent variation of at least three components graphically, mineralogists and petrologists commonly use a *triangular diagram*; such a diagram, with rulings at every 2% division, is shown in Fig. 6.1. This diagram allows us to plot three independent components at the three corners. For the selection of three appropriate components, it is useful to look over Tables 5.1 and 5.2, the chemical analyses of a feldspar and of a monoclinic pyroxene, respectively. The widest variations of components in feldspar are shown by CaO, Na_2O, and K_2O (in molecular percentages) or in the equivalent Ca, Na, and K components (in atomic percentages). In the pyroxene group major chemical variations are shown by CaO, FeO, and MgO (in molecular percentages) or in the equivalent Ca, Fe, and Mg components (in atomic percentages). In general, we may choose any three components in a chemical analysis that serve some specific purpose in the graphical illustration of chemical variation. In the pyroxenes we might have chosen FeO, MgO, and Al_2O_3, or CaO, FeO, and MnO, instead of the three components we did choose. We can select any of the three corners of the triangular diagram for a specific component; however, for the feldspars the top corner is commonly assigned to K_2O, the lower left corner to Na_2O, and the lower right to CaO. For the pyroxenes the upper corner is generally assigned to CaO, the lower left to MgO, and the lower right to FeO.

Once the corners of the diagram have been assigned their chemical components, in atomic or molecular percentages, the question is how to plot chemical analyses in terms of the chosen components. Figure 6.2a shows a triangular diagram with CaO (molecular percentage = 100%) at the top, MgO (molecular percentage = 100%) at the left corner, and FeO (molecular percentage = 100%) at the lower right corner. The component assignment means that a mineral that calculates to have 100 molecular percent CaO in its chemical composition (this would occur in pure calcite with the composition $CaCO_3$, or in pure wollastonite with composition $CaSiO_3$) would plot at the top corner.

In the same way pure enstatite, $MgSiO_3$, would plot at the 100% MgO corner. A member of the orthopyroxene series, which is mainly a continuous solid-solution series between the end-members of $MgSiO_3$ and $FeSiO_3$, would plot somewhere along the lower edge (between MgO and FeO), if indeed its composition is completely devoid of CaO. Let us assume that such an orthopyroxene has a chemical analysis with the following molecular percentages: MgO = 60 weight percent and FeO = 40 weight percent. (See exercise 5 for the steps involved in calculating molecular or atomic percentages for specific elements in a mineral analysis.) Because the lower left-hand corner of Fig. 6.2a is 100% MgO, an analysis with 60% MgO will plot at the 60% mark, 10% *to the left* of the center of the edge between MgO and FeO. We could have used the information of FeO = 40 molecular percent for this same analysis instead. If the analysis contains 40 molecular percent FeO, it must plot farther away from the right-hand corner of the diagram (which is defined as 100% FeO) than from the center point along the MgO–FeO edge. Indeed, it plots 10% to the left of the middle point. This is exactly where we plotted the point on the basis of its MgO content. As such, using *one* molecular percentage value in a two-oxide compound system (e.g., MgO and FeO) suffices; the second, complementary value is redundant in terms of plotting the point.

Figure 6.2*b* shows a composition in which CaO, MgO, and FeO are all major variables. This composition is CaO = 50%, MgO = 20%, FeO = 30% (all in molecular percentages). The question now is how we plot a composition on a triangular diagram that involves all three of the end-member compositions. Notice that the triangular diagrams in Figs. 6.2*a* and *b* have 100% notations at the three corners; they also show percentage graduations, identified by numbers such as 90, 80, and 70, along lines that are parallel to the edges of the triangle. As shown in Fig. 6.2*a*, stacks of such lines (marked with percentages) define directions (perpendicular to the edges of the triangle and away from the corners) in terms of the continually decreasing percentage of the corner component. This means that an analysis with 50 molecular percent of the CaO component plots on the horizontal line marked 50% (this line is halfway between the 100% CaO corner and the 0% CaO along the horizontal edge of the triangle). We now know it plots along this 50% CaO line (see Fig. 6.2*b*), but where along this line will be defined only after we also plot the MgO (or FeO) value of the analysis. The MgO content is 20% (molecular percentage), which will plot along the 20% MgO line. Such a line is one of those that are parallel to the inclined right-hand edge of the triangle, at a distance 80% away from the MgO corner. Where this line intersects the 50% CaO line is where the three-component composition plots. Instead we could have used the value for FeO = 30 molecular percentage. This FeO = 30% line intersects the CaO = 50% line at exactly the same spot as we just plotted the analysis point. Thus, in a three-component system the third component value is not necessary for the location of the graphical position, but it is very useful as a check on the accuracy of the plotted point.

Now that we have addressed the mechanics of the plotting of data on triangular diagrams, let us look at what type of data are commonly plotted. Chemical analyses of minerals (as shown in exercise 5) are generally given in weight percentage values. We could plot the weight percentage values of three components (after normalization to 100%) on a triangular diagram. However, it is a much more common and more meaningful procedure in mineralogy and petrology to plot molecular (or the equivalent atomic) percentages that can be derived from such a weight percent analysis. In accordance with this custom you will be asked to plot molecular percentages of three oxide components, CaO, MgO, and FeO, for a set of pyroxene analyses. (See the assignment in Table 6.1.)

The procedure for obtaining molecular percentages from a mineral analysis given originally in weight percentages was outlined in exercise 5. Please refer back to column 9 in Table 5.2 and the text that describes this column. First, the weight percentage values of the components selected for a graphical display must be converted to molecular proportions (by dividing the weight percentage values by the appropriate weights; see columns 1, 2, 3, and 4 in Table 5.2); second, the values obtained for the molecular proportions (for CaO, MgO, and FeO in this case) must be normalized to 100% (see column 9 in Table 5.2). These normalized values can now be plotted directly to obtain a graphical representation of molecular percentages of, for example, CaO, MgO, and FeO. Other components could have been selected from the analyses, but in the pyroxene group a CaO–MgO–FeO diagram is a very standard representation.

MATERIALS

Table 6.1, which lists 13 partial analyses for members of the pyroxene group of silicates, is the basis of this assignment. An electronic calculator is essential, and to be able to compute the appropriate molecular weights, you must consult a tabulation of atomic weights (e.g., Table 3.2 in *Manual of Mineral Science*, 23rd ed.). The triangular diagram in Fig. 6.1 is needed for the graphical representation.

ASSIGNMENT

Table 6.1 lists partial chemical analyses (in weight percentages) for 13 pyroxenes. The largest variations in chemical components in these analyses are shown by FeO, MgO, CaO, and to a lesser degree Al_2O_3. In a triangular diagram three independent chemical components can be plotted. For this exercise the chosen components are CaO, MgO, and FeO. You could plot these three components *as relative weight percentages*. That is, after having totaled the weight percentages of CaO + MgO + FeO in each analysis and after having normalized this total to 100%, you would have the new normalized values for CaO, MgO and FeO, in relative weight percentages. But as pointed out earlier, it is standard procedure to plot such diagrams *in molecular percentages*.

1. For each analysis select the weight percentage values for FeO, MgO, and CaO and convert these (by dividing with the appropriate molecular weights) to molecular proportions. See Table 5.2, columns 1, 2, 3, and 4 and the text accompanying these conversions. Once you know the values for FeO, MgO, and CaO in terms of molecular proportions, you sum these values, so you can obtain the ratio

$$\frac{\text{molecular proportion}}{\text{total of molecular proportions}} \times 100\%$$

EXERCISE 6

Student Name

FIGURE 6.1 Triangular diagram with percentage lines at 2% intervals.

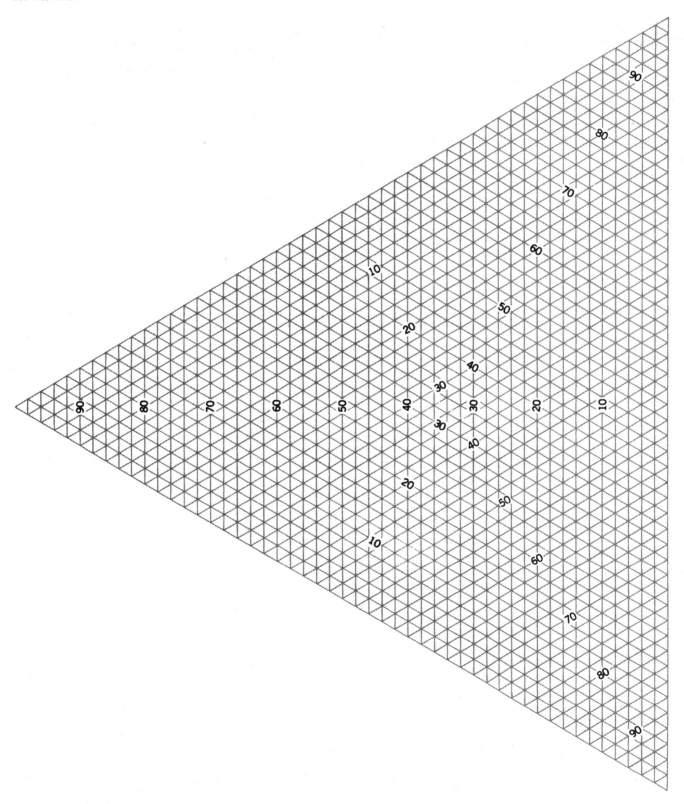

Please make the necessary photocopies of this diagram.

FIGURE 6.2   Illustration of the mechanics of plotting two different pyroxene compositions on a triangular diagram.

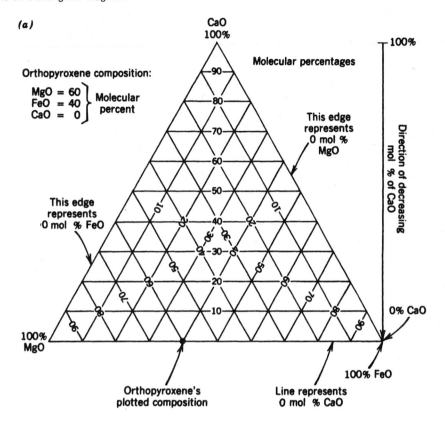

(a)

Orthopyroxene composition:

MgO = 60 ⎫ Molecular
FeO = 40 ⎬ percent
CaO = 0 ⎭

CaO 100%

Molecular percentages

This edge represents 0 mol % MgO

This edge represents 0 mol % FeO

Direction of decreasing mol % of CaO

100%

0% CaO

100% MgO

Orthopyroxene's plotted composition

100% FeO

Line represents 0 mol % CaO

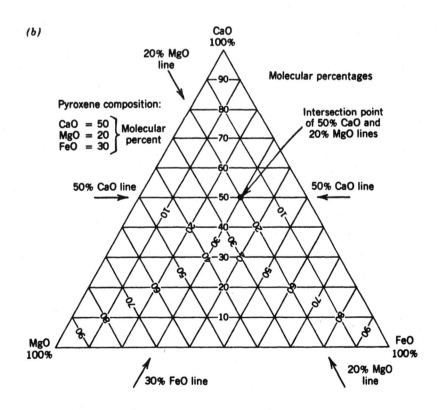

(b)

CaO 100%

20% MgO line

Pyroxene composition:

CaO = 50 ⎫ Molecular
MgO = 20 ⎬ percent
FeO = 30 ⎭

Molecular percentages

Intersection point of 50% CaO and 20% MgO lines

50% CaO line

50% CaO line

MgO 100%

FeO 100%

30% FeO line

20% MgO line

Student Name _____

TABLE 6.1 Partial Analyses of Pyroxenes to be Converted to Molecular Percentages of CaO, MgO, and FeO Before Plotting Them on a Triangular Diagram.

	1	2	3	4	5	6	7	8	9	10	11	12	13
Weight Percentage Values													
SiO_2	59.92	55.94	54.11	52.07	50.26	49.34	54.61	49.81	52.67	46.93	50.02	50.18	44.76
Al_2O_3	0	1.61	1.52	1.70	3.13	0.38	1.87	6.42	1.42	6.61	1.98	0.0	1.70
FeO	0.38	7.15	15.73	22.65	26.54	34.91	0.0	2.53	4.34	6.64	9.72	17.33	19.27
MnO	0	0.19	0.34	0.48	0.76	1.19	0.0	0.06	0.13	0.10	0.83	0.28	6.22
MgO	39.51	32.12	27.03	21.13	16.36	12.96	18.42	16.22	15.59	12.32	10.84	6.87	5.25
CaO	0.32	1.48	1.16	1.55	1.76	0.71	23.14	23.92	23.81	22.35	22.37	23.28	22.20
Molecular Percentage Values													
CaO													
MgO													
FeO													
Total													

and thereby a relative percentage value (in molecular percentage). See column 9 in Table 5.2 for an example of this type of conversion.

2. Normalized molecular percentages for the CaO, MgO, and FeO values should be recorded in the space provided (below the weight percentage columns) in Table 6.1. Clearly the sum of these sets of three values must total to 100%.

3. Plot each of the analyses on the triangular diagram given in Fig. 6.1 or on a photocopy thereof. Make sure that the CaO component is at the top of the triangle, the MgO component at the lower left, and the FeO component at the lower right. Label each analysis point with the analysis number as given in Table 6.1.

4. In order to make your diagram identical to the standard pyroxene composition diagram expressed in molecular components of $CaSiO_3$, $MgSiO_3$, and $FeSiO_3$, write each of these three components next to the appropriate corner of the diagram. Keep in mind that you could have plotted this diagram in terms of percentages of atomic components of Ca, Mg, and Fe (see, for example, the comparative calculations in columns 8 and 9 of Table 5.2). You would have obtained an identical plot had you calculated the molecular percentages of the compound components $CaSiO_3$ ($= CaO \cdot SiO_2$), $MgSiO_3$ ($= MgO \cdot SiO_2$), and $FeSiO_3$ ($= FeO \cdot SiO_2$). The notation in parentheses shows that SiO_2 is a constant common factor to the three silicate components; such plots in terms of atomic Ca, Mg, and Fe, or molecular percentages of CaO, MgO, and FeO, or $CaSiO_3$, $MgSiO_3$, or $FeSiO_3$ are all equivalent and indeed identical.

5. To complete the diagram, plot the compositions of diopside $= CaMgSi_2O_6$ and hedenbergite $= CaFeSi_2O_6$ along the two inclined sides of the triangle. It will help you to plot these compositions if you keep in mind that $CaMgSi_2O_6 = CaSiO_3 + MgSiO_3$. In other words, $CaMgSi_2O_6$ can be expressed in terms of equal amounts of $CaSiO_3$ and $MgSiO_3$; this means that a 50:50% mix of these two components gives you the proper graphical position.

6. Compare your final and clearly labeled diagram with a standard compositional diagram for the pyroxenes (e.g., Fig. 18.15 in *Manual of Mineral Science*, 23rd ed.).

EXERCISE 7

Graphical Representation of Mineral Formulas on Triangular Diagrams

PURPOSE OF EXERCISE

Learning the steps involved in converting a mineral formula into a graphical data point on a triangular diagram.

FURTHER READING AND CD-ROM INSTRUCTIONS

Klein, C. and Dutrow, B., 2008, *Manual of Mineral Science,* **23rd ed., Wiley, Hoboken, New Jersey, pp. 105–108.**

CD-ROM, entitled *Mineralogy Tutorials,* **version 3.0, that accompanies** *Manual of Mineral Science,* **23rd ed., click on "Module I", and subsequently on the button "Graphical Representation".**

Nesse, W. D., 2000, *Introduction to Mineralogy,* Oxford University Press, New York, pp. 72–73.

Perkins, D., 2002, *Mineralogy,* 2nd ed., Prentice Hall, Upper Saddle River, New Jersey, pp. 97, 124, 165–166.

Wenk, H. R. and Bulakh, A., 2004, *Minerals: Their Constitution and Origin,* Cambridge University Press, New York, New York, pp. 256–258.

Background Information: If you have successfully completed the previous two exercises (5 and 6), you will have to learn very little that is new. Exercise 5 should have given you a basic understanding of reasonably complex mineral formulas (as in Table 5.2, for a pyroxene), and exercise 6 gave you the overall approach to graphical representation of chemical data on a triangular diagram. Here, the approach is somewhat different from what you have already learned in that we start with the *formulas* of some common minerals that we wish to plot in a three-component triangular diagram.

As a lead into the exercise, let us look at a relatively simple example of the approach. Figure 7.1 is a triangular diagram in which the three corners represent the three chemical components: S (at the top), Fe (at the lower left corner), and Ni (at the lower right corner). Because the diagram's chemical components are all simple atoms (not compounds such as, e.g., FeO), and because we do *not* wish to plot the weight percentages (see exercise 6), the units of measurement will be *atomic percentages.* On this diagram are plotted the chemical locations for the formulas of eight sulfide minerals whose composition can be described in terms of the three components, Fe, Ni, and S. These are, with name abbreviations in parentheses,

Pyrite (py) – FeS_2

Troilite (tr) – FeS

Pyrrhotite (po) – $Fe_{0.9}S$

Heazlewoodite (hz) – Ni_3S_2

Millerite (ml) – NiS

Polydymite (pdy) – Ni_3S_4

Violarite (vl) – $FeNi_2S_4$

Pentlandite (pn) – $(Fe,Ni)_9S_8$

An overview of these formulas will tell you immediately that (1) the first three sulfides (involving only Fe^{2+} and S) will plot somewhere along the inclined left edge of the triangle, (2) the next three formulas (involving only Ni and S) will plot somewhere along the inclined right edge of the triangle, and (3) the last two formulas of violarite and pentlandite, because their formulas show various ratios of all three components of the diagram, must plot somewhere in the general field of the diagram (away from the edges). The only question that remains is where exactly do any of these formulas plot. Having mineral formulas, with exact subscripts (except for the notation for the mineral pentlandite), makes the answer easy. Let us take the first three minerals in the chemical system Fe–S:

Pyrite has a total of three atoms (1 Fe and 2 S), and Fe constitutes only one of these three; in other words, Fe is 33.3% (on an atomic percentage basis) of the analysis in terms of Fe and S.

Troilite shows a total of two atoms (1 Fe and 1 S), and as such one Fe in troilite constitutes 50% of the total in terms of Fe and S.

Pyrrhotite, with the analysis $Fe_{0.9}S_1$, is deficient by 10% in Fe with respect to the troilite analysis; this means that it is, in a relative sense, more sulfur-rich than troilite, and must therefore plot closer to the S corner of the diagram than troilite. The exact position is a function of the total number of atoms (totaling 1.9) of which 0.9 is Fe and 1 is S. The Fe percentage contribution is $0.9/1.9 \times 100\% = 47.4\%$.

The foregoing values for the atomic percentages of Fe (33.3% Fe in py; 50% Fe in tr; and 47.4% Fe in po) locate the three analyses on the left inclined edge of the diagram in Fig. 7.1. We could have done these atomic percentage conversions for sulfur, instead of iron, and we would have obtained the complementary values, which give the same plot positions as obtained for Fe.

FIGURE 7.1 Graphical representation of sulfide formulas in the system Fe-Ni-S (see text for discussion) and of some coexisting minerals (pyrite, pyrrhotite, and pentlandite).

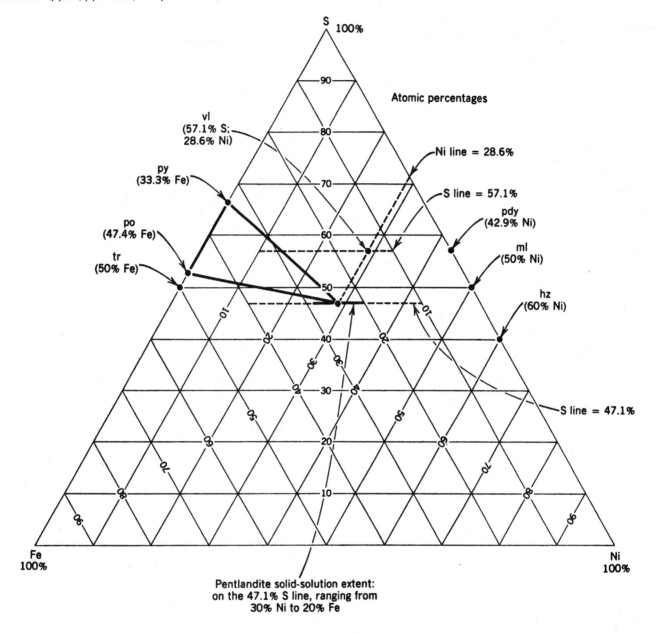

Pentlandite solid-solution extent:
on the 47.1% S line, ranging from
30% Ni to 20% Fe

We can now do the same simple calculations for the next three minerals in the system Ni–S. Notice that *millerite* with the formula NiS must plot on the Ni–S edge, at the same location as FeS on the Fe–S edge (indeed, NiS has 50% Ni and 50% S, halfway along the edge). The other two Ni–S minerals recalculate as follows.

Heazlewoodite, with the formula Ni_3S_2, has 3 Ni atoms out of a total of 5 (Ni + S) atoms. On an atomic percentage basis this recalculates as $3/5 \times 100\% = 60\%$ Ni.

Polydymite, with the formula Ni_3S_4, has 3 Ni atoms in a total of 7 (Ni + S), resulting in a Ni percentage of $3/7 \times 100\% = 42.9\%$ Ni.

These three Ni–S minerals have been plotted in Fig. 7.1 on the basis of the foregoing computed values for the atomic percentages of Ni (50% Ni in ml; 60% Ni in hz; and 42.9% Ni in pdy).

The last two minerals involve all three components, Fe, Ni, and S.

Violarite, with the formula $FeNi_2S_4$, has a total number of atoms of $1 + 2 + 4 = 7$. Fe (in atomic %) represents $1/7 \times 100\% = 14.3\%$; Ni (in atomic %) represents $2/7 \times 100\% = 28.6\%$; and S (in atomic %) represents $4/7 \times 100\% = 57.1\%$.

Pentlandite, with the formula $(Fe, Ni)_9S_8$, gives the total number of atoms for Fe + Ni = 9, and sulfur as 8, giving an overall total of 17. The formula statement implies that there is a chemical substitution (that is, a solid solution) between Fe and Ni in the structure. This solid solution will be represented as a line on the diagram, and the location of this line will be at a sulfur (atomic %) value of $8/17 \times 100\% = 47.1\%$. At temperatures of about 400°C, the extent of Fe–Ni solid solution in pentlandite ranges from about 30 atomic percent Ni to about 20 atomic percent Fe.

The foregoing data allow us to plot the composition of violarite at 14.3 atomic percent Fe, 28.6 atomic percent Ni, and with the complementary value of S at 57.1 atomic percent (a good check on the location of the point). The pentlandite solid solution plots horizontally along the sulfur line of 47.1%. It was noted earlier that the extent of solid solution is limited at both the Fe- and Ni-rich ends, and these limits of 20% Fe and 30% Ni are shown.

This completes the plotting of chemical formulas on a triangular diagram in a reasonably simple sulfide system. Mineralogists and petrologists commonly use such a diagram not only to show the distribution of mineral compositions in a chemical system, but also to record the *coexistence* of pairs (or triplets) of minerals in a hand specimen or thin section. As such, triangular diagrams are very useful in the graphical recording of *coexistences* or *assemblages* of minerals. For example, pyrrhotite and pentlandite commonly occur together in Ni-rich sulfide ores, and pyrite may be present as well. One can express such an association in terms of pairs, as follows:

py coexists with pn

py coexists with po (troilite is found only in meteorites not in terrestrial rocks)

po coexists with pn

These coexistences are shown graphically by *tielines*, lines that connect coexisting minerals. Three such tielines are shown in Fig. 7.1.

The assignment in this exercise will consist of the plotting of mineral formulas in a sulfide system (Cu–Fe–S) and showing tielines between minerals in this system on the basis of reported pairs. An additional part will consist of the plotting of some common mineral compositions in the system SiO_2–MgO–Al_2O_3. The molecular proportions are obtained from the mineral formulas in the same way that atomic proportions are calculated from formulas (see assignment).

MATERIALS

Table 7.1 with a listing a sulfide formulas (for a plot of their compositions in the system Cu–Fe–S) and Table 7.2 with a listing of silicate and oxide formulas (for a plot of their compositions in the system SiO_2–MgO–Al_2O_3) form the basis of this exercise. An electronic calculator is helpful for percentage calculations. Photocopies of the triangular diagram in Fig. 7.2 are needed.

TABLE 7.1 Mineral Formulas in the System Cu–Fe–S.

Pyrite (py)–FeS_2

Pyrrhotite (po)–$Fe_{0.9}S$

Covellite (cv)–CuS

Chalcocite (cc)–Cu_2S

Chalcopyrite (cp)–$CuFeS_2$

Bornite (bn)–Cu_5FeS_4

Coexistences in the above system:

cc-po; cp-po; py-cv; cp-bn; py-cp; py-bn; cc-bn; cv-cc; py-po and cv-bn

ASSIGNMENT

In this assignment you are asked to complete two different triangular plots on the basis of given mineral formulas.

1. The first graphical plot is based on the mineral formulas in Table 7.1. Before plotting these formulas (as outlined in the previous text), mark the top of the diagram as 100% S, the left lower corner as 100% Cu, and the right lower corner as 100% Fe. After all the compositions have been located, plot the coexistences as given in Table 7.1, by connecting pairs of minerals with tielines.

2. The second part of this assignment asks you to plot the mineral formulas in Table 7.2 on a triangular diagram in which SiO_2 is the top corner, MgO the lower left corner, and Al_2O_3 the lower right corner. In the listing in Table 7.2 there are several hydrous mineral formulas (those with OH groups), which can be plotted on the SiO_2–MgO–Al_2O_3 diagram by ignoring their OH component. As in the previous assignment, you must determine the number of Si, Mg, and Al that are recorded in each formula. However, because you are asked to represent the formulas in terms of SiO_2,

TABLE 7.2 Mineral Formulas in the System SiO_2–MgO–Al_2O_3–(H_2O).

Quartz(qz)–SiO_2

Corundum(cor)–Al_2O_3

Periclase(per)–MgO

Spinel(sp)–$MgAl_2O_4$

Kyanite(ky), sillimanite(sill), andalusite(and)–Al_2SiO_5

Forsterite(fo)–Mg_2SiO_4

Enstatite(en)–$MgSiO_3$

Anthophyllite(anth)–$Mg_7Si_8O_{22}(OH)_2$

Pyrope(pyr)–$Mg_3Al_2Si_3O_{12}$

Talc(tc)–$Mg_3Si_4O_{10}(OH)_2$

Serpentine(ser)–$Mg_3Si_2O_5(OH)_4$

Pyrophyllite(pyph)–$Al_2Si_4O_{10}(OH)_2$

Kaolinite(kao)–$Al_2Si_2O_5(OH)_4$

EXERCISE 7

FIGURE 7.2 Triangular diagram with percentage lines at 2% intervals.

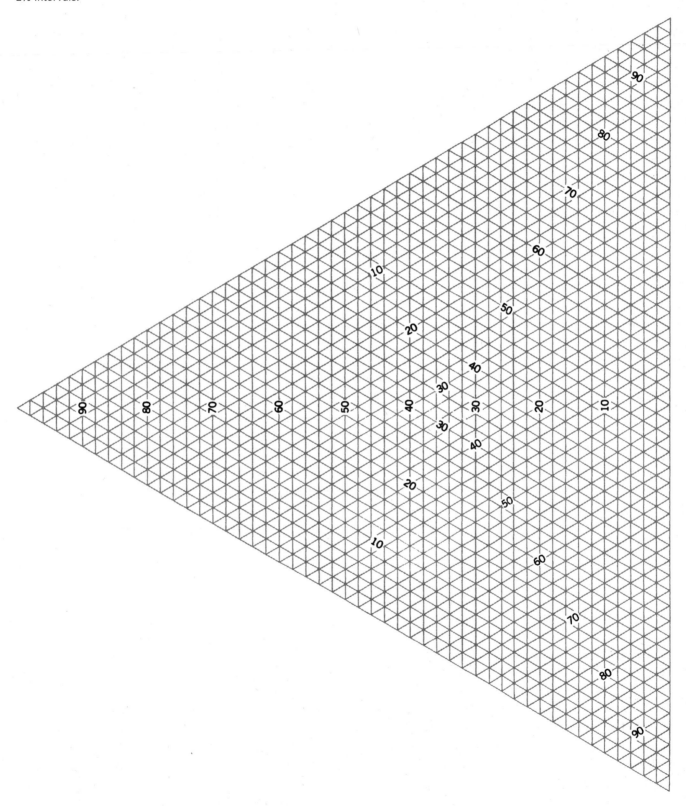

Please make the necessary photocopies of this diagram.

MgO, and Al_2O_3 (note: two atoms of Al per oxide compound), a mineral formula such as that of Al_2SiO_5 contributes one Al_2O_3 and one SiO_2; that is, it plots at 50:50 on the SiO_2–Al_2O inclined right-hand edge. In other words, a mineral formula with only one Al per formula contributes only one-half of one Al_2O_3 component.

In this part of the assignment you are asked to plot only the various compositions listed in Table 7.2. Make sure you label each plotted composition with the appropriate mineral abbreviation. An evaluation of possible mineral assemblages in this diagram will be done in exercise 34.

Symmetry Contents of Letters, Motifs, and Common Objects

PURPOSE OF EXERCISE

Recognition of symmetry elements, and their location, in common shapes, designs, and objects as a preamble to the systematic study of the symmetry content of crystals.

FURTHER READING AND CD-ROM INSTRUCTIONS

Klein, C. and Dutrow, B., 2008, *Manual of Mineral Science*, 23rd ed., Wiley, Hoboken, New Jersey, pp. 109–120.

CD-ROM, entitled *Mineralogy Tutorials*, version 3.0, that accompanies *Manual of Mineral Science*, 23rd ed., click on "Module II", and subsequently on the button "Symmetry Operations".

Nesse, W. D., 2000, *Introduction to Mineralogy*, Oxford University Press, New York, pp. 12–13.

Perkins, D., 2002, *Mineralogy*, 2nd ed., Prentice Hall, Upper Saddle River, New Jersey, pp. 175–178.

Wenk, H. R. and Bulakh, A., 2004, *Minerals: Their Constitution and Origin*, Cambridge University Press, New York, New York, pp. 34–37.

Hargittai I. and Hargittai, M., 1994, *Symmetry, a Unifying Concept*, Shelter Publications Inc., Bolinas, California, p. 222.

Stevens, P. S. 1991, *Handbook of Regular Patterns: An Introduction to Symmetry in Two Dimensions*, MIT Press, Cambridge, Massachusetts, pp. 1–45.

Background Information: Various symmetry elements, and their combinations, can be recognized in everyday life in the motifs of wallpaper patterns, of printed cloth patterns, of wall and floor tiles, in the shape and form of buildings, in letters of the alphabet, and in common objects such as cubes and pyramids. The recognition of the presence of symmetry elements in our surroundings is often enjoyable but also instructive.

In this exercise you will be asked to recognize the presence or absence of symmetry elements in letters of the alphabet, in design motifs, and in some common objects. The symmetry elements that you will be asked to identify are as follows:

a. *Mirror planes* (abbreviated as *m*), which reflect a motif into a mirror image thereof; the presence of a mirror perpendicular to a page is shown by a solid line, for example, ——, or |; and

b. *Rotation axes,* which are imaginary lines about which a motif may be rotated and repeated in its appearance one or more times during a complete (360°) rotation; the notations for rotational symmetry are as follows.

Type of Rotation Axis	Angle of Rotation	Symbol When Axis Is Perpendicular to Page
1-fold rotation = 1	360°	None
2-fold rotation = 2	180°	⬤
3-fold rotation = 3	120°	▲
4-fold rotation = 4	90°	■
6-fold rotation = 6	60°	⬡

A *motif* (or motif unit) is the smallest representative unit of a pattern, as, for example, an arrangement of several dots in printed wallpaper, a flower in printed cloth, and a design in ceramic tiles. In the external shape of crystals, the motif may be a specifically shaped crystal face, and in the internal structure of crystals, the motif may be an atom, ion, or cluster of atoms or ions. Indeed, *a crystal structure consists of motifs* (atoms, ions, or clusters of them) *that are repeated in three dimensions at regular* (= periodic) *intervals.* As such, in two- and three-dimensional patterns (arrays), the motif is translated.

In this exercise we will concern ourselves only with translation-free patterns, namely the symmetry elements inherent in letters, motifs, shapes, and common objects. This is a preamble to the study of symmetry elements as revealed by the external form of well-shaped crystals (exercise 9). You will discover that in three-dimensional objects that show symmetry (see Fig. 8.3) the various symmetry elements that may be present intersect each other at a point, at the center of the object. In exercise 9 you will note this again, namely, that well-formed crystals may show combinations of symmetry elements that intersect in a point, at the center of the crystal. The classification of the various symmetry contents of crystals is, therefore, based on *point groups*, which express the symmetry about a central point. There are 32 of these point groups (see exercise 9).

EXERCISE 8

Student Name

FIGURE 8.1 Letters of the alphabet (and some other symbols) and their symmetry content as a function of type style.

A	B	C	D	E	F	G
H	I	J	K	K	L	M
M	N	N	O	O	P	Q
R	S	T	U	V	W	X
Y	Y	Z	!	:	$	¢

EXERCISE 8

Student Name

FIGURE 8.2 The symmetry content of several patterns.

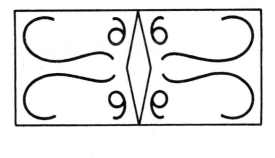

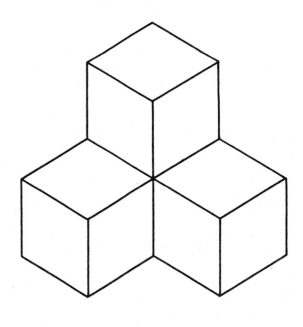

Student Name

FIGURE 8.3 The symmetry content of some common objects.

Unsharpened pencil

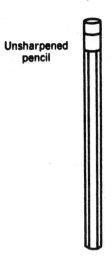

- - - - - - - - - - - -

Rubber eraser

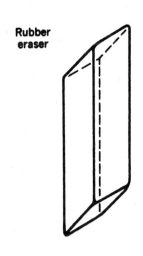

- - - - - - - - - - - -

Equilateral triangle

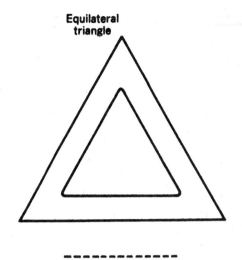

- - - - - - - - - - - -

Chair

- - - - - - - - - - - -

Card table

- - - - - - - - - - - -

45° angle triangle

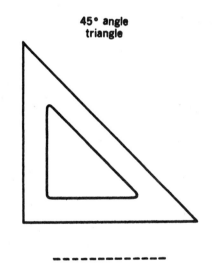

- - - - - - - - - - - -

Cube

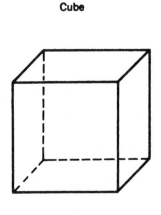

- - - - - - - - - - - -

Baseball (or tennis ball)

- - - - - - - - - - - -

Pyramid

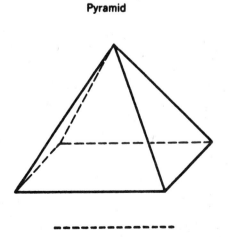

- - - - - - - - - - - -

MATERIALS

The drawings in Figs. 8.1 and 8.2, and the objects depicted in Fig. 8.3. Inspection and handling of the real objects is essential in the evaluation of the symmetry contents of the objects sketched in Fig. 8.3. A colored pencil and a ruler are needed for the location of symmetry elements in the various drawings.

ASSIGNMENT

To recognize and locate the inherent symmetry in the objects depicted in Figs. 8.1 through 8.3.

1. In Fig. 8.1, which depicts letters and several symbols, indicate (using a colored pencil and ruler) where a mirror, and/or a rotation axis is located. Use the proper symbols as noted earlier. Some letters may contain more than one symmetry element, and some may have none. Below each letter, above the dashed line, note the presence of, one or more mirrors by writing 1*m* or 2*m*s; the presence of a rotation axis by writing, e.g., one 2-fold; and the absence of all symmetry by "none." If more symmetry exists, write, e.g., 2*m*s, + one 2-fold, etc.

2. In Fig. 8.2 locate all symmetry elements present by drawing them in with a colored pencil aided by a ruler (do it neatly!). Use the proper symbols as noted earlier. Below each of the figures, note the presence of the number of each of the symmetry elements (e.g., 4*m*s, and/or one 6-fold, etc.) in the space above the dashed line.

3. Figure 8.3 gives sketches of common objects. To evaluate fully their symmetry contents, it is very helpful to study the real objects at home, or in the classroom. This is essential especially for the last three objects, cube, baseball, and pyramid. For each of the objects locate in Fig. 8.3 (with a colored pencil and ruler) the mirror planes and rotational axes that may be present. Below each object give the number of each of the symmetry element(s) and the types that are present (e.g., *m* for mirror, 2-fold for 2-fold rotation, etc.). Two of the objects (cube and baseball) have many symmetry elements, some of which (such as rotoinversion axes) we have not yet introduced. However, try them both; they are a real challenge with respect to their total symmetry content! Please note that a three-dimensional object containing several symmetry elements will have these elements intersect in a point at the center of the object. Because of this, the total symmetry content of a three-dimensional object is commonly referred to as its *point group* symmetry. This is an expression of the total symmetry content of an object about a central, stationary point (see exercise 9).

Symmetry Elements and Their Combinations as Expressed by External Crystal Form: 32 Point Groups

PURPOSE OF EXERCISE

Recognition of symmetry elements in crystals or idealized models of crystals. Such elements are rotation axes: 1, 2, 3, 4, and 6; rotoinversion axes: $\bar{1}$ (= center of symmetry, or i), $\bar{2}$, $\bar{3}$, $\bar{4}$, and $\bar{6}$; and m (mirrors). All these elements can occur singly; however, they can also occur in various combinations. When all possible single elements and combinations thereof are tallied, it turns out that there are 32 possibilities. These symmetry elements plus their combinations are known as the *32 crystal classes*, or *point groups*, and are expressed by the external form (i.e., the morphology) of crystals.

FURTHER READING AND CD-ROM INSTRUCTIONS

Klein, C. and Dutrow, B., 2008, *Manual of Mineral Science*, 23rd ed., Wiley, Hoboken, New Jersey, pp. 109–142.

CD-ROM, entitled *Mineralogy Tutorials*, version 3.0, that accompanies *Manual of Mineral Science*, 23rd ed., click on "Module II", and subsequently on the buttons "Symmetry Operations", and "Crystal Classes".

Nesse, W. D., 2000, *Introduction to Mineralogy*, Oxford University Press, New York, pp. 13–19.

Perkins, D., 2002, *Mineralogy*, 2nd ed., Prentice Hall, Upper Saddle River, New Jersey, pp. 175–195.

Wenk, H. R. and Bulakh, A., 2004, *Minerals: Their Constitution and Origin*, Cambridge University Press, New York, New York, pp. 73–79.

Bloss, F. D., 1994, 2nd printing with minor revisions, *Crystallography and Crystal Chemistry*, Mineralogical Society of America, Chantilly, Virginia, pp. 1–25.

Background Information: In the external form of well-shaped crystals (or in the wooden or plaster or paper models thereof) we can recognize the presence (or absence) of various symmetry elements. Notice that these symmetry elements (when in combination) intersect at one point, namely the center of the crystal, or crystal model. For this reason the 32 combinations of symmetries are known as *point groups* (meaning: symmetries about a stationary point), which are equivalent to the 32 crystal classes.

The symmetry elements that you will be asked to recognize and locate in crystals (or crystal models) are as follows:

a. *Rotation axes.* These are imaginary lines about which a crystal face (or a motif, as discussed in exercise 8) may be rotated and may repeat itself once or several times during a complete rotation (of 360°). Rotation axes are numbered as $n = 1$, $n = 2$, $n = 3$, $n = 4$, and $n = 6$, where n represents a divisor of 360°; $n = 1$ means that the crystal face is repeated every 360°, that is, once per full 360° rotation; similarly, $n = 3$ means that the crystal face is repeated three times during a 360° rotation, that is, every 120° of rotation about the rotation axis. The standard symbols for rotation axes, with their respective angles of rotation, are given in Table 9.1.

b. *Rotoinversion axes.* These are imaginary lines about which a crystal face (or a motif) is rotated as well as inverted, and repeats itself one or more times during a 360° rotation. Rotoinversion axes are numbered as $n = \bar{1}$, $n = \bar{2}$, $n = \bar{3}$, $n = \bar{4}$, and $n = \bar{6}$. The numbers express the same angles of rotation as they do for rotation axes; that is, $n = \bar{1}$ involves a repeat at 360°, $n = \bar{2}$ involves a repeat at 180°, $n = \bar{3}$ is for a 120° repeat, $n = \bar{4}$ is for a 90° repeat, and $n = \bar{6}$ involves a 60° repeat angle. In the operation of rotoinversion, a specific crystal face is rotated a specific number of degrees (as a function of n), but the face is not repeated in the same regular fashion as it is if only rotation is involved. Rotoinversion, as the name implies, is a combination of rotation as well as inversion. For example, for $n = \bar{4}$, there are four 90° rotations combined with inversions (through the center of the crystal) at every 90° angle. In practice, rotoinversion axes are generally more difficult to recognize and locate in a crystal than rotation axes. This will be discussed more fully in the subsequent text that accompanies Fig. 9.2. The standard symbols for rotoinversion axes, with their respective angles of rotation, are given in Table 9.1.

c. *Mirrors.* These are planes that reflect a specific crystal face into its mirror image. Mirrors are identified by the letter m and are shown on drawings by solid lines. In three-dimensional sketches of crystals, mirrors are shown as planes that intersect the crystal (see Table 9.1).

d. *A center of symmetry.* This relates a specific face on a crystal to an equivalent crystal face by inversion through a point. This point is the center of the crystal. A center of symmetry is equivalent to $\bar{1}$, that is, a rotoinversion axis involving 360° of rotation, and inversion. Its presence is not generally identified by a symbol, but it is represented by the letter i, for inversion.

TABLE 9.1   Rotation Axes, Rotoinversion Axes and Mirror Planes

Type of rotation axis	Angle of rotation	Symbol
1–fold rotation = 1	360°	none
2–fold rotation = 2	180°	⬥
3–fold rotation = 3	120°	▲
4–fold rotation = 4	90°	■
6–fold rotation = 6	60°	⬡

Type of rotoinversion axis	Angle of rotation	Letter or symbol
1–fold rotoinversion = $\bar{1}$ [a]	360°	i [a]
2–fold rotoinversion = $\bar{2}$ [b]	180°	m [b]
3–fold rotoinversion = $\bar{3}$ [c]	120°	△
4–fold rotoinversion = $\bar{4}$	90°	◩
6–fold rotoinversion = $\bar{6}$ [d]	60°	⬡

[a] i = inversion, which is equivalent to a center of symmetry.

[b] m = mirror; m is used instead of $\bar{2}$ in the description of the symmetry of crystals.

[c] $\bar{3}$ is equivalent to a 3–fold rotation axis in combination with a center of symmetry (i).

[d] $\bar{6}$ is equivalent to a 3–fold rotation axis with a mirror perpendicular to it; expressed as 3/m.

Mirrors

When seen edge on (that is, when they are perpendicular to the page)

When in a perspective sketch; intersection line is one of 2–fold rotation

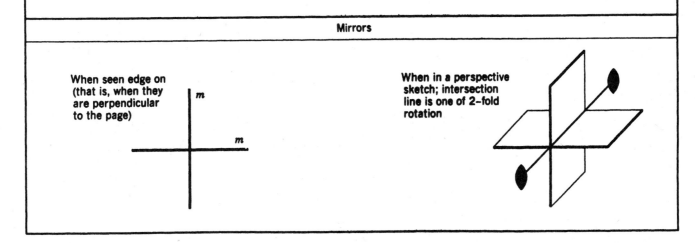

FIGURE 9.1 The determination of the presence of rotational axes and mirror planes in a crystal (see text for discussion).

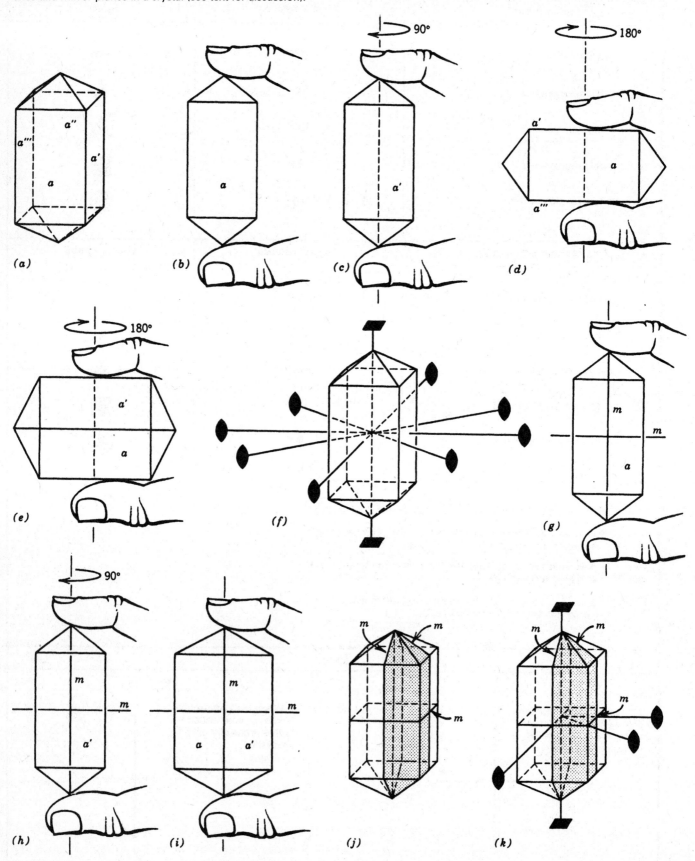

At this stage it will be helpful to illustrate how one determines the presence, or absence, of these four symmetry elements (that is, rotation axes, rotoinversion axes, mirror planes, and a center of symmetry) in a crystal (or crystal model). The various drawings in Fig. 9.1 show the observations and steps that are necessary to evaluate the total symmetry content of a specific crystal. Figure 9.1*a* shows a perspective drawing (also known as a clinographic projection) of a specific crystal; this might represent a wooden or plaster crystal model in the laboratory. Figure 9.1*b* is a sketch of how such a crystal (model) is held between the thumb and index finger to determine the presence of a possible *rotation axis* parallel to its longest dimension. Figure 9.1*c* illustrates the identical appearance of the crystal upon 90°, 180°, 270°, and 360° rotations. The imaginary axis between index finger and thumb is therefore one of $n = 4$, or 4-fold rotation. Figure 9.1*d* shows the same crystal held between thumb and index finger, but with the axis of crystal elongation horizontal. Now the index finger is centered on the top rectangular face (*a'*) and the thumb on the opposite bottom rectangular face (*a'''*). The identical appearance can be repeated only upon a 180° rotation (about the imaginary axis between thumb and index finger), which indicates the presence of a 2-fold axis ($n = 2$) between the center of the top and the center of the bottom face. An identical 2-fold rotation is seen when the fingers hold the crystal between the faces marked *a* and *a''*; these two faces are at 90° to the faces marked *a'* and *a'''*. In other words, there are two sets of 2-fold rotation axes; each is perpendicular to a set of parallel vertical crystal faces (see Fig. 9.1*a*). There are additional 2-fold rotation axes that can be located between the opposite (vertical) edges of the crystal. In Fig. 9.1*e* the crystal is held between the centers of the opposite edges; rotation of 180°, when the crystal is held in this fashion, produces an identical appearance, from which one concludes the presence of a 2-fold rotation axis. Because there are two sets of opposite, vertical edges (see Fig 9.1*a*), there are two horizontal 2-fold rotation axes between the edges. Now that all rotational symmetry elements have been located (there are no rotoinversion axes in the crystal shown in Fig. 9.1*a*), their positions can be shown as in Fig. 9.1*f*.

Let us now see where and how many *mirrors* can be located. In Fig. 9.1*g* a vertical and horizontal mirror are noted, and upon 90° rotation of the crystal in this same orientation (see Fig. 9.1*h*), an additional vertical mirror is seen. Additional vertical mirrors are found with the elongation of the crystal vertical and when viewing it along the vertical edges (Fig. 9.1*i*). There are two such vertical mirror planes between the opposite edges of the crystal. Figure 9.1*j* shows the original crystal with the location of three of the mirror planes (the horizontal mirror and two vertical mirrors, each of which is perpendicular to the vertical faces *a*, *a'*, *a''*, and *a'''*). There are, in all, five mirror planes, the two additional ones lying between the opposite edges of the crystal; these two additional mirrors

have not been shown to reduce the complexity of the figure. Figure 9.1*k* shows the same three mirror planes and their locations in the crystal as well as several (not all) of the rotation axes. Notice that the rotation axes are coincident with the intersection lines between the various mirror planes.

Although the presence of rotation axes and mirrors can be fairly easily recognized in the drawings of crystals (see Fig. 9.6), the recognition of *rotoinversion axes* is really possible only on well-formed crystals or crystal models. If one inspects the footnote to the tabulation of rotoinversion axes in Table. 9.1, one notes that there is only one rotoinversion axis, $\bar{4}$, which has not been expressed in terms of *i*, *m*, *m* and inversion, or a rotation perpendicular to a mirror (3/*m*). Here we will illustrate the recognition of $\bar{4}$ in a crystal. Figure 9.2*a* shows a perspective drawing of a crystal whose four faces are isosceles triangles (meaning that two of the sides of each of the triangles are equal in length); all four of these triangular faces are related by a vertical 4-fold rotoinversion axis. Figure 9.2*b* shows the appearance of face *a* when held between index finger and thumb. Upon a 90° rotation the crystal looks as in Fig. 9.2*c*. The difference between Figs. 9.2*b* and *c* is the fact that in Fig. 9.2*b* the crystal face appears as an inverted isosceles triangle but in Fig. 9.2*c* it is the same triangle but right side up. If, in this same 90° rotational position, we wish to see an isosceles triangular face in the inverted position (equivalent to the drawing in Fig. 9.2*b*), we must invert the crystal, as shown in Fig. 9.2*d*. In this figure the crystal was inverted through a 180° vertical rotation of the wrist; the thumb is now at the top and the index finger at the bottom of the crystal. In this orientation, the appearance of face *a''* is identical to that of face *a* in Fig. 9.2*b*. We have now completed a 90° rotation and an inversion at the 90° angle. Continued rotations of 90° and inversions will prove to you that the vertical axis of elongation of this specific crystal is not just one of 4-fold rotation, but one of 4-fold rotoinversion. Figure 9.2*e* shows all the rotoinversional and rotational symmetry of the crystal ($\bar{4}$ is vertical, and there are two 2-fold rotation axes in the horizontal plane between the centers of opposite edges of the crystal). Figure 9.2*f* shows the presence of two mutually perpendicular mirror planes as well.

The only symmetry element that we have not discussed in terms of its recognition in crystals (or crystal models) is a *center of symmetry* (or *inversion*, *i*). In a crystal, or crystal model, the easiest element to recognize is the center of symmetry, $\bar{1}$, or *i*. With the crystal, or crystal model, laid down on any face on a table top, see whether a face of equal size and shape, but inverted, is present in a horizontal position at the top of the crystal (or crystal model). If this is so, the crystal contains a center of symmetry. Thus, the presence of a center of symmetry can be easily determined in a physical object (crystal, or crystal model), but it is much harder to recognize in the drawings of crystals as in Fig. 9.6.

FIGURE 9.2 The determination of the presence of a rotoinversion axis in a crystal (see text for discussion).

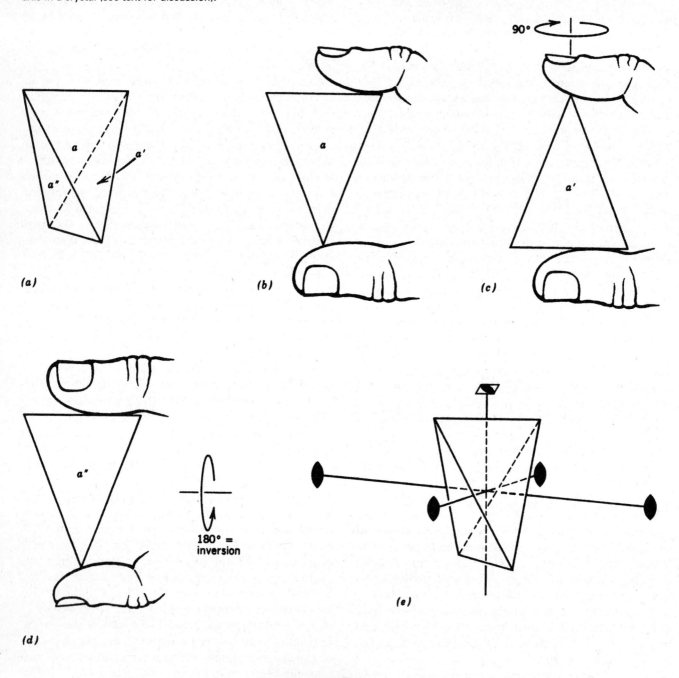

(a)

(b)

(c)

(d)

180° = inversion

(e)

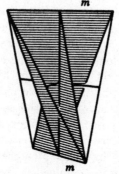

(f)

MATERIALS

Wooden or plaster models of crystals representing various crystal classes and a pad of plain paper for possible sketches. Wooden or plaster models represent highly idealized crystals. They are made in this idealized fashion in order to show physical similarity (e.g., equal size and shape) for faces (of a form) that are symmetry related. If no models are available, a number of paper models are provided in Fig. 9.5, at the very back of this laboratory manual; these can be cut out, glued, and studied (see instructions in assignment). Even if wooden or plaster models are available in the classroom, the paper cutout models can be taken home and used as independent study materials. Furthermore, a representative number of crystal drawings are given in Fig. 9.6. Such drawings, known as *clinographic projections*, can be used to locate symmetry elements, directly onto the drawing of the crystal. This method is by no means as instructive as the use of wooden or plaster models; if, however, such are not available, the drawings are helpful.

ASSIGNMENTS

A. With wooden (or plaster) models, or crystals available

1. Select several different wooden (or plaster) models; a suite of different types is probably assigned by the instructor.

2. Locate all symmetry elements present in each model (see "some helpful hints" below). Summarize the symmetry elements as noted in the column of Table 9.2, entitled "symmetry content." List your findings in Table 9.3.

3. From the entries under "symmetry content" derive the notation that is internationally accepted for such symmetry content. That is, derive the *Hermann–Mauguin* (or *international*) notation from the column "symmetry content" (see Table 9.2 for reference). This system of symbolic symmetry notations will be explained further in exercise 10.

4. As an optional part of this assignment you may wish to make a relatively neat and large three-dimensional sketch of each model you are to study and describe. You can make a reasonably successful sketch only after you have familiarized yourself with the shape of the model and all its faces. The sketch should be fairly large so you can locate on it the symmetry elements, as requested in item 2. In addition to, or in place of a sketch of the model, you may find it helpful to make some drawings of, for example, the front view, or of the top view, and so on. Such drawings will help in locating symmetry elements. Examples of how you might represent aspects of two isometric crystals, one a cube, the other a tetrahedron, are given in Fig. 9.3.

By the way, you may find that such free-form sketches are stretching your artistic skills. If you really do not enjoy such sketches, you can use the three-dimensional drawings that are given in Fig. 9.6.

Some Helpful Hints

The easiest element to identify is the center of symmetry $\bar{1}$ (or i). Lay the model on any face and see whether a face of equal size and shape, but inverted, is present in a horizontal position at the top of the model. If this is so, the crystal contains a center of symmetry.

Next, look for a principal rotation axis, that is, a rotation axis with rotational symmetry greater than 2. If one 3-fold or one 4-fold axis is present, you can determine its crystal system (see Table 9.2). If three 4-fold axes are present, the crystal is in yet another, higher symmetry, crystal system (see Table 9.2). Hold the principal axis you may have found vertically (between thumb and index finger) and examine the plane perpendicular to it. This plane may be a mirror or it may not. It may, instead, contain one or

TABLE 9.2   The 32 Crystal Classes and Their Symmetry

Crystal System	Crystal Class	Symmetry Content
Triclinic	1	none
	$\bar{1}$	i
Monoclinic	2	$1A_2$
	m	$1m$
	$2/m$	$i, 1A_2, 1m$
Orthorhombic	222	$3A_2$
	$mm2$	$1A_2, 2m$
	$2/m2/m2/m$	$i, 3A_2, 3m$
Tetragonal	4	$1A_4$
	$\bar{4}$	$1\bar{A}_4$
	$4/m$	$i, 1A_4, m$
	422	$1A_4, 4A_2$
	$4mm$	$1A_4, 4m$

TABLE 9.2 *(continued)*

Crystal System	Crystal Class	Symmetry Content
	$\bar{4}\,2m$	$1\bar{A}_4$, $2A_2$, $2m$
	$4/m2/m2/m$	i, $1A_4$, $4A_2$, $5m$
Hexagonal[a]	3	$1A_3$
	$\bar{3}$	$1\bar{A}_3$ ($= i + 1A_3$)
	32	$1A_3$, $3A_2$
	$3m$	$1A_3$, $3m$
	$\bar{3}\,2/m$	$1\bar{A}_3$, $3A_2$, $3m$
		$(1\bar{A}_3 = i + A_3)$
	6	$1A_6$
	$\bar{6}$	$1\bar{A}_6$ ($= 1A_3 + m$)
	$6/m$	i, $1A_6$, $1m$
	622	$1A_6$, $6A_2$
	$6mm$	$1A_6$, $6m$
	$\bar{6}m2$	$1\bar{A}_6$, $3A_2$, $3m$
		$(1\bar{A}_6 = 1A_3 + m)$
	$6/m2/m2/m$	i, $1A_6$, $6A_2$, $7m$
Isometric	23	$3A_2$, $4A_3$
	$2/m\,\bar{3}$	$3A_2$, $3m$, $4\bar{A}_3$
		$(1\bar{A}_3 = 1A_3 + i)$
	432	$3A_4$, $4A_3$, $6m$
	$\bar{4}\,3m$	$3\bar{A}_4$, $4A_3$, $6m$
	$4/m\bar{3}2/m$	$3A_4$, $4\bar{A}_3$, $6A_2$, $9m$
		$(1\bar{A}_3 = 1A_3 + i)$

[a] In this table all crystal classes (point groups) beginning with 6, $\bar{6}$, 3, and $\bar{3}$ are grouped in the hexagonal system. In earlier editions of the *Manual of Mineralogy* the hexagonal system was divided into the hexagonal and the rhombohedral divisions. The use of these two subdivisions, as based on the presence of 6 or $\bar{6}$ versus 3 or $\bar{3}$ axes in the morphological symmetry of a crystal, results in confusion when subsequent X-ray investigations show a specific crystal with, for example, 32 symmetry to be based on a hexagonal lattice. This is the case in low quartz, which allows morphological symmetry 32 but is based on a primitive hexagonal lattice, resulting in a space group $P3_12$ (or $P3_22$).

The hexagonal system can, however, be divided on whether the lattice symmetry is hexagonal ($6/m2/m2/m$) or rhombohedral ($\bar{3}2/m$). This results in the following groupings.

Hexagonal (hexagonal lattice division)		Rhombohedral (rhombohedral lattice division)
$6/m2/m2/m$		$\bar{3}\,2/m$
$\bar{6}m2$	$\bar{3}\,2/m$	$3m$
$6m$	$3m$	$3\,2$
$6\,2\,2$ and	$3\,2$	$\bar{3}$
$\bar{6}/m$	$\bar{3}$	3
6	3	
6		

SOURCE: From *Manual of Mineralogy*, 21st ed., 1993, John Wiley and Sons, p. 39.

Student Name _____

TABLE 9.3   Record of Symmetry Content of Crystal Models

Model Number	Symmetry Content	Crystal Class Notation (Hermann–Mauguin)

Please make photocopies of this form if more copies are needed.

FIGURE 9.3 Clinographic projection (three-dimensional "sketches") of (*a*) a cube and (*d*) a tetrahedron. Various views (*b, c*) of the cube, and (*e, f, g*) of the tetrahedron.

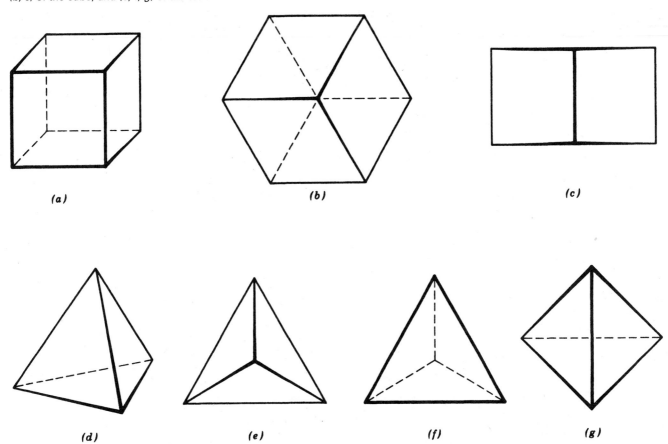

several sets of horizontal 2-fold axes. These axes may or may not have mirrors perpendicular to them. In the symbolic representation of symmetry, an axis with a mirror perpendicular to it is shown (e.g., for a 3-fold axis) as 3/*m*.

If you find no principal axis, you may find either one 2-fold axis, several mutually perpendicular 2-fold axes, only a mirror, or a 2-fold axis with a mirror perpendicular to it.

The combination of the foregoing observations leads you toward the establishment of the correct symbol for the point group (= crystal class). Figure 9.4 is a very handy scheme (a "flow sheet") for the determination of such point group symmetry. More is said about these symmetry symbols (known as **Hermann–Mauguin,** or international notation) in exercise 10, especially Table 10.3.

B. In the absence of wooden (or plaster) models, or in addition to assignment A

1. Use the common and simple crystal forms in Fig. 9.5 as one part of this exercise. (The crystal patterns for Fig. 9.5 are printed on heavy paper and appear at the end of

the book.) Cut these figures out along the external solid lines. Score, but do not cut through, the dashed lines. Carefully fold inward along the scored lines. Fasten with a good-quality glue (Duco or rubber cement is very successful). Alternatively, hold the model together with transparent adhesive tape.

2. The paper models can be evaluated in the same sequence of steps as was outlined in the previous section (A) under the heading "with models available."

3. As these paper models are your own, and not restricted to study in a classroom or laboratory, they are very useful for further study at home.

4. Use the clinographic projections of some commonly observed forms as given in Fig. 9.6. These illustrations are meant as a basis for the insertion of the various symmetry elements that their shapes imply. Locate such symmetry elements through the center of the drawing; use standard symbols for rotation axes and mirrors, as noted in Table 9.1 and in Figs. 9.1 and 9.2.

FIGURE 9.4 Scheme for the assignment of the symbolic point group notation on the basis of the symmetry content of a crystal. See Exercise 10 for further discussion of this symmetry notation. (By permission of B. M. Loeffler, Colorado College, Colorado Springs, Colorado.)

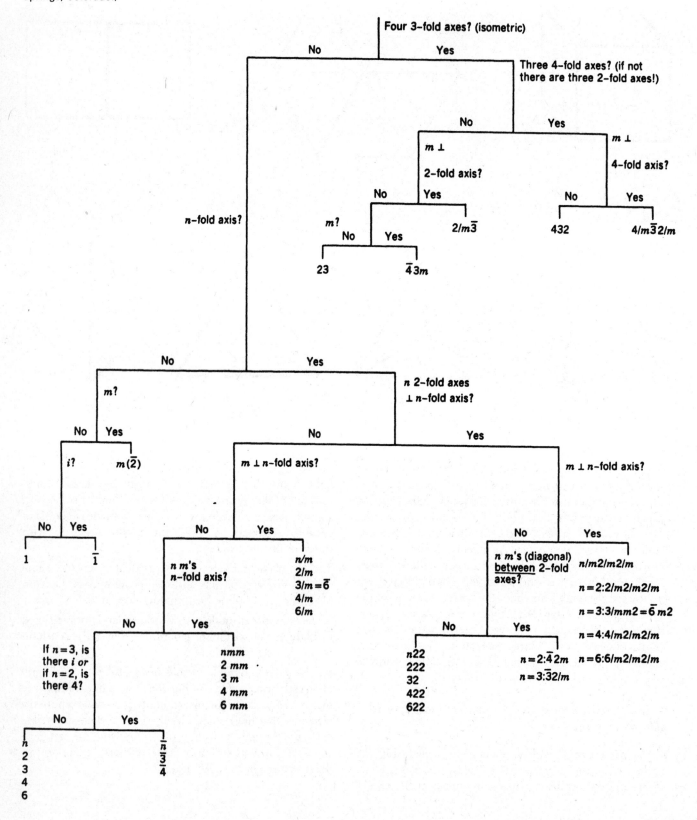

Student Name

FIGURE 9.6   Clinographic projections of some commonly observed crystal forms.

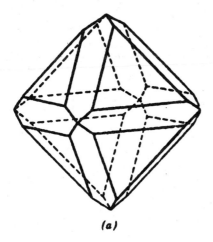

(a)

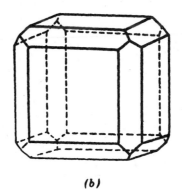

(b)

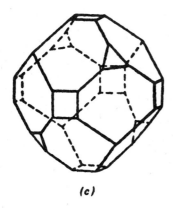

(c)

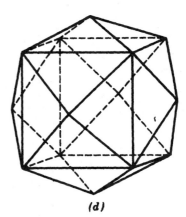

(d)

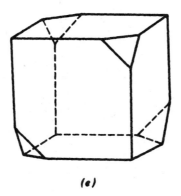

(e)

FIGURE 9.6 *(continued)*

Student Name

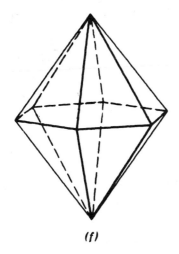

(f)

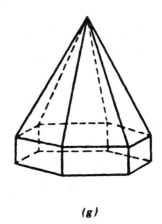

(g)

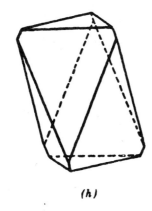

(h)

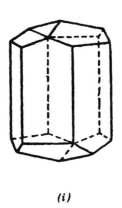

(i)

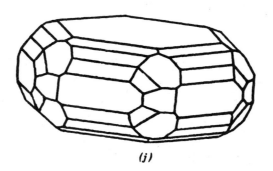

(j)

FIGURE 9.6 *(continued)*

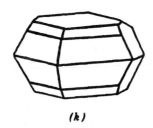

(k)

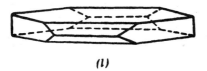

(l)

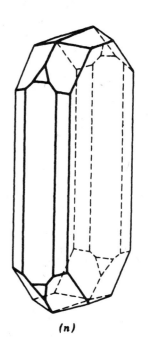

(n)

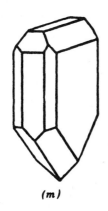

(m)

Student Name

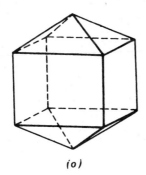

(o)

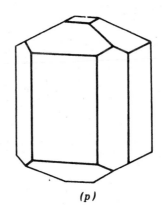

(p)

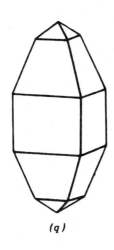

(q)

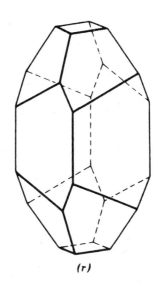

(r)

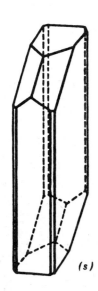

(s)

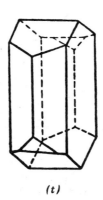

(t)

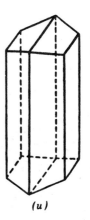

(u)

Common Crystal Forms, Choice of Coordinate Axes, and Symmetry Notation of Crystals

PURPOSE OF EXERCISE

Learning to identify crystal forms, the symmetry inherent in them, and to orient crystals according to their appropriate coordinate axes. All these aspects of crystals involve an understanding of the symmetry notation known as the *Hermann–Mauguin*, or *international notation*.

FURTHER READING AND CD-ROM INSTRUCTIONS

Klein, C. and Dutrow, B., 2008, *Manual of Mineral Science*, **23rd ed., Wiley, Hoboken, New Jersey, pp. 120–131 and 134–142.**

CD-ROM, entitled *Mineralogy Tutorials*, **version 3.0, that accompanies** *Manual of Mineral Science*, **23rd ed., click on "Module II", and subsequently on the button "Crystal Classes".**

Nesse, W. D., 2000, *Introduction to Mineralogy*, Oxford University Press, New York, pp. 24–33.

Perkins, D., 2002, *Mineralogy*, 2nd ed., Prentice Hall, Upper Saddle River, New Jersey, pp. 187–205.

Wenk, H. R. and Bulakh, A., 2004, *Minerals: Their Constitution and Origin*, Cambridge University Press, New York, New York, pp. 66–79.

Bloss, F. D., 1994, 2nd printing with minor revisions, *Crystallography and Crystal Chemistry*, Mineralogical Society of America, Chantilly, Virginia, pp. 26–47.

Background Information: A *crystal form* (in short, *form*) consists of a group of symmetry-related crystal faces. The number of symmetrically equivalent faces in a form depends on the symmetry elements present that relate them, and on the orientation of the form with respect to the symmetry elements. On wooden or plaster models, the faces that belong to a specific form are made identical in size and shape. This simplifies the task of recognizing all faces that belong to the same form. In real crystals this may not be the case; however, they should show identical luster, etch pits, or striations. *Forms* are said to be *open* when they do not enclose space and *closed* when they do. For example, a form consisting of two parallel faces (see Fig. 10.1), a pinacoid (or parallelohedron), does not enclose space, whereas a cube does. The shapes and names of 48 forms (or 47, depending on the classification) are given in Fig. 10.1 and Table 10.1, respectively. The most commonly used nomenclature for forms in the

English-speaking world is that of Groth–Rodgers, which contains 48 form names (33 for nonisometric forms, and 15 more for isometric forms). The internationally recommended system lists 47 forms (32 for nonisometric forms, and 15 more for isometric forms). The reason for the differences is that the dome and sphenoid (see Table 10.1, entries 3 and 4) are both named dihedron in the internationally recommended system. The logic of the derivation of the various names (etymology) based on Greek words in the internationally recommended nomenclature is especially attractive and easy to learn (see footnote to Table 10.1).

The recognition of one or several forms on a crystal allows the observer to orient the crystal on the basis of its inherent symmetry content; that is, the total amount of symmetry in a crystal (e.g., numbers of rotation axes and mirrors, as discussed in exercise 9) dictates how a crystal should be conventionally oriented (when holding it between thumb and index finger). For example, certain rotation axes are, by convention, equivalent to specific directions in a coordinate axis system (see Tables 10.2 and 10.3).

Coordinate (reference) *axes* in a crystal are usually chosen parallel to crystal edges. Commonly these directions are coincident with symmetry directions. Information on the choice of coordinate axes, appropriate to the inherent symmetry content of the crystal, is given in Fig. 10.2 and Table 10.2.

As noted before, the choice of one of the coordinate axis systems illustrated in Fig. 10.2 is a function of the total symmetry content inherent in the crystal. If the crystal has no or relatively low overall symmetry, it may be part of the triclinic or monoclinic systems, whereas if it shows the highest possible symmetry it would undoubtedly be classified as part of the isometric system. There are six *crystal systems: triclinic, monoclinic, orthorhombic, tetragonal, hexagonal,* and *isometric*, each of which groups a number of point groups (or crystal classes) on the basis of common symmetry characteristics.

Before one can decide on the appropriate coordinate axes for a specific crystal, one must derive its total symmetry context (as was done in exercise 9). This symmetry content is best expressed in a symbolic notation known as the *Hermann–Mauguin*, or *international notation*. In exercise 9 symmetry content was evaluated, and indeed the

FIGURE 10.1 The 48 (or 47) different crystal forms and some of their symmetry elements. (From *Manual of Mineral Science,* 23rd ed., pp. 139–142.)

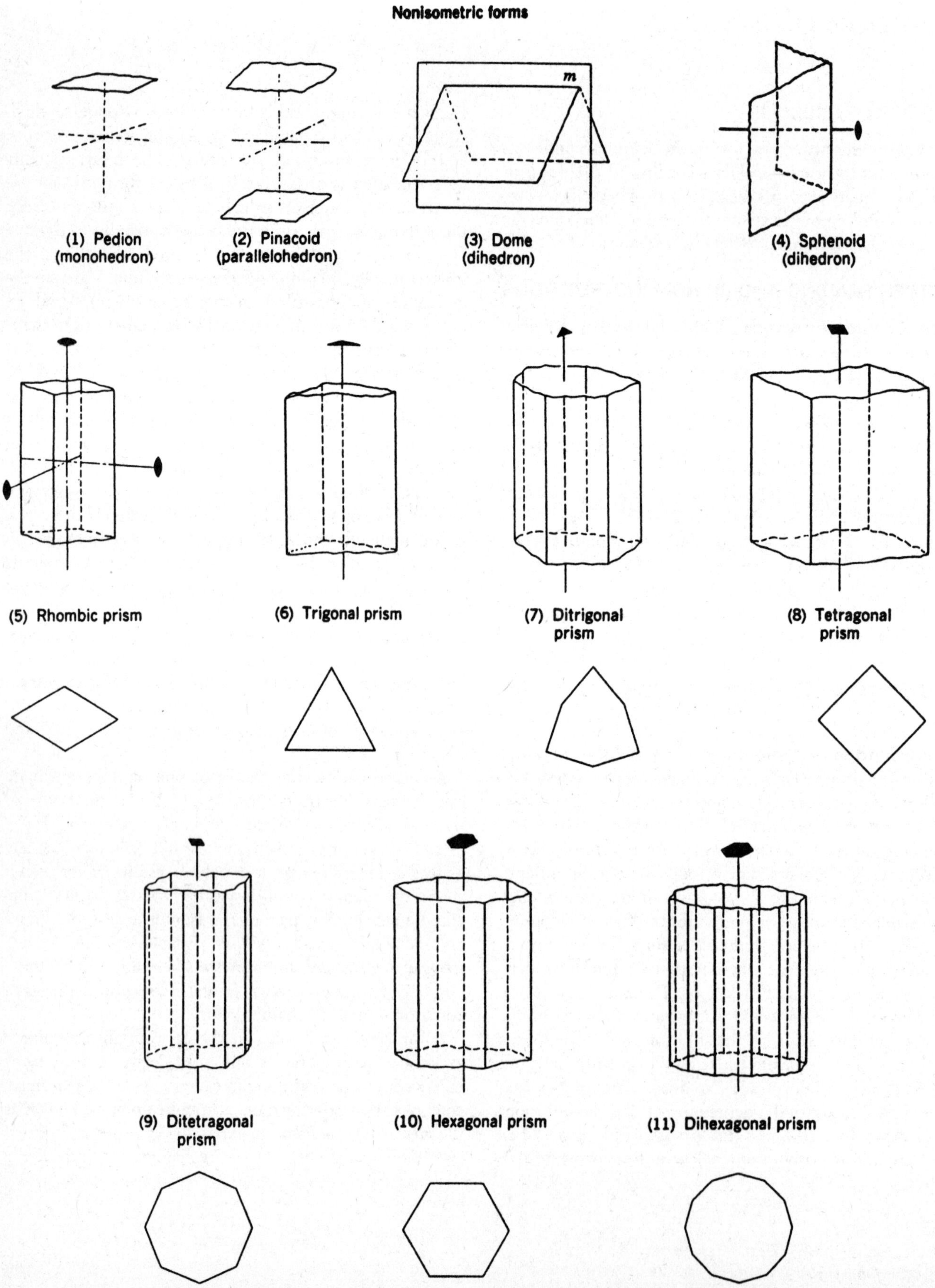

Nonisometric forms

(1) Pedion (monohedron)

(2) Pinacoid (parallelohedron)

(3) Dome (dihedron)

(4) Sphenoid (dihedron)

(5) Rhombic prism

(6) Trigonal prism

(7) Ditrigonal prism

(8) Tetragonal prism

(9) Ditetragonal prism

(10) Hexagonal prism

(11) Dihexagonal prism

FIGURE 10.1   *(continued)*

Nonisometric forms (cont'd.)

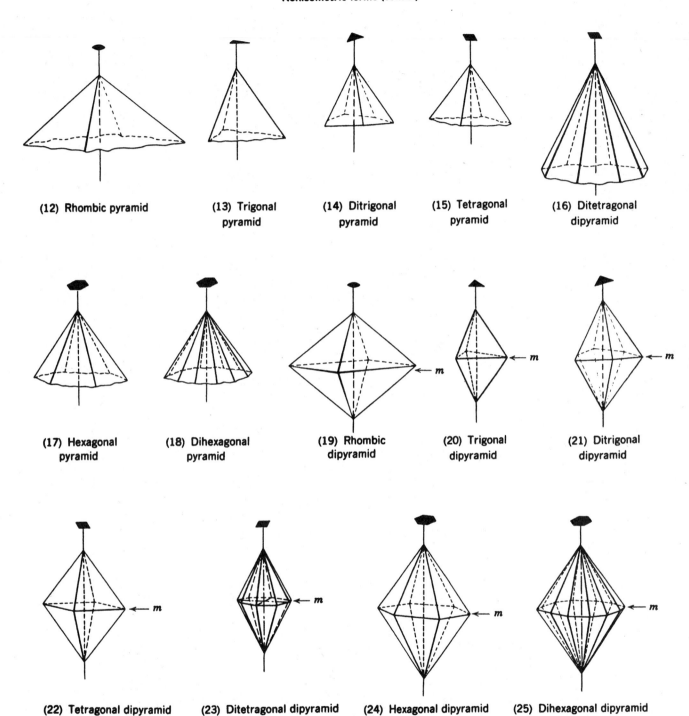

(12) Rhombic pyramid

(13) Trigonal pyramid

(14) Ditrigonal pyramid

(15) Tetragonal pyramid

(16) Ditetragonal dipyramid

(17) Hexagonal pyramid

(18) Dihexagonal pyramid

(19) Rhombic dipyramid

(20) Trigonal dipyramid

(21) Ditrigonal dipyramid

(22) Tetragonal dipyramid

(23) Ditetragonal dipyramid

(24) Hexagonal dipyramid

(25) Dihexagonal dipyramid

FIGURE 10.1   *(continued)*

Isometric forms

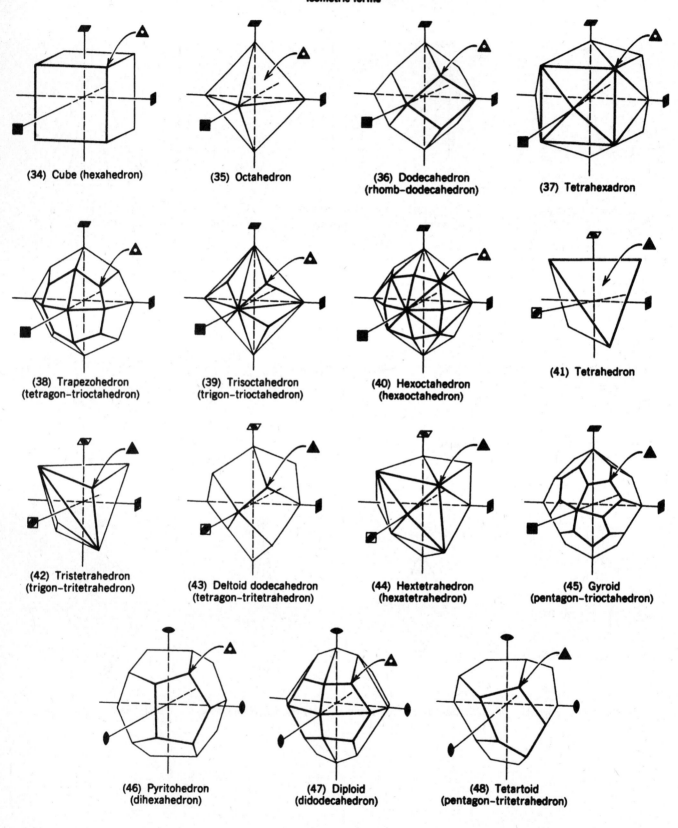

(34) Cube (hexahedron)

(35) Octahedron

(36) Dodecahedron
(rhomb-dodecahedron)

(37) Tetrahexadron

(38) Trapezohedron
(tetragon–trioctahedron)

(39) Trisoctahedron
(trigon–trioctahedron)

(40) Hexoctahedron
(hexaoctahedron)

(41) Tetrahedron

(42) Tristetrahedron
(trigon–tritetrahedron)

(43) Deltoid dodecahedron
(tetragon–tritetrahedron)

(44) Hextetrahedron
(hexatetrahedron)

(45) Gyroid
(pentagon–trioctahedron)

(46) Pyritohedron
(dihexahedron)

(47) Diploid
(didodecahedron)

(48) Tetartoid
(pentagon–tritetrahedron)

TABLE 10.1 The Names of the 33 (or 32) Different Types of Nonisometric Crystal Forms and of the 15 Isometric Crystal Forms

Name According to System of Groth–Rogers[a]	Number of Faces[b]	Internationally Recommended Name (after Fedorov Institute)
1. *Pedion*	1	*Monohedron*
2. *Pinacoid*	2	*Parallelohedron*
3. *Dome*	2	*Dihedron*
4. *Sphenoid*	2	*Dihedron*
5. Rhombic prism	4	Rhombic prism
6. Trigonal prism	3	Trigonal prism
7. Ditrigonal prism	6	Ditrigonal prism
8. Tetragonal prism	4	Tetragonal prism
9. Ditetragonal prism	8	Ditetragonal prism
10. Hexagonal prism	6	Hexagonal prism
11. Dihexagonal prism	12	Dihexagonal prism
12. Rhombic pyramid	4	Rhombic pyramid
13. Trigonal pyramid	3	Trigonal pyramid
14. Ditrigonal pyramid	6	Ditrigonal pyramid
15. Tetragonal pyramid	4	Tetragonal pyramid
16. Ditetragonal pyramid	8	Ditetragonal pyramid
17. Hexagonal pyramid	6	Hexagonal pyramid
18. Dihexagonal pyramid	12	Dihexagonal pyramid
19. Rhombic dipyramid	8	Rhombic dipyramid
20. Trigonal dipyramid	6	Trigonal dipyramid
21. Ditrigonal dipyramid	12	Ditrigonal dipyramid
22. Tetragonal dipyramid	8	Tetragonal dipyramid
23. Ditetragonal dipyramid	16	Ditetragonal dipyramid
24. Hexagonal dipyramid	12	Hexagonal dipyramid
25. Dihexagonal dipyramid	24	Dihexagonal dipyramid
26. Trigonal trapezohedron	6	Trigonal trapezohedron
27. Tetragonal trapezohedron	8	Tetragonal trapezohedron
28. Hexagonal trapezohedron	12	Hexagonal trapezohedron
29. *Tetragonal* scalenohedron	8	*Rhombic* scalenohedron
30. *Hexagonal* scalenohedron	12	*Ditrigonal* scalenohedron
31. Rhombohedron	6	Rhombohedron
32. Rhombic *disphenoid*	4	Rhombic *tetrahedron*
33. Tetragonal *disphenoid*	4	Tetragonal *tetrahedron*
Total number of forms = 33		Total number of forms = 32

Isometric Crystal Forms

34. Cube	6	Hexahedron
35. Octahedron	8	Octahedron
36. Dodecahedron	12	Rhomb-dodecahedron[c]
37. Tetrahexahedron	24	Tetrahexahedron
38. Trapezohedron	24	Tetragon-trioctahedron
39. Trisoctahedron	24	Trigon-trioctahedron
40. Hexoctahedron	48	Hexaoctahedron
41. Tetrahedron	4	Tetrahedron
42. Tristetrahedron	12	Trigon-tritetrahedron
43. Deltoid dodecahedron	12	Tetragon-tritetrahedron
44. Hextetrahedron	24	Hextetrahedron
45. Gyroid	24	Pentagon-trioctahedron
46. Pyritohedron	12	Dihexahedron
47. Diploid	24	Didodecahedron
48. Tetartoid	12	Pentagon-tritetrahedron

[a]The terminology in the first column is a compromise of generally accepted names used in the English-speaking world. The internationally recommended system of nomenclature is based on the generally accepted terms for geometric shapes with adjectives derived from Greek words. The adjectives give the shape of the cross section of the ideal geometric form, either at midheight or near an apex. In the rhombic prism, for example, the cross section of the prism is rhomb-shaped. The Greek prefixes for 1, 2, 3, 4, 5, 6, 8, and 12, are *mono, di, tetra, penta, hexa, octa,* and *dodeca;* the word *gonia* means angle, as in hexagon; *hedron* means face, as in polyhedron; *rhomb,* a planar figure with four equal sides and opposite angles equal (two acute, two obtuse); *prism,* three or more faces intersecting in parallel edges; *pyramid,* three or more faces whose edges intersect in a point; *trapezohedron,* a form with faces that are trapezoids (four-sided faces in which no two sides are parallel but two adjacecent sides are equal); and *scalenohedron,* in which faces are scalene triangles (triangles in which no two sides are equal).

[b]The number of faces of each form is given. The numbers on the left refer to Fig. 10.1. In this listing the eight synonymous terms are in italics.

[c]In the internationally recommended scheme a prefix describes the shape (as in rhomb) or the number of edges (as in trigon).

SOURCE: From *Manual of Mineralogy,* 21st ed., 1993, John Wiley and Sons, p. 47.

TABLE 10.2   Crystal Systems and Their Coordinate Axes

Crystal System	Coordinate Axes
Triclinic	No symmetry constraints on choice of axes ($a \neq b \neq c$; $\alpha \neq \beta \neq \gamma$)[a]
Monoclinic	If 2-fold axis is present, set this as b; if only a mirror is present, choose b perpendicular to it ($a \neq b \neq c$; $\alpha = \gamma = 90°$; $\beta > 90°$)
Orthorhombic	The three mutually perpendicular directions about which there is binary symmetry (2 or m) are chosen as the axial directions, a, b, and c ($a \neq b \neq c$; $\alpha = \beta = \gamma = 90°$)
Tetragonal	The unique 4-fold axis is chosen as c; the two a axes are in a plane perpendicular to c ($a \neq b$, which results in $a_1 = a_2$; $\alpha = \beta = \gamma = 90°$)
Hexagonal	The unique 6-fold (or 3-fold) axis is chosen as c; the three a axes are in a plane perpendicular to c ($a = b$, which results in $a_1 = a_2 = a_3$; $\alpha = \beta = 90°$; $\gamma = 120°$)
Isometric	The three 4-fold rotation axes (when present) are chosen parallel to the three coordinate axes (a_1, a_2, a_3); if 4-fold rotation axes are absent, locate the four 3-fold axes at 54°44′ to the a axes (the 3-fold directions are parallel to diagonal directions from corner to corner in a cube) ($a_1 = a_2 = a_3$; $\alpha = \beta = \gamma = 90°$)

[a]The notation $a \neq b$ means that the a axis is nonequivalent to the b axis. Similarly, $\alpha \neq \beta$ means that the angle α is not equivalent to the angle β. When a specific axis turns out to be equivalent to another axis, $a = b$ results, which crystallographers rename as $a_1 = a_2$

TABLE 10.3   Relationship of Hermann–Mauguin Symbols to Coordinate Axes in Crystals

Point Group (crystal class)	Crystal System	Symmetry Constraints on Hermann-Mauguin Notations
$1, \bar{1}$	Triclinic	No symmetry constraints on location of coordinate axes
$2, m, 2/m$	Monoclinic	2-fold = b ("second setting"); mirror is the a–c plane
$222, mm2,$ $2/m2/m2/m$	Orthorhombic	2-fold axes coincide with coordinate axes in the order a, b, c.
$4, \bar{4}, 4/m,$ $422, 4mm,$ $\bar{4}2m, 4/m2/m2/m$	Tetragonal	4-fold axis (unique) is c axis; second symbol (if present) refers to both a_1 and a_2 axial directions; third symbol (if present) refers to directions at 45° to the a_1 and a_2 axes. *Example:* 422: 4 = c; first 2 means 2-fold axes along a_1 and a_2; second 2 means two more 2-fold axes along diagonal directions.
$6, \bar{6}, 6/m,$ $622, 6mm, \bar{6}m2,$ $6/m2/m2/m,$ $3, \bar{3}, 32, 3m, \bar{3}2m$	Hexagonal	6-fold axis (or 3-fold axis) is c axis; second symbol (if present) refers to three axial directions (a_1, a_2, and a_3); and third symbol (if present) refers to directions at 30° to the a_1, a_2, a_3 axes. *Example:* $\bar{6}m2$; 6 is coincident with the c axis; m's present along the three axial directions (a_1, a_2, a_3) and 2-fold rotations occur along the directions half-way to the crystallographic axes (a_1, a_2, a_3)
$23, 2/m\bar{3},$ $432, \bar{4}3m,$ $4/m\bar{3}2/m$	Isometric	The first entry refers to the three crystallographic axes (a_1, a_2, a_3); the second symbol refers to four directions at 54° 44′ to the crystallographic axes (these directions run from corner to corner in a cube); the third symbol (if present) refers to six directions that run from edge to edge in the cube. Example: $4/m\bar{3}2/m$; all three a axes are axes of 4-fold rotation, with mirrors perpendicular to them; the "corner to corner" directions (in a cube) $\bar{3}$ (there are four such directions); the "edge to edge" directions (in a cube) are 2-fold rotations (there are six such directions) with mirrors perpendicular to them.

FIGURE 10.2 Choice of coordinate axes among the six crystal systems. The three directions are labeled *a, b, c,* unless symmetry makes them equivalent. The coordinate axis system used in crystallography is a "right-handed" system, which means that if the positive end of *a* is turned toward the positive end of *b,* a screw motion can be imagined, with the advance of the screw toward the positive end of *c.* In the drawings of the coordinate system the positive ends of the axes are oriented as follows: *a* toward the observer, *b* toward the right, and *c* to the top. Where there is equivalency in the axes, *a* = *b* becomes $a_1 = a_2$, *a* = *b* = *c* becomes $a_1 = a_2 = a_3$, and so forth.

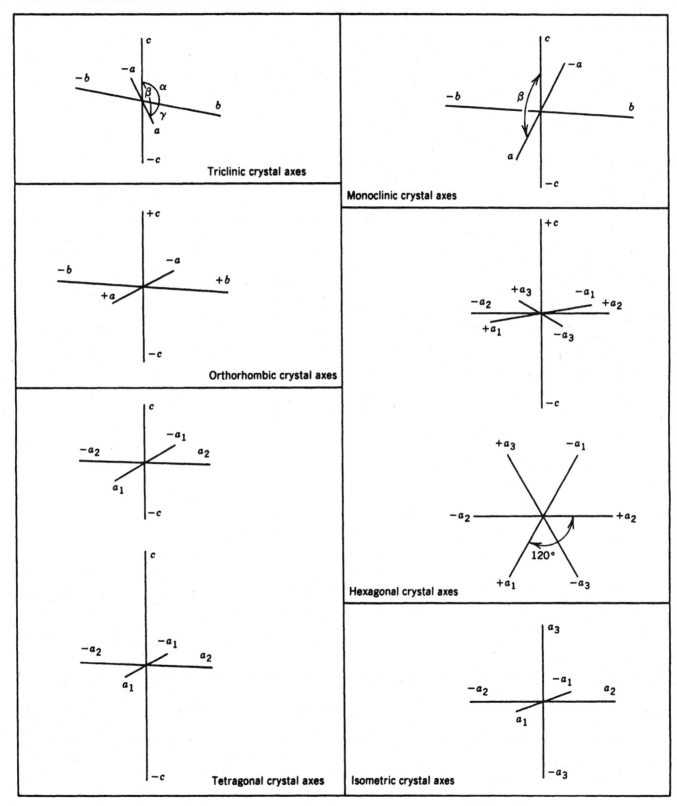

Hermann-Mauguin notation was shown in Table 9.2 as well as Fig. 9.4. However, at that time no explicit instructions were given to enable you to convert the overall symmetry content of a crystal into this notation. We will do this now.

The overall symmetry content of a crystal "pigeonholes" it in groupings of crystal systems (six of these), and these systems in turn dictate the type of coordinate axes choices.

If a crystal contains only $\bar{1}$, or 1 (= no symmetry, identity only), it belongs to the triclinic system; in this system there are no symmetry constraints on coordinate axes (= axial directions); see Fig. 10.2.

If, however, a crystal (or model) shows only $2/m$, m, or 2 as its symmetry content, it is part of the monoclinic system. The description of this system is most commonly based on choosing the b axis as the 2-fold axis, or choosing the b axis direction as perpendicular to the mirror (in point group m). This orientation is referred to as the "second setting" in the monoclinic system. (The first setting, with the c axis as the 2-fold direction, is very infrequently used in the literature.) If indeed your crystal (or model) has monoclinic symmetry, you must also decide on the directions of the a and c axes and the size of the β angle (the b axis is fixed, as was noted earlier; see also Table 10.2). Figure 10.3 illustrates a typical monoclinic crystal, upon which appropriate coordinate axis choices and β angle are superimposed. Once the 2-fold axis (= b) is located in this example, it is clear that the inclined and vertical edges make an angle of 128° (or 52°) with each other. The β angle is chosen as the *obtuse* angle (larger than 90°); thus

$\beta = 128°$. The a and c axes are chosen at this angle of intersection, parallel to pronounced edges on the crystal. Such evaluations of the symmetry content and the subsequent deduction of the proper choice of coordinate axes is a little more straightforward in the orthorhombic, tetragonal, hexagonal, and isometric systems because in these there are clear symmetry constraints on several unique axial directions (see Table 10.2). Please refer back, at this time, to Table 9.2, where the column labeled "Crystal Class" shows the Hermann–Mauguin notation of all 32 crystal classes. These Hermann–Mauguin symbols may contain a single item (e.g., as in triclinic or monoclinic systems) or two, or three items. Each item symbolizes the symmetry along a certain direction (e.g., in $2/m$, the 2-fold rotation axis = b, which is perpendicular to a mirror in the a–c plane). In higher symmetry crystal systems unique rotational symmetry axes (see Table 10.2) are chosen as coordinate axes. The remaining directions in the Hermann–Mauguin symbol may be additional coordinate directions. As an example, for the tetragonal crystal class (point group), $4/m2/m2/m$, the unique 4-fold axis = c, the second entry with 2 refers to the two a axial directions with mirrors perpendicular to them, and the third entry refers to 2-fold rotation axes in positions diagonal to the a axes.

By combining the symmetry elements deduced from your model with the types of constraints given in Table 10.2, you are gradually establishing the correct symbolism for the point group (crystal class) notation. A résumé of the conventions that relate to the Hermann–Mauguin system of notations is given in Table 10.3.

FIGURE 10.3 Choice of coordinate axes and β angle in a crystal with monoclinic symmetry ($2/m$).

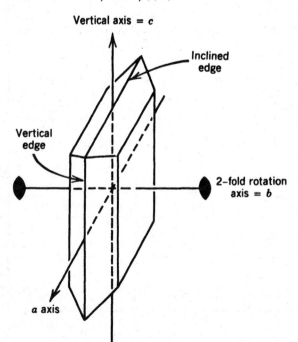

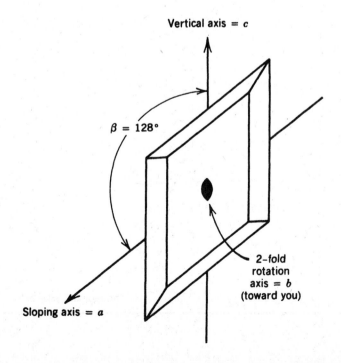

MATERIALS

Wooden or plaster crystal models with various form developments. If such models are not available, use the illustrations in Fig. 10.4 as the basis for this assignment.

ASSIGNMENTS

A. *A selection of wooden or plaster crystal models is available*

1. Select several wooden (or plaster) models; a suite of different types is probably assigned by the instructor.

2. On each of your models identify all the faces that belong to one and the same form. If a model shows more than one form, extend mentally the faces of the most prominent form (the one with the largest faces) and visualize what this form would be if it occurred alone. Then, visualize the extended shape of the second most pronounced form, and so forth. Consult Fig. 10.1 and Table 10.1. In crystal drawings faces of the same form are commonly identified by the same letter, which is helpful in the recognition of faces belonging to the same form. Crystal faces may also be identified by their Miller index, again showing which faces belong to what form. However, Miller indices are not discussed until exercise 11. Record your model number, the form names, and form type (open vs. closed) in Table 10.4.

3. The recognition of forms facilitates your assessment of the total symmetry content of the crystal (or model). All of the 32 possible symmetries (the 32 *point groups* or *crystal classes*) are grouped into six crystal systems. The coordinate axes, and their orientations, within the six crystal systems, are tabulated and illustrated in Fig. 10.2

and Table 10.2, respectively. In the fifth column of Table 10.4 enter the names of the crystal systems to which each of your models belong, and record their point group symmetry in Hermann–Mauguin notation (you may wish to refer to Fig. 9.4 in exercise 9 to help with the derivation of point groups).

4. In the last column of Table 10.4 note which symmetry elements coincide with which of the coordinate (axial) directions.

B. *An additional exercise to assignment A using Fig. 10.4*

1. In all, or a selection of, the crystal drawings in Fig. 10.4 write the appropriate letter on all the faces that compose the same form. In each of the drawings you will note that letters are given for specific faces; however, a letter is given only once. In this assignment you are asked to write the given letter on all equivalent faces.

2. In Table 10.5 record the letter given for the form, and the number of faces that make up the form.

3. Complete the various columns labeled "open or closed form," and "name of form" (refer to Fig. 10.1 and Table 10.1).

4. Establish, from the drawings in Fig. 10.4, the overall point group symmetry of each crystal, and record this as well as the name of the crystal system in Table 10.5. It would be best to record the point group (crystal class) symmetry in Hermann–Mauguin notation.

5. On the crystal drawings of Fig. 10.4, sketch the locations of the positive ends of the axes. That is, $+a_1$, $+a_2$, $+a_3$, or $+a_1$, $+a_2$, and $+c$, or $+a$, $+b$, $+c$, etc., as appropriate. Along these axial directions locate any symmetry axes (using standard symbols) that may be present.

TABLE 10.4 Results on Form Recognition and Location of Coordinate Axes, Using Wooden (or Plaster) Crystal Models

Model Number	Number of Faces on Form	Open or Closed	Name of Form	Crystal System and Point Group	Notes on Location of Coordinate Axes

Please make photocopies if more tables are required to record your results.

FIGURE 10.4 Some common forms on crystals.

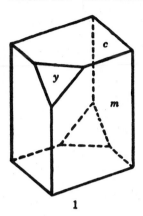

1

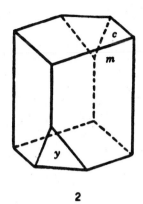

2

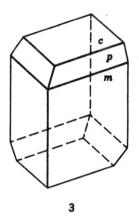

3

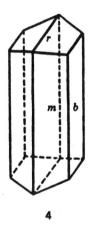

4

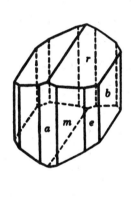

5

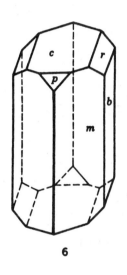

6

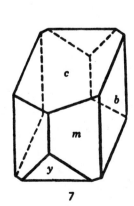

7

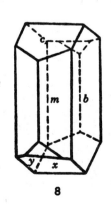

8

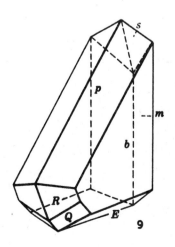

9

FIGURE 10.4 *(continued)*

Student Name

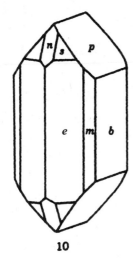

10

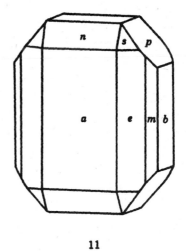

11

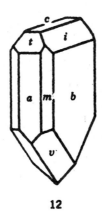

12

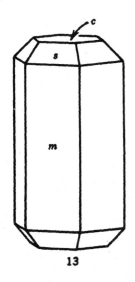

13

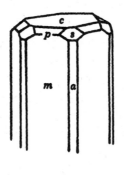

14

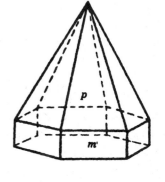

15

FIGURE 10.4 *(continued)*

16

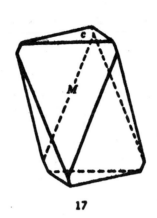

17

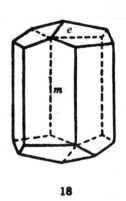

18

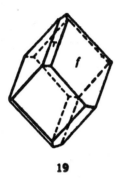

19

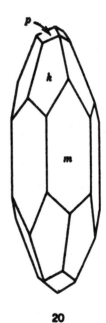

20

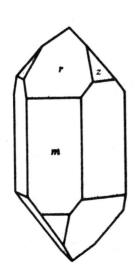

21

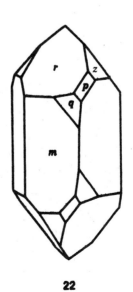

22

FIGURE 10.4 *(continued)*

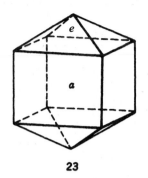

23

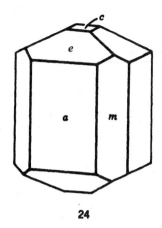

24

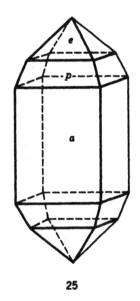

25

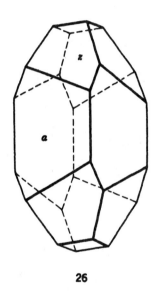

26

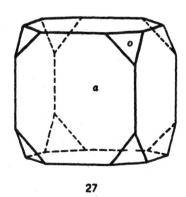

27

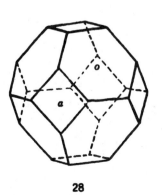

28

FIGURE 10.4   *(continued)*

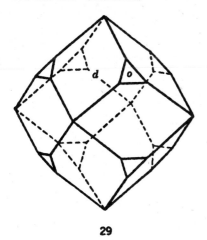

29

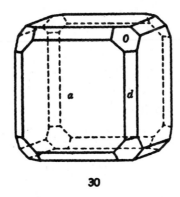

30

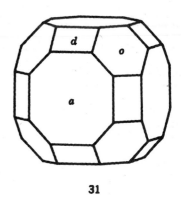

31

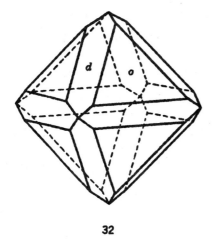

32

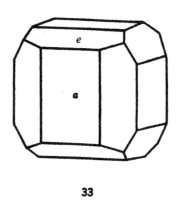

33

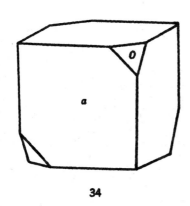

34

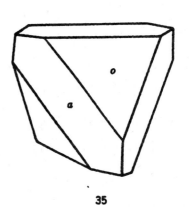

35

TABLE 10.5 Results from the Study of the Crystal Drawings in Fig. 10.4

Crystal Number	Letter of form	Number of Faces on Form	Open or Closed	Name of Form	Crystal System and Point Group

Please make photocopies if more tables are required to record your results

Indexing of Planes and Directions in a Lattice and on Crystals

PURPOSE OF EXERCISE

Determination of intercepts and Miller indices (a) for planes in a crystal lattice, (b) for faces on crystals, and (c) for zonal directions in crystals.

FURTHER READING AND CD-ROM INSTRUCTIONS

Klein, C. and Dutrow, B., 2008, *Manual of Mineral Science*, 23rd ed., Wiley, Hoboken, New Jersey, pp. 131–137.

CD-ROM, entitled *Mineralogy Tutorials*, version 3.0, that accompanies *Manual of Mineral Science*, 23rd ed., click on "Module II", and subsequently on the button "Miller Indices".

Nesse, W. D., 2000, *Introduction to Mineralogy*, Oxford University Press, New York, pp. 19–24.

Perkins, D., 2002, *Mineralogy*, 2nd ed., Prentice Hall, Upper Saddle River, New Jersey, pp. 244–258.

Wenk, H. R. and Bulakh, A., 2004, *Minerals: Their Constitution and Origin*, Cambridge University Press, New York, New York, pp. 44–50.

Bloss, F. D., 1994, 2nd printing with minor revisions, *Crystallography and Crystal Chemistry*, Mineralogical Society of America, Chantilly, Virginia, pp. 48–54.

Background Information

a. Planes in a lattice. The definition of a lattice is "an imaginative pattern of points (or nodes)—in three-dimensions—in which every point (or node) has an environment that is identical to that of any other point (or node) in the pattern. A lattice has no specific origin as it can be shifted parallel to itself" (*Manual of Mineral Science*, 2008, p. 145). In such a lattice three coordinate axes are selected parallel to the unit cell edges of the lattice. The standard type of axis notation is shown in Fig. 11.1, with the positive end of the a axis forward, the positive end of the b axis toward the right, and the positive end of the c axis vertical. The translations along the axes are t_1 along a, t_2 along b, and t_3 along c. Figure 11.1 shows such translations for the nodes along the axial directions (note that the coordinate axes are not necessarily at right angles to each other, and that the translations along the three axes are not necessarily different).

Once a coordinate system has been chosen for a specific lattice, the orientation of any plane (made up of nodes in the lattice) can be uniquely described in terms of the *intercepts* of such a plane on the coordinate axes (that is, in terms of its intersections t_1, t_2, and t_3). Examples of the intercepts of various planes with respect to the coordinate

axes of a lattice are shown in Fig 11.2. However, instead of intercept values (which can be determined directly by inspection of a lattice, as in Fig. 11.2), *Miller indices* are generally used for the definition of the orientation of a plane in a lattice. Miller indices are the reciprocals of the intercepts of a plane, with the axis notation omitted, but reported such that the first digit reflects a, the second b, and the third c. When a plane is parallel to one of the three axes, its "intercept" (that is, its parallellism) is noted as infinity, by the symbol ∞. With reference to Fig. 11.2.

	Plane	Intercepts	Reciprocals	Miller Index
(a)	C	$\infty\, a, \infty\, b, 1\, c$	$\frac{1}{\infty}, \frac{1}{\infty}, \frac{1}{1}$	(001)
(b)	X	$2\, a, \infty\, b, 3\, c$	$\frac{1}{2}, \frac{1}{\infty}, \frac{1}{3} \times 6$	(302)
(c)	A	$1\, a, 2\, b, 1\, c$	$\frac{1}{1}, \frac{1}{2}, \frac{1}{1} \times 2$	(212)
	B	$1\, a, 1\, b, 1\, c$	$\frac{1}{1}, \frac{1}{1}, \frac{1}{1}$	(111)
	C	$\frac{1}{2}\, a, 1\, b, \frac{1}{2}\, c$	$\frac{1}{\frac{1}{2}}, \frac{1}{1}, \frac{1}{\frac{1}{2}}$	(212)

In the foregoing table the column headed "*Reciprocals*" contains a number of fractions. These are eliminated by multiplying with a common factor (e.g., 6 or 2) in the derivation of the Miller index. It should be noted that two planes in Fig. 4.2c are parallel (namely, planes A and C) and these, although having different intercepts, turn out to have the same Miller index (212), implying their parallelism.

Although the Miller indices in the table are merely a sequence of three digits, it is conventional to report them inside parentheses for the identification of a specific plane, e.g., (001), (302), (212), and (111); read oh oh one, three oh two, etc.

In order to understand further the relationship of a digit in Miller index notation to intercepts on coordinate axes, turn to Fig. 11.3. The plane shown has intercepts $3t_1$ along a, ∞ along b, and $1t_3$ along c, resulting in $3a$, $\infty\, b$, $1c$. The reciprocals of these intercepts are $\frac{1}{3}, \frac{1}{\infty}, \frac{1}{1}$, which when multiplied by 3, become the Miller index (103). In this Miller index the largest digit (3) reflects the intercept with the smallest multiple of $t(1t_3)$, whereas the smallest digit (1) reflects the larger multiple of $t(3t_1)$. The 0 digit is the reciprocal of ∞, and as such a zero in a Miller index always means intersection at infinity for the axis specified by the zero in the index.

Four digits are used to describe the orientation of a plane when the lattice (or crystal) belongs to the hexagonal system (with a unique 6- or 3-fold axis). In this system (see exercise 10, Fig. 10.2) four coordinate axes are used; three horizontal axes of equal lengths, at 120° to each other, labeled a_1, a_2, and a_3, and a uniquely defined (6- or 3-fold) vertical c axis. Figure 11.4 shows a plane that intersects all four coordinate axes in a hexagonal lattice. Its intercepts are $1t_1$, $1t_2$, $-\frac{1}{2}t_3$, $1t_4$, with t_1, t_2, and t_3 representing the translations along a_1, a_2, and a_3, and t_4 the translation along c. The reciprocals of these intercepts are: $\frac{1}{1}, \frac{1}{1}, \frac{1}{-\frac{1}{2}}, \frac{1}{1}$, which results in $(1\,1\,\bar{2}\,1)$; read as one, one, minus two, one. This four-digit index is known as the *Bravais-Miller index.* The third digit in this symbol is always equal to the negative sum of the first two, because one axis is redundant. However, the presence of three horizontal axes is the direct result of the presence of a $\bar{6}, 6, \bar{3}$, or 3 axis in the hexagonal system.

b. *Faces on a crystal.* In a lattice, as explained earlier, one can count the number of intercepts of a specific plane along coordinate axes. In a crystal, or crystal model, one can (after the coordinate axes are chosen) determine the general orientation of a plane with respect to these axes, but one can estimate only *relative* values for the intercepts. How then does one go about assigning intercepts, or Miller indices, to the faces of crystals? On the basis of the overall symmetry of the crystal (or model), one first locates the appropriate coordinate axes (see exercise 10). When these have been selected, identify the *largest face* on the crystal that cuts all three (a, b, c) axes and assign it the (estimated) intercepts of $1a$, $1b$, $1c$. Such a face is referred to as the *unit face* and is given the Miller index of (111). Once the identity of such a unit face is established, the Miller (or Bravais-Miller) indices of all other faces are fixed. In the hexagonal system select as the unit face a large face that intersects the c axis and has equal intercepts on the

FIGURE 11.1 Standard type of axis notation for three coordinate axes (a, b, and c) at right angles to each other. Translations along these axes are shown by the spacings of the nodes and are designated as t_1, t_2, t_3.

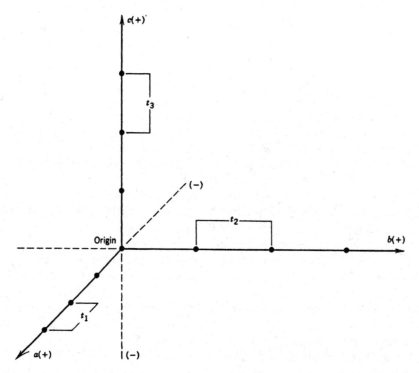

FIGURE 11.2 Intercepts for several planes with various orientations to the coordinate axes.

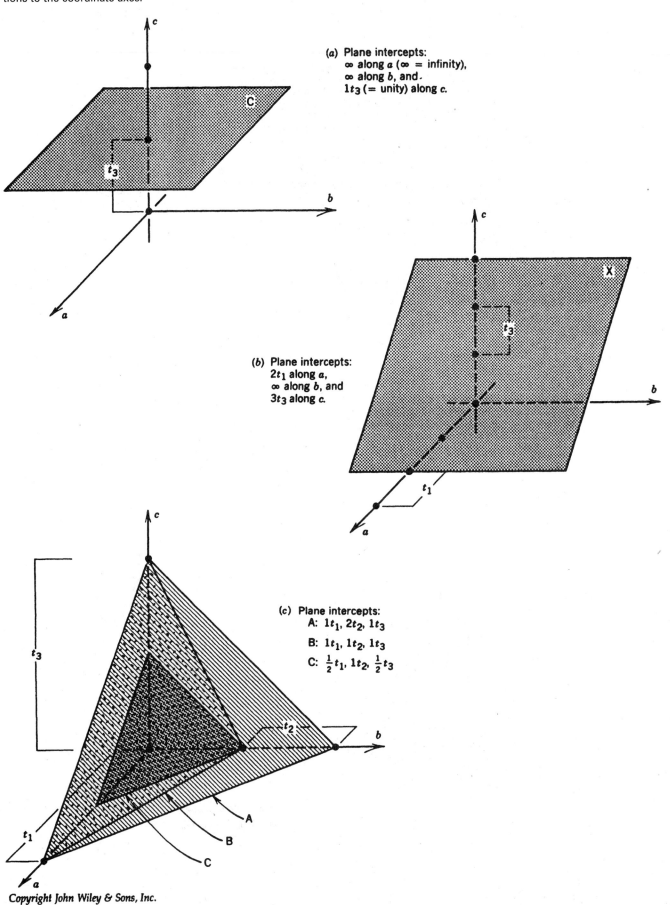

(a) Plane intercepts:
∞ along a (∞ = infinity),
∞ along b, and
$1t_3$ (= unity) along c.

(b) Plane intercepts:
$2t_1$ along a,
∞ along b, and
$3t_3$ along c.

(c) Plane intercepts:
A: $1t_1$, $2t_2$, $1t_3$
B: $1t_1$, $1t_2$, $1t_3$
C: $\frac{1}{2}t_1$, $1t_2$, $\frac{1}{2}t_3$

FIGURE 11.3 Relationship of digits in Miller index notation to multiples of intercepts; see text for discussion.

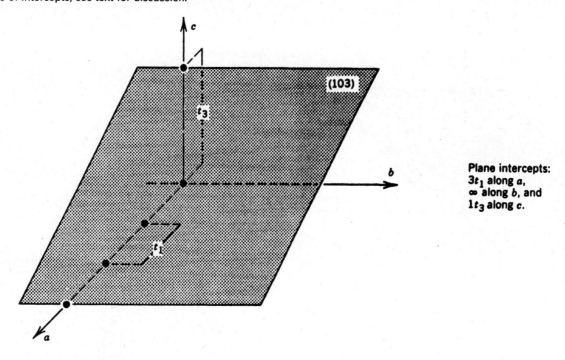

(103)

Plane intercepts:
$3t_1$ along a,
∞ along b, and
$1t_3$ along c.

FIGURE 11.4 (a) Perspective view of coordinate axes in the hexagonal system, and a plane intersecting all four axes. (b) Plan view of the horizontal axes showing the horizontal trace of the plane.

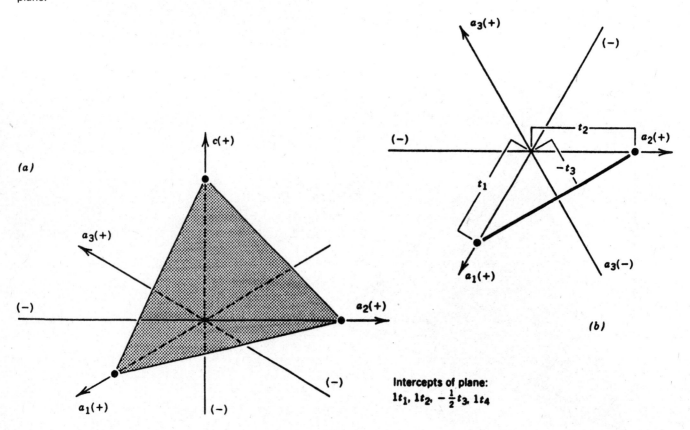

(a)

(b)

Intercepts of plane:
$1t_1$, $1t_2$, $-\frac{1}{2}t_3$, $1t_4$

a_1 and $-a_3$ axes. This unit face (see Fig. 11.5) will thus be parallel to the a_2 axis, and its Bravais-Miller index is ($10\bar{1}1$). This is a common rhombohedral face in, for example, calcite and quartz.

When it is impossible to obtain the *absolute* intercept values for crystal faces, a more general symbol of the Miller (or Bravais-Miller) index is commonly used. Such a general symbol is (hkl) where h, k, and l are, respectively, variables representing the reciprocals of rational but undefined intercepts along the a, b, and c axes; similarly, in the hexagonal system the ($hkil$) Bravais-Miller symbol represents h, k, i, and l as rational but undefined intercepts along the a_1, a_2, a_3, and c axes. Less general symbols would be ($hk0$), ($0kl$), ($h0l$) with the crystal faces cutting only two of the coordinate axes (0 referring to an intercept of ∞; $\frac{1}{\infty} = 0$). Even more specifically oriented crystal faces might have indices such as (100), (010), (001), or (0001), where only one axis is cut by the plane (that is, it is parallel to two other axes, or the three horizontal axes as in the hexagonal system).

In exercise 10 we discussed crystal forms and noted that faces that belong to the same form are commonly identified by the same letter on crystal drawings (see Fig. 10.4). Instead of using letters, one can use the Miller index of each specific crystal face that is part of a form to label all the faces in the form. This is also commonly done in crystal drawings. However, for referring to a specific crystal form, it is conventional to use only one of the form's several Miller indices, namely the Miller index symbol with positive digits enclosed in braces. For example, a side pinacoid consists of two parallel faces (010) and ($0\bar{1}0$); the form symbol for this pinacoid is {010}. An octahedron in the isometric system has eight faces with the following eight Miller indices: (111), ($\bar{1}11$), ($1\bar{1}1$), ($11\bar{1}$), ($\bar{1}\bar{1}1$), ($\bar{1}1\bar{1}$), ($1\bar{1}\bar{1}$), and ($\bar{1}\bar{1}\bar{1}$). The form symbol for this octahedron becomes {111}. The Miller index notation for a form that has a completely general orientation with respect to coordinate axes would be {hkl}.

Forms are commonly described in two general categories: *general forms* and *special forms*. A crystal face (hkl), which is part of the form {hkl}, is oriented in the most general position with respect to the symmetry elements of the crystal. Accordingly, the {hkl} form is the *general form*. On the other hand, another crystal face may have a relatively specialized position with respect to the inherent symmetry elements of the crystal; such a form is known as a *special form*. In the $4/m\bar{3}2/m$ crystal class (point group) of the isometric system, the general form {hkl} is represented by the hexoctahedron (containing 48 faces); in this same point group the cube {001} is a special form with only six faces. In most point groups the general form has a larger number of crystal faces than any of the possible special forms in the same point group.

c. *Zones on crystals.* On examination of a number of crystals, crystal models or crystal drawings, one observes that a prominent feature on most of them is that the faces are so arranged that the intersection edges of several of them are parallel. See Fig. 11.6a, where four vertical faces intersect parallel to the vertical direction, the c axis. Such a group of faces constitutes a *zone*, and the common direction of edge intersections is the *zone axis*. Zone axes can be expressed by symbols similar to Miller indices but distinguished from them by being enclosed in square brackets such as [hkl], [001], or [0001]. The distinction should be noted that Miller indices represent planes and zone symbols represent directions.

When coordinate crystal axes are selected, it is common to choose them parallel to the zone axes of three major zones. In orthogonal crystal systems with mutually perpendicular coordinate crystal axes, there appears to be a simple relation between zone symbols and Miller indices. That is, [100], [010], and [001] are respectively perpendicular to (100), (010), and (001) (see Fig. 11.6b). However, it is obvious in the monoclinic system that [001] ($=c$) cannot be perpendicular to (001) because (001), which is parallel to the $a-b$ plane, makes an angle with c greater than 90°. (See Figs. 11.6c and d.) Here [100] is the zone axis of faces (010) and (001) and is parallel to their line of intersection; [010] is the zone axis of (100) and (001); and [001] is the zone axis of faces (100) and (010) and is parallel to their intersection edge.

Zonal relations are clearly shown on the stereographic projection and are most useful in locating poles of crystal faces (see exercise 12).

MATERIALS

(a) One or several lattices (made of balls and metal connecting rods), in which various planes are outlined by variously colored balls. If such are not available, use the lattice drawings in Fig 11.7. (b) Wooden or plaster models of crystals. If these are not available, use the paper models constructed in exercise 9, Fig. 9.5 and/or Fig. 10.4, as well as Fig. 11.8.

ASSIGNMENTS

A. *With lattices available in the laboratory; alternatively, use Fig. 11.7*

1. Choose a set of coordinate axes for the lattice and determine the intercepts on these axes for various planes (planes that may be outlined by balls of various colors); if such a lattice is not available, use Fig. 11.7.

2. Determine the intercepts of each of the marked planes.

3. Obtain the Miller indices for the same planes.

4. Report your observations in Table 11.1.

FIGURE 11.5 (*a*) Perspective view of coordinate axes in the hexagonal system, and a plane defined as the unit plane. (*b*) Plan view of the three horizontal axes showing the horizontal trace of the plane.

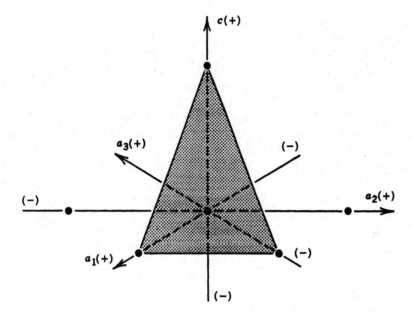

Intercepts of plane:
$1t_1$, ∞t_2, $-1t_3$, $1t_4$

(a)

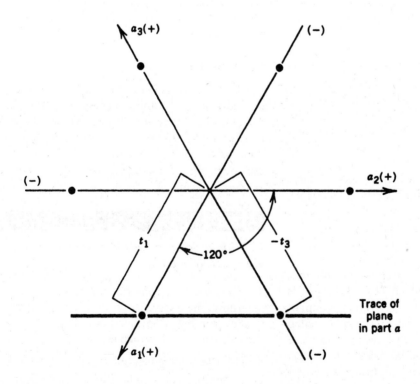

Intercepts of the above plane
on the horizontal axes:
$1t_1$, ∞t_2, $-1t_3$

(b)

FIGURE 11.6 (*a*) Tetragonal crystal with vertical faces parallel to [001]. (*b*) Vertical cross section through crystal in part *a*, showing zones [001] and [010]. (*c*) Monoclinic crystal. (*d*) Vertical cross section through crystal in part *c*; notice that here zone axes and perpendiculars to crystal faces are in different directions.

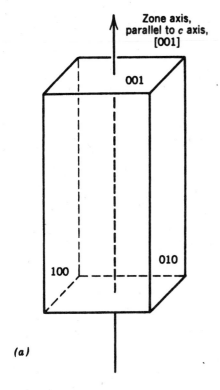

(*a*)

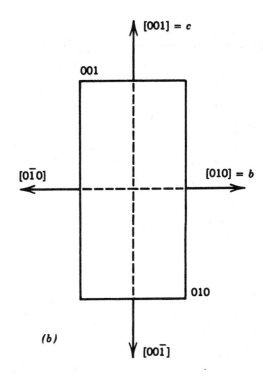

(*b*)

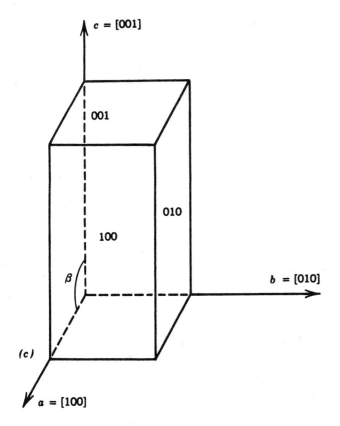

(*c*)

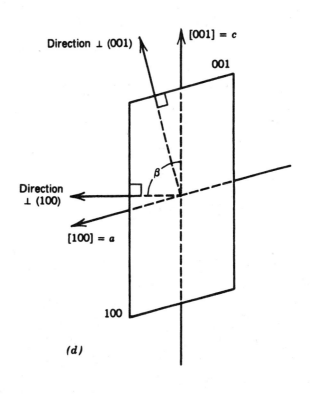

(*d*)

B. *With wooden crystal models available in the laboratory*

1. Determine the overall symmetry (crystal class) of the model.

2. Orient it according to the appropriate coordinate axes (see exercise 10).

3. Define a *unit face*, if such is present on the crystal [this is the (111) face]; if this is not present, the faces of your crystal (model) will have indices such as $(hk0)$, $(0kl)$, $(h00)$, $(00l)$, and (hkl).

4. Index all other faces on your crystal. Notice that in the foregoing "background information" we discussed only the indexing in the positive part of the coordinate system (e.g., along the + end of a, the + end of b, and the + end of c). A crystal will also have faces that intersect at the negative ends of a, b, and c. As such the total listing of the indices of faces on a crystal must include not only 100, but also $\bar{1}00$ (read: minus one, oh, oh); 310, but also $3\bar{1}0$ or $\bar{3}10$, etc. This should remind you that all the faces of a *form* have very similar Miller (or Bravais-Miller) indices.

5. Report your findings in Table 11.2.

C. *In the absence of wooden crystal models, or as an alternative to assignments A and B, use Fig. 9.6 and/or Fig. 10.4*

1. Figures 9.6 and 10.4 both depict common crystal forms and combinations. The forms are completely unidentified in Fig. 9.6, but they are given letters in Fig. 10.4. Using Fig. 9.6 as a base, locate the appropriate Miller indices on the crystal faces.

2. Using Fig. 10.4 give the appropriate Miller index for each of the forms identified by a letter. For example, write next to the crystal drawing $m = \{110\}$, etc.

D. *Using Fig. 11.8 as a basis for Miller index notation*

1. The external outline of an orthorhombic crystal, in an $a-b$ section, is shown in Fig. 11.8. This outline is superimposed on a net of lattice nodes in which the origin and the $+a$ and $+b$ axes are identified. All the crystal faces are parallel to the c axis, which is vertical and comes out of the page toward you, from the point of origin. As such, all the faces, as well as the one set of crystal planes (dashed lines), are of the $(hk0)$ type. On the right-hand side of the figure list the intercepts for each of the faces.

2. In the bottom part of the figure convert these to Miller indices for each face.

3. Lastly, give the Miller index notation for each of the forms, in the lower right part of the figure.

Student Name

FIGURE 11.7 Two lattices with five planes (A, B, C, D, E) and four planes (F, G, H, I), respectively. For clarity nodes are shown only along edges of the lattice. Record your results in the lower half of Table 11.1.

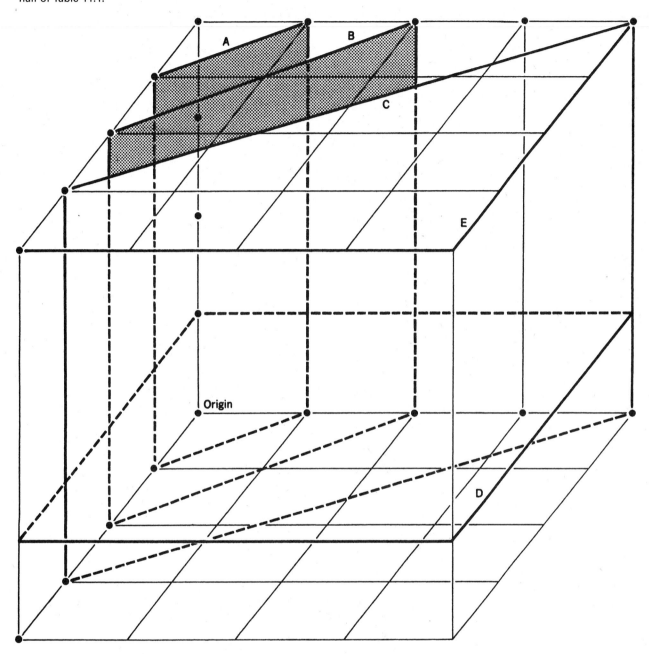

FIGURE 11.7 (continued)

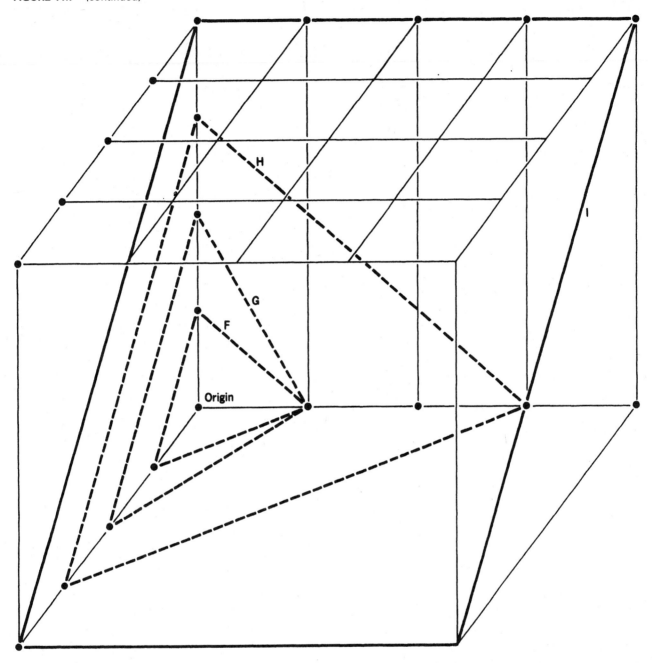

Origin

F

G

H

I

EXERCISE 11

FIGURE 11.8 Horizontal cross section through an orthorhombic crystal, superimposed on a regular pattern of lattice nodes. Record intercepts and Miller indices for faces on the crystal, and record form index symbols at the bottom of the page.

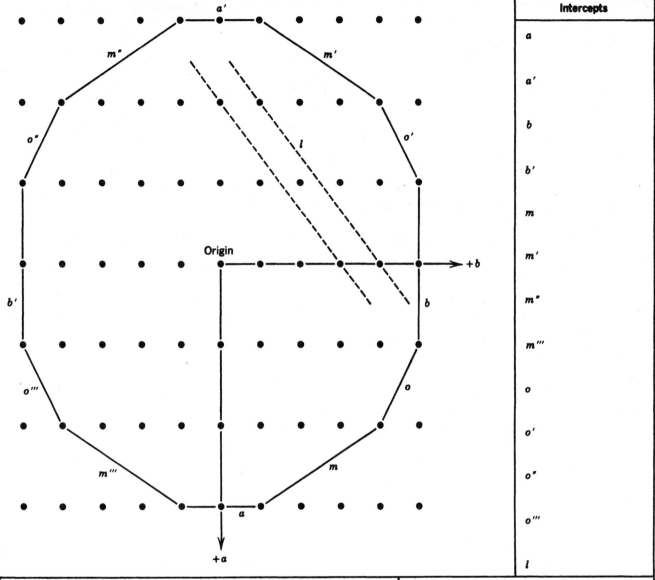

Intercepts
a
a'
b
b'
m
m'
m"
m'''
o
o'
o"
o'''
l

Miller indices		Form index symbol
a	o	a
a'	o'	b
b	o"	m
b'	o'''	o
m	l	l
m'		
m"		
m'''		

TABLE 11.1 Planes in a Lattice Model that has been provided by the instructor.

Coordinate Axes (Their Lengths and Angles Between Them)	Intercepts of Planes (Identify Plane by Color or Letter)	Miller Indices of the Same Planes

With reference to the lattice in Fig. 11.7, identify the coordinate axes on the figure and tabulate your observations.

Intercepts of Planes	Miller Indices of the Same Planes
A.	
B.	
C.	
D.	
E.	
F.	
G.	
H.	
I.	

TABLE 11.2   Faces and Directions in Crystal Models provided by the instructor.

Overall Symmetry (Crystal Class of Crystal)	Miller Index for Each Crystal Face
	Faces
	a.
	b.
Coordinate Axes (Their Lengths and Angles Between Them)	c.
	d.
	e.
	f.
	etc.
Crystal Sketch (optional)[a]	**Define Obvious Zones in Crystal by Direction Index Symbols:**

[a]For reference it may be useful to sketch your crystal and to give letters to the various forms. If more of these forms are needed, please make photocopies.

EXERCISE 12

Stereographic Projection

PURPOSE OF EXERCISE

Representation of a three-dimensional object (crystal) and its symmetry content in a two-dimensional drawing, through stereographic projection.

FURTHER READING AND CD-ROM INSTRUCTIONS

Klein, C. and Dutrow, B., 2008, *Manual of Mineral Science*, 23rd ed., Wiley, Hoboken, New Jersey, pp. 169–181.

CD-ROM, entitled *Mineralogy Tutorials*, version 3.0, that accompanies *Manual of Mineral Science*, 23rd ed., click on "Module II", and subsequently on the button "Stereographic Projection."

Perkins, D., 2002, *Mineralogy*, 2nd ed., Prentice Hall, Upper Saddle River, New Jersey, p. 196.

Wenk, H. R. and Bulakh, A., 2004, *Minerals: Their Constitution and Origin*, Cambridge University Press, New York, New York, pp. 57–67.

Bloss, F. D., 1994, 2nd printing with minor revisions, *Crystallography and Crystal Chemistry*, Mineralogical Society of America, Chantilly, Virginia, pp. 72–98.

Background Information: In previous exercises you have seen many examples of clinographic projections of crystals. At times you have been asked to make perspective sketches of crystals similar to these clinographic projections. If free-hand sketching is not your favorite pastime, you will be glad to know there are other ways of presenting three-dimensional information about crystals on a two-dimensional page; one such way is *stereographic projection*. A stereographic projection is the reduction of a sphere onto a page, by a method in which all angular relations that can be defined on a sphere are preserved on the page. This is essential for the representation of symmetry. You are referred to one of several texts for the derivation of a stereographic net (see references above) from a spherical representation. Here it is assumed that you know how a stereographic net was constructed, because only if this is clear to you will you be able to manipulate the projection of crystal faces and directions on a stereographic net.

A stereographic net of 10-cm radius, which forms the basis of all aspects of this exercise, is printed in Fig. 12.1 Tear Fig. 12.1 out and mount it on thin poster board. The exact center of the net should be pierced, from the back, by the sharp point of a thumbtack. This point will function as the pivot about which a sheet of tracing paper (as an overlay) can be rotated.

The process of plotting crystallographic orientations (of planes and directions) is best illustrated in various steps with reference to Fig. 12.2.

a. Trace the outer circle of the net (called the primitive) on the tracing paper.

b. Make a reference mark at the positive end of the y direction (or b axis direction; at the exact "east side" of the east–west great circle) of the overlay. This point is the $\phi = 0°$ position.

c. You may also wish to mark $\phi = 90°$, as well as $\phi = -90°$, at the "south" and "north" poles of the net on the overlay, respectively.

d. You are now ready to plot ϕ and ρ angles on the overlay as given to you from a reference, or as measured by you directly using a contact goniometer on a crystal model (exercise 13). The ϕ angles are plotted in clockwise (+) or counterclockwise (−) directions. A step-by-step sequence for the plotting of ϕ and ρ angles is given in Figs. 12*a* to *c*.

e. The assignment in this exercise consists of the plotting of already measured ϕ and ρ angles. These ϕ and ρ angles represent the *poles* of specific crystal faces. The pole of a face is a direction that is perpendicular to the face.

MATERIALS

A stereographic net mounted on thin board, translucent paper, a thumbtack, a ruler, and a soft pencil.

ASSIGNMENT

1. Take the ϕ and the ρ angles listed in Table 12.1 and plot all faces. Figure 12.3 depicts many of the faces you are asked to plot. First, on the primitive count from $\phi = 0°$ to the required number of degrees for ϕ. Make a mark on the tracing paper (count clockwise for positive angles, counterclockwise for negative angles). The pole of the face, depending on the ρ value of the face, must lie somewhere between the center of the stereonet and the mark you just made for the appropriate ϕ. To locate the pole, rotate the tracing paper until the appropriate ϕ mark overlies the east–west line and then measure ρ from the center out. To plot the next face return the $\phi = 0°$ reference mark to the due east direction. Keep repeating the foregoing procedure until all face poles are plotted. If poles are required for faces in the northern as well as southern hemisphere, indicate poles in the upper part of the projection by • and in the lower part by ○ (put this ○ about the •).

2. After all poles for the faces have been marked on your tracing, *plot the symmetry elements of the point group* shown in the heading of Table 12.1.

3. Subsequently generate the poles of all the other faces (e.g., $00\bar{1}$, $0\bar{1}0$, etc.) that must be present on the crystal as a result of its symmetry. You will have generated forms that consist of sets of symmetrically related faces.

4. Mark the position of the a_1, a_2, and c axes and give the Miller index for each of the poles on your tracing paper. Note that poles that superimpose (for northern and southern hemisphere projection) should have two sets of Miller indices. Refer to pages 174 to 181 and especially Figs. 8.10 and 8.11 in *Manual of Mineral Science* (2008), for worked examples and the final expected results.

5. A very useful feature of a stereographic net is that the angular distance between any two points can be easily determined. An arc connecting two points is always part of a great circle. By rotation of the tracing paper you can make any two points lie on the same great circle, and you can then measure between the two poles by counting off the small circles crossed.

Graphically measure the angles between several face poles as requested in Table 12.2. Trace the great circle on which each pair of poles lies, and write along it the measured angle along the great circle, between the poles specified. Report your angular measurements in Table 12.2.

6. Observe some zonal relationships. *A group of crystal faces whose intersection edges are parallel are said to define a zone.* Zones are important in revealing the symmetry of a group of face poles and in helping to indicate the positions of faces that you may have accidentally missed in plotting. *It follows from the definition of a zone that all the faces in a zone lie on the same great circle.* The zone axis is then the pole to that great circle. In Table 12.2 you are asked to define several zonal relations. Fill out the information requested on the basis of the stereographic plot.

Student Name

FIGURE 12.1 Stereographic net with a 10-cm radius.

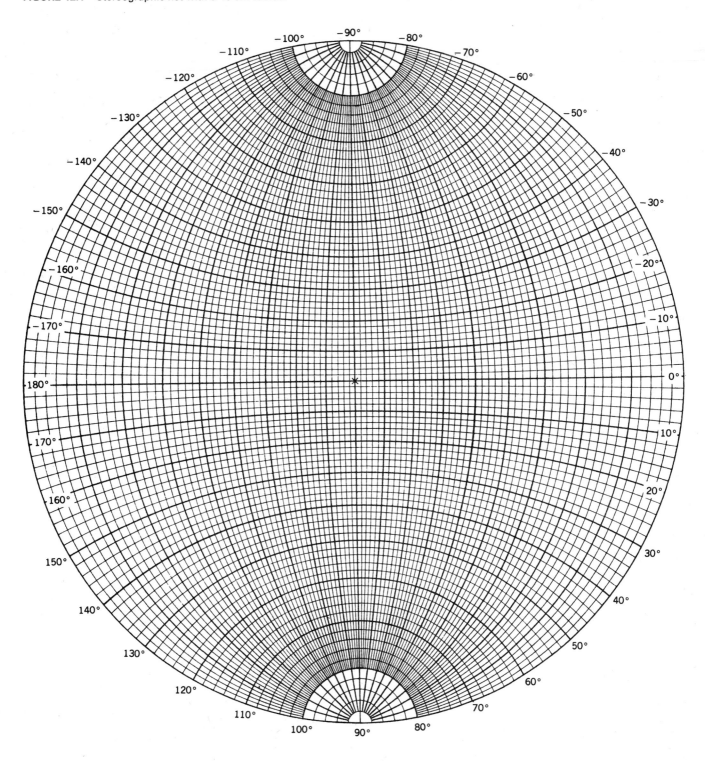

Please make photocopies of this form if more copies are needed.

FIGURE 12.2 (*a*) Illustration of the use of a stereographic net (mounted on thin board), pierced in its center by a thumbtack from the back and overlaid by a somewhat transparent paper (tracing paper or onionskin paper). The primitive circle, as well as the locations of $\phi = 0°$, $\phi = 90°$, $\phi = -90°$, must always be marked on the transparent overlay before any angles are plotted. (*b*) To project the pole of a plane with $\phi = 30°$, and $\rho = 60°$, we plot the angle $\phi = 30°$, on the primitive (*x*) in a clockwise direction from $\phi = 0°$. (*c*) The direction of $\phi = 30°$ has been rotated to coincide with the E-W direction and the angle $\rho = 60°$ can be measured directly along the vertical great circle. The black dot is the pole of the crystal face with $\phi = 30°$, $\rho = 60°$. (This is Fig. 8.8 in *Manual of Mineral Science*, 23rd ed., p. 176.)

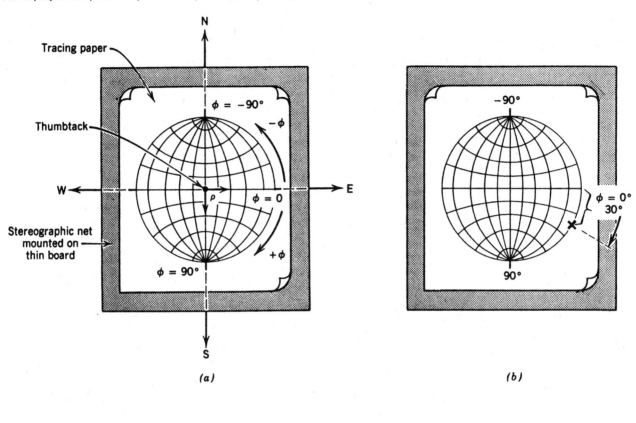

(a)   (b)

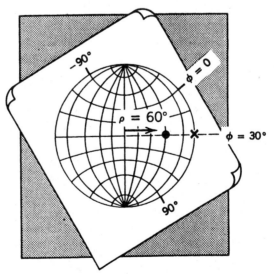

(c)

TABLE 12.1 ϕ and ρ Angles for Crystal Forms on Hausmannite ($MnMn_2O_4$) with Point Group Symmetry $4/m2/m2/m$

Forms[a]	Miller index	ϕ	ρ
c	{001}	...	0°
a	{010}	0°	90°
m	{110}	45°	90°
p	{011}	0°	58°34′
s	{013}	0°	28°36′
n	{021}	0°	73°
e	{112}	45°	49°10′
k	{121}	26°34′	74°43′
h	{136}	18°26′	40°46$\frac{1}{2}$′
r	{123}	26°34′	50°39′

[a]See Fig. 12.3 for illustrations of several of these forms.

SOURCE: Data from *Dana's System of Mineralogy*, vol. 1, C. Palache, M. Berman, and C. Frondel, seventh ed, 1944, Wiley, New York, p. 713.

FIGURE 12.3 (*a, b*) Clinographic projections of two different hausmannite crystals. (From *Dana's System of Mineralogy*, vol. 1, C. Palache, M. Berman, and C. Frondel, seventh ed., 1944, Wiley, New York, p. 137.)

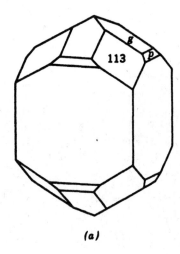

(a)

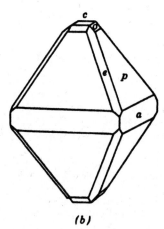

(b)

TABLE 12.2 Report on Results of Angular Measurements, Zone Axes, and Number of Faces per Form, Using the Stereographic Projection of the Data in Table 12.1

A. Angular Measurements Between Face Poles

$(001) \wedge (021) =$

$(013) \wedge (011) =$

$(112) \wedge (110) =$

$(021) \wedge (121) =$

$(136) \wedge (112) =$

$(011) \wedge (123) =$

$(201) \wedge (211) =$

B. Give ϕ and ρ Angles for Several Zone Axes

$(001), (013), (011), (021), (010)$ define $\phi =$ ————, ρ ————

$(001), (112), (110)$ define $\phi =$ ————, ρ ————

$(021), (121), (100)$ define $\phi =$ ————, ρ ————

C. Forms, Their Total Number of Faces, and Their Type

Form	Number of Faces	Special(S) or General(G)
c, {001}		
a, {010}		
e, {112}		
r, {123}		
n, {136}		

Contact Goniometer Measurements of Interfacial Angles and the Plotting of these Angles on a Stereographic Net

PURPOSE OF EXERCISE

Transferring the information on symmetry, forms, and interfacial angles—as expressed by the morphology of a crystal—into a two-dimensional representation (projection) that reveals true symmetry. In such a representation all directions and symmetry elements pass through a point. This, then, leads to the graphical representation of *point groups* (crystal classes).

FURTHER READING

Klein, C. and Dutrow, B., 2008, *Manual of Mineral Science*, 23rd ed., Wiley, Hoboken, New Jersey, pp. 174–181.

Perkins, D., 2002, *Mineralogy*, 2nd ed., Prentice Hall, Upper Saddle River, New Jersey, p. 196.

Wenk, H. R. and Bulakh, A., 2004, *Minerals: Their Constitution and Origin*, Cambridge University Press, New York, New York, pp. 57–67.

Bloss, F. D., 1994, 2nd printing with minor revisions, *Crystallography and Crystal Chemistry*, Mineralogical Society of America, Chantilly, Virginia, pp. 72–98.

Background Information: This assignment is a sequel to exercises 11 and 12. Manipulation of a stereographic net (as discussed in exercise 12) is a skill that you must acquire before you can tackle this assignment. The indexing of crystal faces (discussed in exercise 11) is required in order to identify the specific crystal face that you are plotting on the stereographic net.

Exercise 12 provided a table of carefully measured ϕ and ρ angles for a specific crystal; these measurements were obtained originally using an optical instrument known as a reflecting goniometer (see Fig. 1.13, *Manual of Mineral Science*, 23rd ed., 2008). Because such angular measurements can be derived with high precision, the data for ϕ and ρ angles are usually reported in terms of degrees, minutes, and seconds of degrees. Such ϕ and ρ angles are generally obtained from small, well-formed crystals (ranging from several centimeters to millimeters in size) and require considerable skill in the operation of a reflecting goniometer.

A much simpler, but also much less accurate, measurement technique is using a *contact goniometer* on large crystals or on wooden (or plaster) models of crystals. They should be big enough to hold easily and to allow alignment of a specific set of faces against the moving arm of the measuring device. A contact goniometer and the angle it measures—the *interfacial angle*—are shown in Fig. 13.1. The interfacial angle as shown in Fig. 13.1b is the angle between the *poles* (perpendiculars) to the two faces in question (the angle between the two faces being 180° − 40° = 140°). As such the 40° angle measured by the contact goniometer equals the interfacial angle, which is the supplemental angle to 140°.

If you are given (or have selected) a crystal (model) with well-developed (001) and (010) faces, you can refer all your measurements of other faces on the crystal to (010), for which $\phi = 0$, and to (001), for which $\rho = 0$. As such you could tabulate your measurement results in terms of ϕ and ρ, but this generally is not the most practical way to measure and record angular measurements on crystals. Rather, it is simpler to measure and plot a new pole from some previously plotted pole.

As an example, let us make interfacial angle measurements on the orthorhombic crystal of olivine shown in Fig. 13.2a. Its overall symmetry is 2/m2/m2/m. The coordinate axes (a, b, and c) are chosen accordingly (see Table 10.3). In the upper right octant of the crystal (where a, b, and c are all positive) the only face that intersects all three axes is the inclined triangular face that is selected as the unit face (111). The Miller indices for all other faces (they are given only for the positive octant) follow from the selection of the inclined face as (111). The face (100) is perpendicular to a; (010) is perpendicular to b; (001) is perpendicular to c; (110) cuts a and b and is parallel to c, etc. Notice that the inclined face that cuts b and c, and is parallel to a, is given the index (021). This Miller index is the result of noting that the b intercept of this inclined plane is half the distance along b of the b intercept of the (111) plane.

Once you have a clear picture of all that is depicted in Fig. 13.2a, you can go ahead with the measurement and plotting of interfacial angles. Figure 13.2b shows a partial vertical (b–c) section of the crystal in (a). The angles that one measures in this section with a contact goniometer are the 49° interfacial angle between (001) and (021) and the 41° interfacial angle between (021) and (010). Similar measurements can be made in the vertical a–c section for

FIGURE 13.1 (*a*) Contact goniometer. (*b*) Schematic enlargement of part *a*, showing the measurement of the internal angle α. (From *Manual of Mineral Science,* 23rd ed., p. 177.)

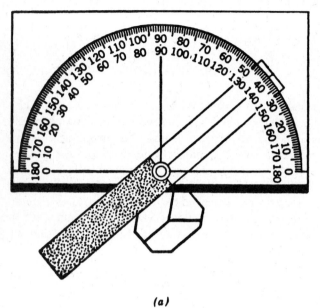

(*a*)

(*b*)

FIGURE 13.2 (*a*) Orthorhombic crystal showing various forms. (*b*) Partial vertical cross section (along *b–c*). (*c*) Some interfacial angles plotted in the upper right positive octant. (*d*) A complete stereographic projection of the crystal in part *a*, with Miller indices for the upper half only.

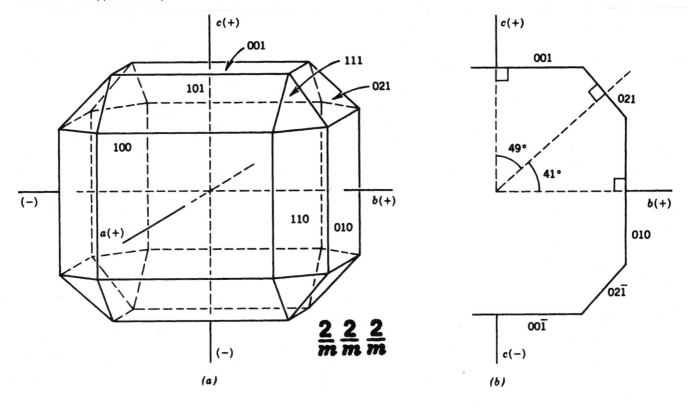

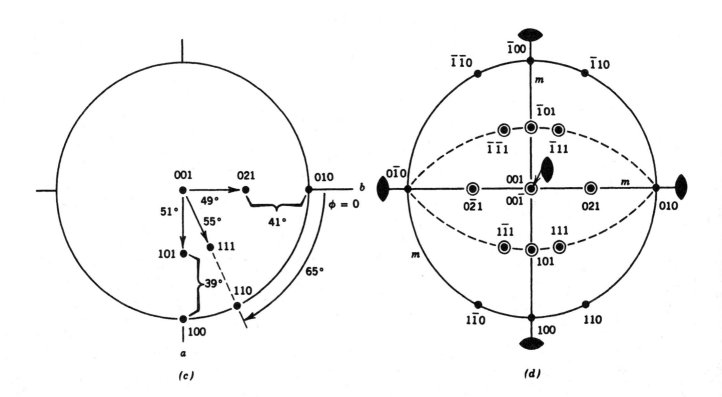

angles between (001) and (101) and/or (101) and (100). The (001) Λ (101) angle turns out to be 51°, and the (100) Λ (101) angle 39°. In a horizontal section the (010) Λ (110) angle can be measured as 65°, and the (110) Λ (100) angle as 25°. We have now measured all interfacial angles in the positive octant except for the angle between, for example, (001) and (111). This turns out to be 55°, measured along a vertical section perpendicular to (110) at a ϕ of 65° [the 65° angle for (010) Λ (100)]. Figure 13.2c shows the plotted positions of the above angles in the positive octant of the crystal in Fig. 13.2a. After all the symmetry elements in $2/m2/m2/m$ have been plotted on the same stereographic net, the poles to all the other faces on the crystal can now be generated (by the symmetry elements) instead of continuing to measure the position of the remaining faces with a contact goniometer. Figure 13.2d shows the location of all face poles and the Miller indices of faces only in the upper half ($+c$) of the crystal. The negative indices are omitted for reasons of clarity. As such, Fig. 13.2d is the completed stereographic projection of the clinographic representation of the orthorhombic crystal in Fig. 13.2a.

Let us briefly look at some possible complications in the measurement and plotting of interfacial angles for crystals (a) without the development of (001), with $\rho = 0°$, and/or (010), with $\phi = 0°$, and (b) with monoclinic symmetry (and therefore a nonorthogonal system of coordinate axes).

a. Figure 13.3a shows an orthorhombic crystal (antlerite), which lacks (010) as well as (001). The Miller indices for faces in the upper octant are shown. The small face (r) is the only one that cuts all three axes (a, b, and c) and as such is designated as (111). It is really very simple to plot the faces of the major forms on this crystal even though two reference faces such as (010) and (001) are missing. Figure 13.3b shows a horizontal cross section (an a–b section) of the crystal in Fig. 13.3b as well as the interfacial angle for (110) Λ ($\bar{1}10$), which turns out to be 110°. Half this angle is the ϕ angle for (110) at 55°, which can be plotted in the positive (clockwise) direction from $\phi = 0°$ (= b axis). This is equivalent to subtending the complete 110° angle symmetrically about the b axis; this then plots the poles of (110) as well as ($\bar{1}10$). Similarly, we can look at the crystal in Fig. 13.3a in a vertical (b–c section) and measure the interfacial angle between (011) and ($0\bar{1}1$), which turns out to be 53°. Half this angle (26°30′) is the ρ angle of (011). The only face left to plot in the upper right quadrant is (111). It has the same ϕ angle as (110) (all faces with the same h and k in their indices have the same ϕ angle), and although we cannot measure its ϕ angle, we can measure its complement, that is, the interfacial angle between (110) and (111). This angle is found to be 48°30′. Turn the projection so that the pole of (110) lies at the end of a vertical great circle (e.g., along the east–west diameter of the net) and plot the pole of (111)

$48\frac{1}{2}°$ in from the primitive. Turn the overlay back to the zero position and trace a great circle through the pole of (011). As would be expected, this great circle also passes through the pole of (111) (see Fig. 13.3c).

An alternative way for the plotting of the pole of (111) is as follows. Measure the interfacial angle (111) Λ ($11\bar{1}$) = 53°. Half this angle is $26\frac{1}{2}°$, the same as the ρ angle of (011). Face (111) therefore must lie on the great circle drawn through the pole (011) in the position $\phi = 0$. Measure (011) Λ (111) = about 32°. Count 32° forward on this great circle to locate the pole of (111) (see Fig. 13.3c).

If you have difficulty with great circles, you should consult any of the references in exercise 12. However, let us review some aspects of zones and great circles. Any two faces determine a zone, and a great circle can be drawn between their poles. Poles of faces with k and l of their indices the same, for example, (011), (111), (211), and so on all lie in a zone. Similarly, poles of faces with h and l of their indices the same lie in a zone. In orthogonal systems all faces with the same h and k of their indices have the same ϕ and lie in a horizontal zone with their poles on a vertical great circle.

All the faces in the positive octant are now located, and when the $2/m2/m2/m$ symmetry is taken into account, all other faces in the other seven octants can be generated. A final stereographic projection of all faces in this antlerite crystal is given in Fig. 13.3d.

b. Figure 13.4a shows a monoclinic crystal of the pyroxene jadeite with symmetry $2/m$. Monoclinic symmetry demands that the coordinate axes are not all at 90° to each other. There is a β angle (in this case of 107°) between the + ends of a and c. The inclination of a is determined by the angle the well-developed (001) plane makes with +c. The (001) plane is parallel to the a and b axes. This plane is not normal to c in the monoclinic and triclinic systems. This is shown in the a–c cross section of the crystal in Fig. 13.4b. This figure also shows that, in such a nonorthogonal crystal, the pole to the (001) face is not the same as the direction of the c axis; indeed the two directions differ by 17° (107° − 90° = 17°)

The following interfacial angles have been determined:

$$(010) \Lambda (110) = 44°$$
$$(100) \Lambda (001) = 73°$$
$$(\bar{1}00) \Lambda (\bar{1}01) = 76°$$

Plotted on the primitive in Fig. 13.4c are the poles of (010) at $\phi = 0°$; (100) at $\phi = 90°$; and (110) at $\phi = 44°$. Because (001) and (100) lie in the same zone, with $\phi = 90°$, the pole of (001) can be plotted 73° in from the primitive along the north–south vertical great circle [ρ (001) = 90° − 73° = 17°]. In like manner the pole of ($\bar{1}01$), which lies in the same zone but with $\phi = -90°$, can be plotted 76° in from the primitive along the north–south vertical great circle [ρ ($\bar{1}01$) = 90° − 76° = 14°].

FIGURE 13.3 (*a*) Orthorhombic crystal showing various forms. (*b*) Partial horizontal cross section (along *a–b*). (*c*) Projection of some face poles on the basis of interfacial angles; see text for discussion. (*d*) A complete stereographic projection of the crystal in part *a*, with Miller indices for the upper half only.

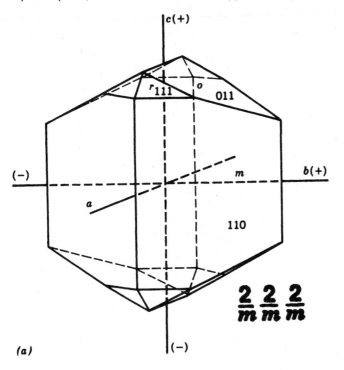

(a)

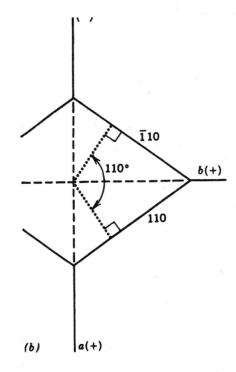

(b)

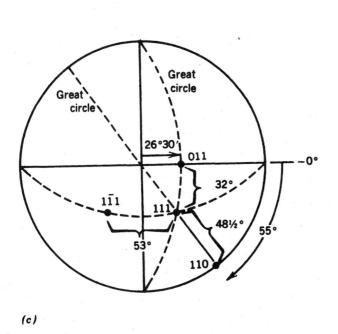

(c)

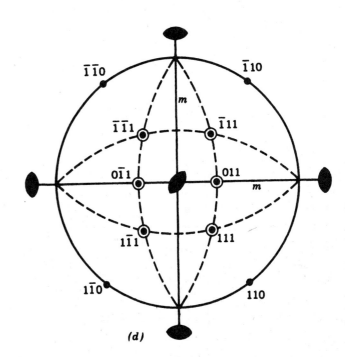

(d)

FIGURE 13.4 (a) Monoclinic crystal with various forms. (b) Vertical cross section (along *a–c*). (c) Some interfacial angles plotted in the upper half of the crystal; see text for discussion. (d) A complete stereographic projection of the crystal in part (a), with Miller indices for all faces.

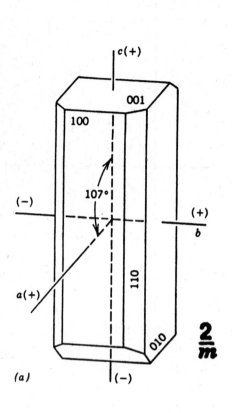

(a)

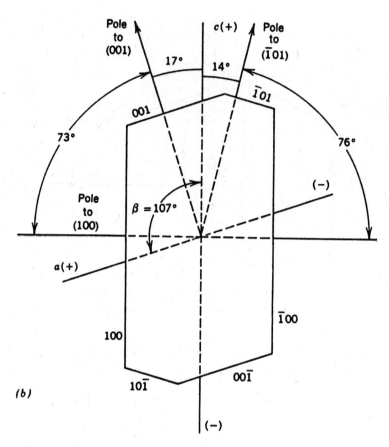

(b)

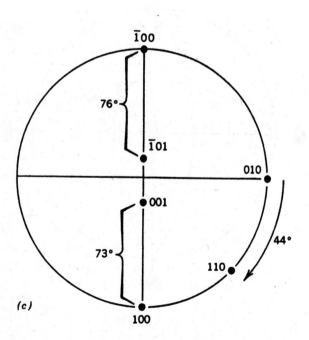

(c)

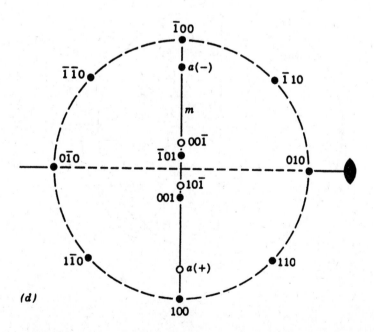

(d)

The only other information that is needed and is not immediately obvious is the location of the *a* axis and its piercing points through the equatorial plane (as in a spherical projection). Because by convention the + end of the *a* axis dips below the equatorial plane in a sphere (see Fig. 13.4*b*) and the − end of *a* is inclined above this equatorial reference plane, the ends of this axis (be it + or −) will plot 17° [as defined by the (001) face] inward from the circumference (from 107° − 90° = 17°). The + end is conventionally shown by a small circle, and the − end by a solid dot, analogous to the convention used for faces at the top (•) and faces at the bottom (○) of a crystal. These piercing points for the *a* axis are located in Fig. 13.4*d*. A complete stereogram generated by the 2/*m* symmetry acting on the faces already plotted in Fig. 13.4*c* is shown in Fig. 13.4*d*. (For further discussion of projecting monoclinic crystals, as well as a complete example of such a projection, see *Manual of Mineral Science*, 23rd ed., pp. 179–181.)

MATERIALS

A stereographic net (please photocopy Fig. 12.1) mounted on thin board (see exercise 12), translucent paper, a thumbtack, a ruler, a soft pencil, and a contact goniometer; also one or several wooden (or plaster) models of crystals. If such models are not available, you can use the paper cutout models from exercise 9 provided you were indeed careful in their construction. If a contact goniometer is not available, you can improvise by using a protractor and a separate plastic ruler.

ASSIGNMENT

1. Place a sheet of tracing paper on the top of the mounted stereonet and keep it in place with the aid of a thumbtack (in the exact center of the net and pierced from the back of the net). This allows for rotation of the tracing paper about the center of the net.

2. Trace the outer circle of the net on the tracing paper.

3. Make a reference mark at $\phi = 0°$, that is, the positive end of the *y* direction (or *b* axis), on the east side of the east–west great circle.

4. Determine the overall symmetry (point group) of the crystal model you selected. Identify the symmetry elements and plot them on the stereonet.

5. Select coordinate axes in accordance with the symmetry (see Table 10.3).

6. Orient your crystal with the *c* axis vertical, *b* east–west, and *a* toward you.

7. Make contact goniometric measurements of the most obvious faces in the upper right–hand part of the crystal. Measure your interfacial angles with respect to (010) and (100) if these are present. If these are lacking, interfacial angles must be measured and plotted somewhat differently (see the "background information").

In making measurements of interfacial angles between two crystal faces, *you must hold the protractor plate at right angles to the edge between faces and normal to both faces!* It may be helpful to hold the crystal with the contact goniometer in place, toward a light source (window or lamp). When the edges of the protractor plate and the ruler are in good contact with the crystal faces, all light (depending on the smoothness of the faces) will be cut out.

8. Plot the angles and their face poles for faces in the upper half, quarter, or octant of the crystal on the stereographic net.

9. Identify these faces with their Miller indices.

10. Generate all the remaining faces of the crystal in projection by letting the symmetry elements (located earlier on the stereonet) act upon the faces already plotted (in the positive octant).

11. Give Miller indices to all face poles so generated.

12. Make sure the elements of symmetry and the coordinate axes are well identified on your stereonet. Once this is done, the assignment for a specific crystal has been completed.

13. If there are additional crystals to be projected, work through points 1 through 12 for each.

Development of the {111} or {11$\bar{2}$1} Form in Several Point Group Symmetries

PURPOSE OF EXERCISE

Further understanding of the interdependence of form and point group symmetry (crystal class).

FURTHER READING AND CD-ROM INSTRUCTIONS

Klein, C. and Dutrow, B., 2008, *Manual of Mineral Science*, 23rd ed., Wiley, Hoboken, New Jersey, pp. 125–142.

CD-ROM, entitled *Mineralogy Tutorials*, version 3.0, that accompanies *Manual of Mineral Science*, 23rd ed., click on "Module II", and subsequently on the button "Crystal Classes".

Nesse, W. D., 2000, *Introduction to Mineralogy*, Oxford University Press, New York, pp. 24–58.

Perkins, D., 2002, *Mineralogy*, 2nd ed., Prentice Hall, Upper Saddle River, New Jersey, pp. 187–203.

Wenk, H. R. and Bulakh, A., 2004, *Minerals: Their Constitution and Origin*, Cambridge University Press, New York, New York, pp. 66–78.

Bloss, F. D., 1994, 2nd printing with minor revisions, *Crystallography and Crystal Chemistry*, Mineralogical Society of America, Chantilly, Virginia, pp. 99–139.

Background Information: This exercise will draw on materials presented in exercise 9 (symmetry elements and their combinations), exercise 10 (recognition of some of the common crystal forms), and exercise 12 (stereographic projection). As such, no new concepts are introduced.

This assignment will develop further understanding of the relationship of crystal symmetry (ranging from low, as in the triclinic system, to high, as in the isometric system) to the shape and complexity of a form. See Fig. 14.1 for a résumé of stereograms of the symmetry contents of the 32 point groups. In these diagrams the points represent the face poles of the general form.

MATERIALS

The various pages that compose Fig. 14.2, ruler, protractor, and soft pencil. (Because in this exercise specific angular measurements are not given or required, a stereographic net is not needed; however, the general concept of stereographic projection of face poles is basic to this exercise.)

ASSIGNMENT

1. Using the various pages that compose Fig. 14.2, beginning with stereogram 1, plot the total symmetry content of the point group onto the equatorial circle of the stereogram. Use standard graphical symbols as shown in Fig. 14.1. Please refer back to Table 10.3 for the various conventions implied by the Hermann–Mauguin notation.

2. Locate coordinate axes, and label them clearly on the stereogram.

3. Locate a (111) face in the upper right front octant of the stereogram, and on the basis of the symmetry elements already plotted, develop the complete {111} form.

4. Index all face poles in the stereogram.

5. List the number of faces comprising the form.

6. State whether the form is open or closed.

7. Is this a special or a general form?

8. Give the form name (see Fig. 10.1, or consult any of the references listed).

9. Do this sequentially for all entries (1 through 14).

For the hexagonal point groups (8 through 11) the Bravais–Miller index for the face that cuts three of the four coordinate axes at unity (and the remaining axis at $-\frac{1}{2}$) is (11$\bar{2}$1). In these three hexagonal point group assignments you will, therefore, develop the {11$\bar{2}$1} form instead of the {111} form as in the other five crystal systems.

In isometric point group assignments (12 through 14), you may find it helpful to use a stereonet (from exercise 12) to locate a number of the inclined symmetry elements (mirror planes, as well as 3- and 2-fold rotation axes). On the stereographic net locate the planes at 45° (along great circles) to the a_1, a_2, and a_3 axes. These contain the $\bar{3}$ axes at the intersection points of pairs of such inclined mirror planes. The four 2-fold rotation axes occur along the intersections of these inclined planes and vertical (axial) sections (see Figs. 6.18 and 6.23, *Manual of Mineral Science*, 23rd ed.). Transfer the stereonet information to the smaller circles on the assignment pages.

NOTE: *The {111} form can be the same form as {hkl} in the triclinic, monoclinic, and orthorhombic systems. However, in the tetragonal, hexagonal, and isometric systems the {111} and {hkl} differ from each other.*

FIGURE 14.1 Graphical representation of the distribution of motif units compatible with the symmetry elements of each of the 32 crystal classes (point groups). The points (representing possible motif units) in these diagrams are equivalent to the face poles of the general forms. For all crystal classes, excepting the triclinic, there are two circular diagrams, with the left-hand diagram showing the distribution of motif units (equivalent to face poles of the general form), and the right-hand diagram illustrating the symmetry elements consistent with these motif units (or poles). The motif units above the page are equivalent to those below the page, but they are differentiated by dots (above the page) and circles (below the page). The symbols for the symmetry elements are given at the top left corner of the diagram. The presence of a center of symmetry is not shown by any symbol; its presence can be deduced by the arrangement of motif units. Instead of 2̄ the symbol for a mirror (m) is used. The diagrams for the monoclinic system are shown in what crystallographers refer to as the "second setting," with m vertical (perpendicular to the page) and the 2-fold axis in an east–west location. Monoclinic symmetry can also be shown by setting the 2-fold rotation axis perpendicular to the page, and orienting the mirror parallel to the page; this is referred to as the "first setting." (From *Manual of Mineral Science*, 23rd ed. p.186 and 187.)

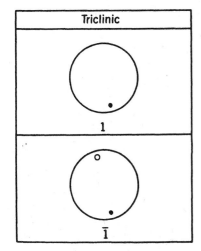

Written symbols	Graphical symbols
1	None
2	⬥
3	▲
4	◆
6	⬢
m	—
1̄ (≡center)	} *See caption
2̄ (≡m)	
3̄ (≡3 plus center)	▲
4̄	◆
6̄ (≡3/m)	⬢

Motif above page •
Motif below page ○

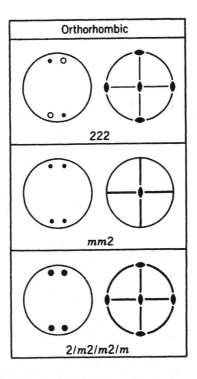

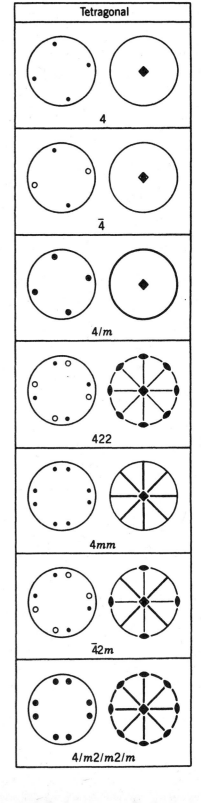

FIGURE 14.1 (continued)

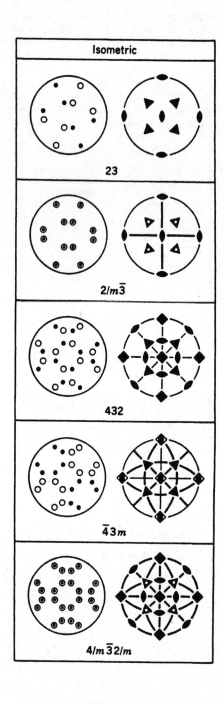

FIGURE 14.2 Stereographic assignments

Student Name

1: *m*

Plot (111) to develop {111} form

Number of faces: _____

Open or closed: _____

Special or general: _____

Name: _____

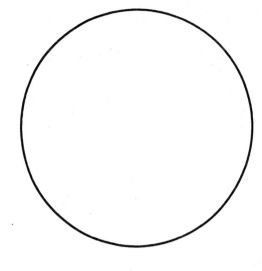

2: 2/*m*

Plot (111) to develop {111} form

Number of faces: _____

Open or closed: _____

Special or general: _____

Name: _____

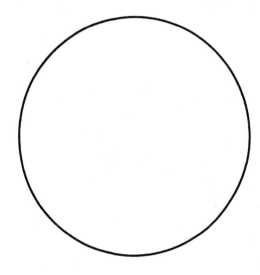

3: 222

Plot (111) to develop {111} form

Number of faces: _____

Open or closed: _____

Special or general: _____

Name: _____

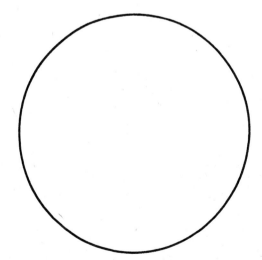

FIGURE 14.2 *(continued)*

4: 2/m2m2/m

Plot (111) to develop {111} form

Number of faces: _____

Open or closed: _____

Special or general: _____

Name: _____

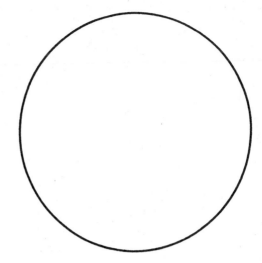

5: $\overline{4}2m$

Plot (111) to develop {111} form

Number of faces: _____

Open or closed: _____

Special or general: _____

Name: _____

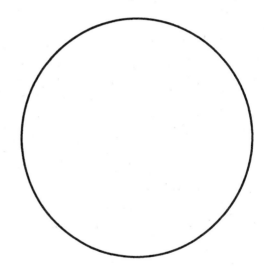

6: 4/m2/m2/m

Plot (111) to develop {111} form

Number of faces: _____

Open or closed: _____

Special or general: _____

Name: _____

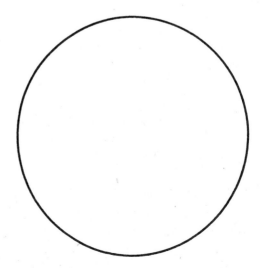

FIGURE 14.2 *(continued)*

Student Name _____

7: 4/m

Plot (111) to develop {111} form

Number of faces: _____

Open or closed: _____

Special or general: _____

Name: _____

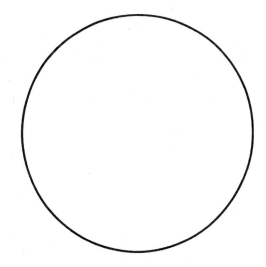

8: 6mm

Plot (11$\bar{2}$1) to develop {11$\bar{2}$1} form

Number of faces: _____

Open or closed: _____

Special or general: _____

Name: _____

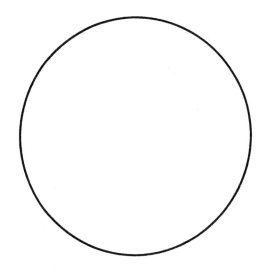

9: 6/m

Plot (11$\bar{2}$1) to develop {11$\bar{2}$1} form

Number of faces: _____

Open or closed: _____

Special or general: _____

Name: _____

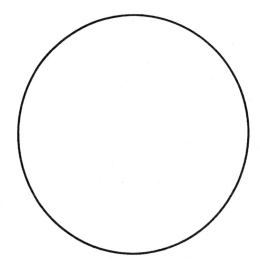

FIGURE 14.2 *(continued)*

10: $\bar{3}2/m$

Plot (11$\bar{2}$1) to develop {11$\bar{2}$1} form

Number of faces: _____

Open or closed: _____

Special or general: _____

Name: _____

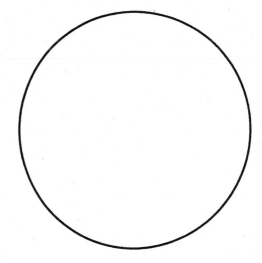

11: $\bar{6}$m2

Plot (11$\bar{2}$1) to develop {11$\bar{2}$1} form

Number of faces: _____

Open or closed: _____

Special or general: _____

Name: _____

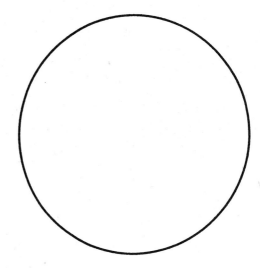

12: 23

Plot (111) to develop {111} form

Number of faces: _____

Open or closed: _____

Special or general: _____

Name: _____

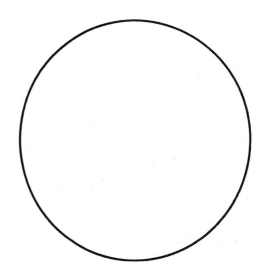

Student Name

13: 432

Plot (111) to develop {111} form

Number of faces: _____

Open or closed: _____

Special or general: _____

Name: _____

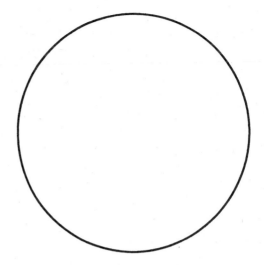

14: 4/m32/m

Plot (111) to develop {111} form

Number of faces: _____

Open or closed: _____

Special or general: _____

Name: _____

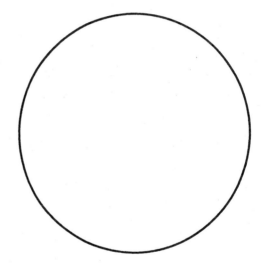

Development of Additional Forms on the Basis of Miller Index and Point Group Symmetry

PURPOSE OF EXERCISE

Further integration of your knowledge of symmetry elements (and point groups), coordinate axes, stereographic projection, Miller index notation, and form development.

FURTHER READING AND CD-ROM INSTRUCTIONS

Klein, C. and Dutrow, B., 2008, *Manual of Mineral Science*, 23rd ed., Wiley, Hoboken, New Jersey, pp. 182–208.

CD-ROM, entitled *Mineralogy Tutorials*, version 3.0, that accompanies *Manual of Mineral Science*, 23rd ed., click on "Module II", and subsequently on the button "Crystal Classes".

Nesse, W. D., 2000, *Introduction to Mineralogy*, Oxford University Press, New York, pp. 24–58.

Perkins, D., 2002, *Mineralogy*, 2nd ed., Prentice Hall, Upper Saddle River, New Jersey, pp. 187–203.

Wenk, H. R. and Bulakh, A., 2004, *Minerals: Their Constitution and Origin*, Cambridge University Press, New York, New York, pp. 66–78.

Bloss, F. D., 1994, 2nd printing with minor revisions, *Crystallography and Crystal Chemistry*, Mineralogical Society of America, Chantilly, Virginia, pp. 99–139.

Background Information: Again, as in exercise 14, no new concepts need to be introduced. Here the goal is to integrate your understanding of point group symmetry, stereographic projection, Miller index notation, and form development in a more general way than in the previous exercise. You will need to draw upon your knowledge of materials in exercise 9 (symmetry elements and their combinations), exercise 10 (recognition of the most common crystal forms), exercise 12 (stereographic projection), and exercise 14 (development of the {111} form).

MATERIALS

The four pages of Fig. 15.1, ruler, protractor, and soft pencil. A stereographic net is not specifically required, although you may wish to refer to one (in exercise 12) for specific angular relationships.

ASSIGNMENT

1. Using the four pages of Fig. 15.1, beginning with entry 1*a*, plot the total symmetry content of the specific point group onto the equatorial circle of the stereogram. Use standard graphical symbols as shown in Fig. 14.1, and refer to Table 10.3 for the various conventions implied by the Hermann–Mauguin notation.

2. Locate coordinate axes and label them clearly on the stereogram.

3. For each of the four selected point groups plot three forms ({010}, {110}, and {*hkl*}; or {10$\bar{1}$0}, {10$\bar{1}$1}, and {*hkil*} (in the hexagonal system), by plotting one face in the positive octant of the coordinate system, and generating all other faces on the basis of the symmetry elements already plotted.

4. For the sake of clarity, each individual form should be plotted on a separate stereogram circle.

5. Index only the face poles that belong to the upper half of the crystal. (Doing more would unnecessarily clutter the stereogram.)

6. List the number of faces making up the form.

7. State whether the form is open or closed.

8. Give the form name (see Fig. 10.1, or any of the references listed above).

EXERCISE 15

Student Name

FIGURE 15.1 Stereographic assignments

2/m

1*a*: Form {010}

Number of faces: ____________________

Open or closed: ____________________

Name: ____________________

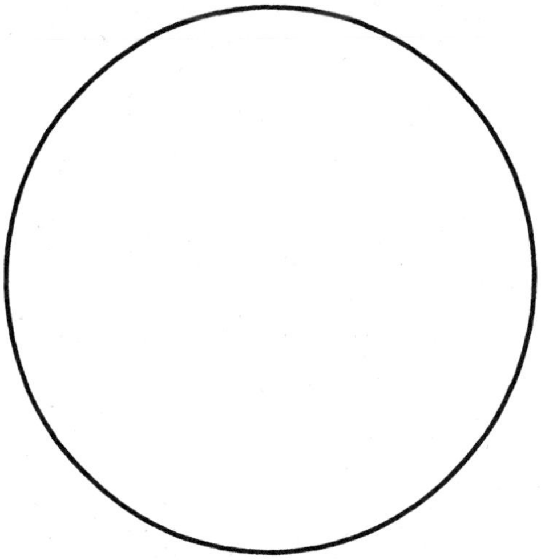

1*b*: Form {110}

Number of faces: ____________________

Open or closed: ____________________

Name: ____________________

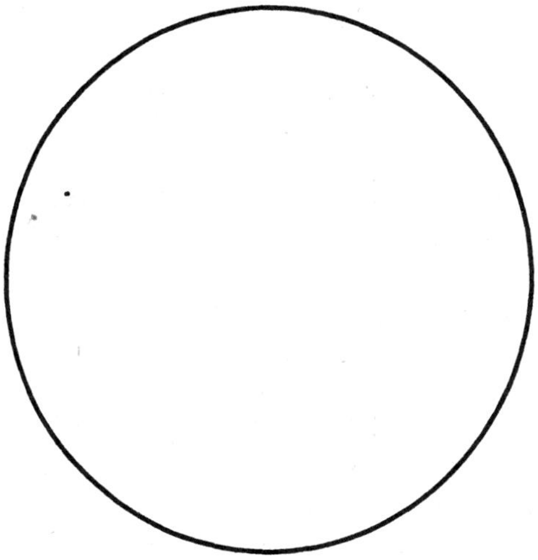

1*c*: Form {*hkl*}, i.e., general form

Number of faces: ____________________

Open or closed: ____________________

Name: ____________________

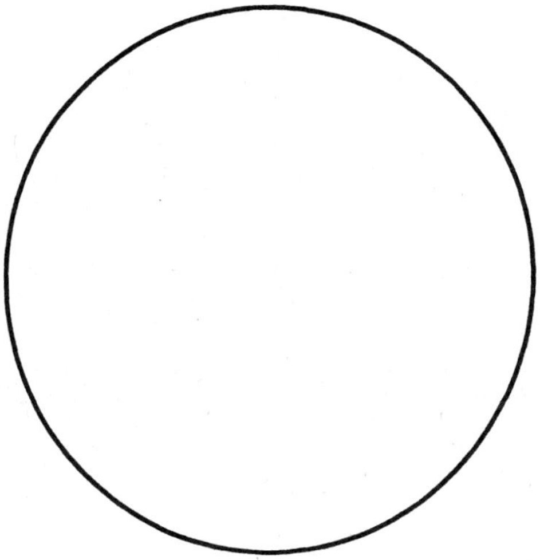

FIGURE 15.1 (*continued*)

Student Name

$$2/m2/m2/m$$

2a: Form {010}

    Number of faces: _____

    Open or closed: _____

    Name: _____

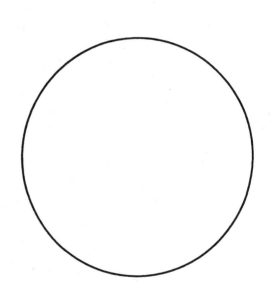

2b: Form {110}

    Number of faces: _____

    Open or closed: _____

    Name: _____

2c: Form {*hkl*}, i.e., general form

    Number of faces: _____

    Open or closed: _____

    Name: _____

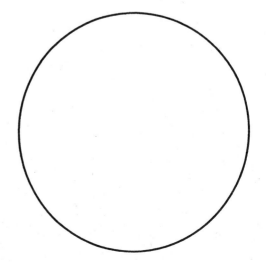

EXERCISE 15

____
Student Name

FIGURE 15.1 *(continued)*

6/m2/m2/m

3*a*: Form $\{10\bar{1}0\}$

Number of faces: ______________________

Open or closed: ______________________

Name: ______________________

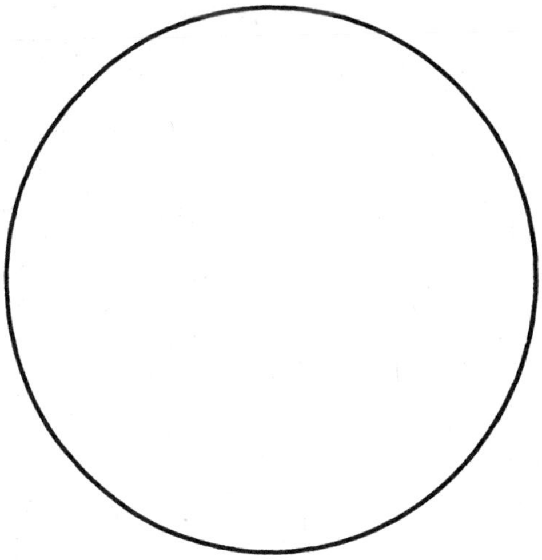

3*b*: Form $\{10\bar{1}1\}$

Number of faces: ______________________

Open or closed: ______________________

Name: ______________________

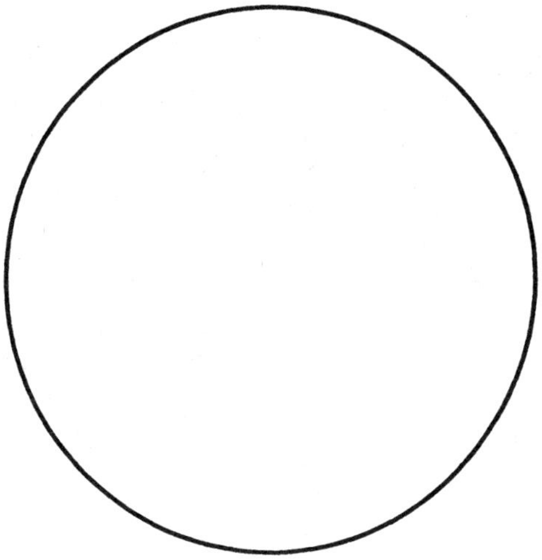

3*c*: Form $\{hk\bar{i}l\}$, i.e., general form

Number of faces: ______________________

Open or closed: ______________________

Name: ______________________

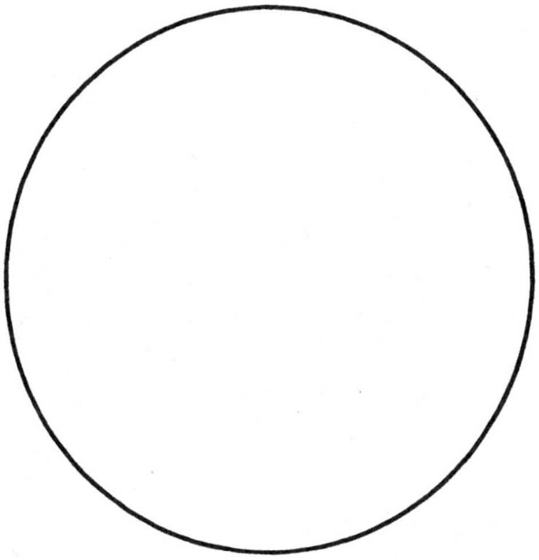

Student Name

FIGURE 15.1 (continued)

$$4/m\bar{3}2/m$$

4a: Form {010}

Number of faces: _____

Open or closed: _____

Name: _____

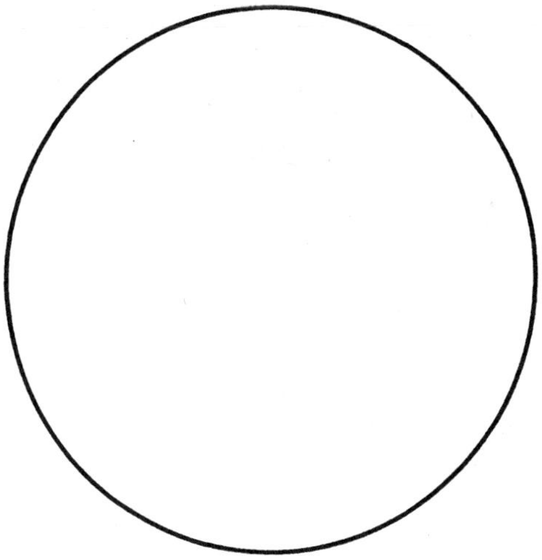

4b: Form {110}

Number of faces: _____

Open or closed: _____

Name: _____

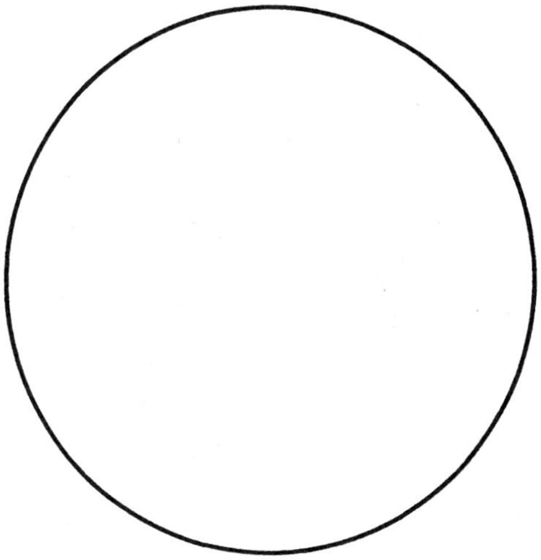

4c: Form {hkl}, i.e., general form

Number of faces: _____

Open or closed: _____

Name: _____

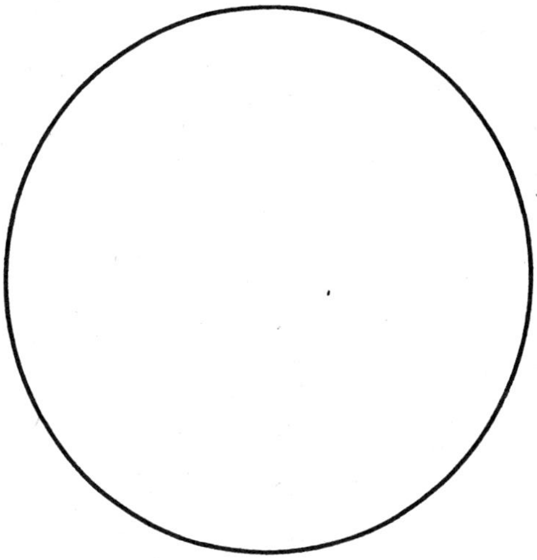

Axial Ratio Calculations

PURPOSE OF EXERCISE

To gain familiarity with axial ratios (1) by determining them graphically from stereographic projections, (2) by determining them trigonometrically from ϕ and ρ angles, and (3) by using them to plot faces on a stereographic net.

FURTHER READING

Klein, C. and Dutrow, B., 2008, *Manual of Mineral Science*, **23rd ed., Wiley, Hoboken, New Jersey, pp. 156–164.**

Nesse, W. D., 2000, *Introduction to Mineralogy*, Oxford University Press, New York, p. 23.

Bloss, F. D., 1994, 2nd printing with minor revisions, *Crystallography and Crystal Chemistry*, Mineralogical Society of America, Chantilly, Virginia, pp. 54–56.

Background Information: Axial ratios express the relative length(s) of one or two coordinate axes with respect to an axis whose length is taken to be unity (= 1). For example, the orthorhombic ($2/m2/m2/m$) mineral barite has the axial ratios of $a:b:c = 1.627:1:1.310$. It is standard procedure to take the length of the b axis as unity and to express the relative lengths of the a and c axes with respect to b axis = 1. In barite the relative length of the a axis is 1.63 times that of b, whereas c is 1.31 times that of b. This example is for an orthorhombic mineral, and it should be noted that a ratio of axes cannot be given in the isometric system, for in this system all three axes are by definition the same length. In the tetragonal and hexagonal systems, the ratio is given between the two axes of different lengths, a and c. In the orthorhombic system, as noted, the ratios are $a:b (= 1):c$. In the nonorthogonal systems, monoclinic and triclinic, the ratios are defined as for orthorhombic, but the values for the nonorthogonal angles are given as well. The following are examples.

Tetragonal:	$a:c = 1:0.906$ (for zircon)
Hexagonal:	$a:c = 1:0.996$ (for beryl)
Orthorhombic:	$a:b:c = 0.993:1:0.340$ (for stibnite)
Monoclinic:	$a:b:c = 0.663:1:0.559$; $\beta = 115°50'$ (for orthoclase)
Triclinic:	$a:b:c = 0.636:1:0.559$; $\alpha = 94°20'$, $\beta = 116°34'$, $\gamma = 87°39'$ (for albite)

As should be clear from this listing, any graphical or trigonometric calculation of axial ratios in nonorthogonal systems is a great deal more involved (because of the β, or the α, β, and γ angles) than in orthogonal systems. In

order to convey the principles that underlie axial ratios, examples and assignments in this exercise will be restricted to orthogonal systems.

By now you should be asking "Where do such ratios come from?" Prior to the discovery of X-rays and the subsequent application in 1912 of X-ray diffraction techniques to the study of crystal structures, axial ratios were determined by crystallographers using optical goniometers for the measurements of the angular coordinates of face poles. Now X-ray diffraction techniques are always used to measure the absolute (*not the relative*) lengths of coordinate axes. An X-ray study of sulfur in 1960 reported measurements of $a = 10.47$ Å, $b = 12.87$ Å, and $c = 24.49$ Å When these direct measurements are used to determine the ratios, as follows; $a/b:b/b = 1:c/b$, they result in $0.8135:1:1.9029$. These numbers are in remarkably close agreement with the axial ratios reported in 1869 for the mineral sulfur: $a:b:c = 0.8131:1:1.9034$.

a. *A worked example for determining axial ratios graphically, using stereographic projection.* For graphical construction based on the stereogram, a circle of any radius representing the primitive may be used, but to reduce the overall size of the construction it should be relatively small. Used here is a circle of 1-inch radius in conjunction with inch graph paper. Thus the units of circle radius and graph paper are the same. The smallest unit on the graph paper that can be read directly is 0.1 inch.

This example applies to orthorhombic minerals and is based on the interfacial angle measurements of the mineral barite (point group: $2/m2/m2/m$), which are

$$(100) \wedge (110) = 58°28'$$
$$(010) \wedge (210) = 50°49'$$
$$(100) \wedge (310) = 28°31'$$
$$(001) \wedge (011) = 52°43'$$
$$(001) \wedge (101) = 38°51'$$
$$(001) \wedge (111) = 57°01'$$

Figure 16.1 is a clinographic projection of a barite crystal that shows several of the faces we will be plotting. This figure also lists the equivalence of form letters and their Miller indices.

In Fig. 16.2, using a protractor, we locate the poles of (110), (210), (310), (100), and (010) on the circles. In Fig. 16.2a a tangent is drawn to the circle at the pole of (110)—this tangent is parallel to the trace of the (110) crystal face. The tangent cuts the a axis at A and the b axis at B.

FIGURE 16.1 Clinographic projection of a barite crystal with various forms. Several of the faces of these forms are used in the calculations in Fig. 16.2.

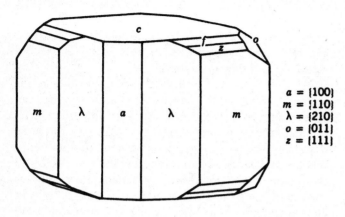

$$a = \{100\}$$
$$m = \{110\}$$
$$\lambda = \{210\}$$
$$o = \{011\}$$
$$z = \{111\}$$

The axial ratio $a:b = $ OA:OB, for the (110) pole, can be measured as 19.5:12.0 on the graph paper (see Fig. 16.2a). Dividing by 12.0, gives an axial ratio statement of $a:b = 1.625:1.$*

In Fig. 16.2b the interfacial angle (010) Λ (210) = 50°49′ is plotted as well as the (210) pole. At this point a tangent is drawn to the circle. This tangent cuts the *a* axis at A′ (distance = 13.0 divisions) and the *b* axis at B′ (distance = 16.0 divisions). Because $a:b = h$ (OA′):k (OB′), the *h* and *k* of the (210) pole must be considered. With $h = 2$, and $k = 1$ factored in, the ratio $a:b = 26.0:16.0$ becomes 1.625:1 when both numbers are divided by 16.0 (thus setting $b = 1$).

A third example of an (*hk0*) face is given in Fig. 16.2c. The pole of (310) is plotted with reference to (100); the interfacial angle is 28°31′. The tangent to the circle cuts the *a* axis at 11.5 units (at OA″) and the *b* axis at 21.2 units (at OB″). Using these unit distances, and also considering that $h = 3$ and $k = 1$ in (310), we find that the measured $a:b = 34.5:21.2$, which reduces to $a:b = 1.627:1$. We now have three statements for the ratio $a:b$, namely 1.625:1; 1.625:1; and 1.627:1. The average for these is $a:b = 1.63:1$.

Clearly all three statements in Figs. 16.2a, b, and c should have been the same; that is, 1.630:1, as published for barite. The small variations and deviations are due to the imprecision of plotting angles and tangents on a small scale. A 10-cm circle using a centimeter graph paper would have been more precise, but it was not used in order to allow for three constructions on one page.

The interfacial angles (001) Λ (011) = 52°43′ and (001) Λ (101) = 38°51′ can be used to obtain *b:c*, and *a:c* ratios respectively. Figures 16.2d and e show constructions for these face poles and interfacial angles, similar to those shown in Figs. 16.2a, b, and c. However, here the sections

are vertical, in contrast to the horizontal views of Figs. 16.2a, b, and c. In Fig. 16.2d the graphical plot of the (011) face leads to an axial ratio for $b:c = 1:1.32$. Using this ratio of *b:c* and combining it with the average of earlier obtained ratios of $a:b = 1.63$, we have the complete axial ratio statement for barite, which is

$$a:b:c = 1.63:1:1.32.$$

Figure 19.2e illustrates how an *a–c* section and the interfacial angle for (001) Λ (101) = 38°51′ allow for an independently obtained *a:c* ratio. Clearly, this ratio, $a:c = 1:0.81$, is in agreement with the foregoing complete axial ratio statement (1.32/1.63 = 0.81).

It is of interest to compare the axial ratio statement, as obtained graphically from interfacial angle measurements, with the directly measured lengths for *a*, *b*, and *c* axes (of barite), as obtained by X-ray diffraction techniques. Such measurements give $a = 8.87$ Å; $b = 5.45$ Å; and $c = 7.14$ Å. When these direct measurements are expressed as ratios with respect to *b*, they result in $a:b:c = 1.6275:1:1.3101$. Our results are in good agreement with these published values.

We could have obtained a complete axial ratio statement for the mineral barite by using interfacial angle measurements for just one face, namely a (*hkl*)-type face. Here follows an illustration of the derivation of axial ratios based on the measurement of a (111) face. Earlier it was noted that (001) Λ (111) = 57°01′; this is the ρ angle of (111). We also need a ϕ angle for (111). The required ϕ angle can be obtained from the measurement for the (110) face (see also Fig. 16.2a). Because (110) and (111) [as well as (001)] define the same zonal direction, as shown by the parallel edges of their intersections, their face poles will lie on the same great circle in a stereographic projection. All three poles lie on a vertical great circle with $\phi = 31°32′$. In Fig. 16.2a the angle (100) Λ (110) = 58°28′ was given. The ϕ angle for (110) = 90° −58°28′ = 31°32′.

Therefore, for (111), $\phi = 31°32′$ and $\rho = 57°01′$. We have already measured the *a:b* ratio as based on angular measurements of (110). This is illustrated in Fig. 16.3a. In Fig. 16.3a the poles of (110), (111), and (001) are plotted on the same vertical great circle. In order to locate the ρ of (111) on this diagram, we need first to locate the ρ angle on a standard stereographic net—for example, with radius = 10 cm—and subsequently to transfer the distance between face poles (001) and (111), properly scaled to the smaller circle. The ρ angle is 57°01′, which is 57° for all practical purposes and measures at 5.5 cm from the origin on a stereographic net with a 10-cm radius. This distance is scaled as 0.55 times 1 inch on the circle of 1-inch radius in Fig. 16.3a. The location of (110) gave rise to the unit measurements of $a:b = 19.5:12.0$.

For clarity of illustration, the face pole of (111) is shown in a vertical section in Fig. 16.3b. This vertical section is taken along the line of the trace of the great circle shown in Fig. 16.3a (along OT). The ρ of (111) is 57°01′, which

---

*Note also that the angle OBA = 58°28′ and that tan 58°28′ = *a/b*. Thus the *a/b* ratio can be determined trigonometrically as well. The tangent of 58°28′ = 1.6297 (which rounds off to 1.63).

FIGURE 16.2   Graphical determination of axial ratios. See text for discussion.

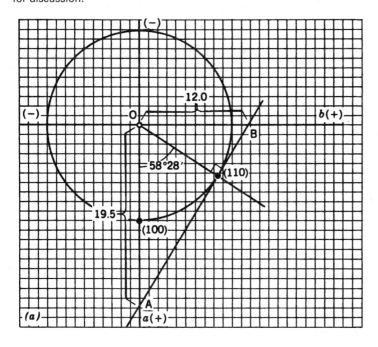

Scale

0   10

$(100) \wedge (110) = 58°28'$

$a : b = OA : OB$
$a : b = 19.5 : 12.0$
Setting $b = 1$, by dividing by 12, gives
$a : b = 1.625 : 1$

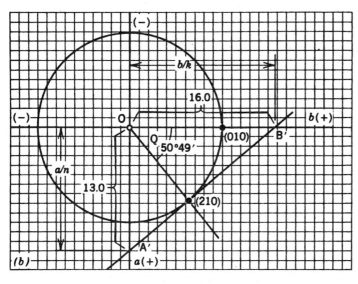

$(010) \wedge (210) = 50°49'$

$a : b = h(OA') : k(OB')$
$a : b = 2(13.0) : 1(16.0)$
$a : b = 26 : 16$
Setting $b = 1$, by dividing by 16, gives
$a : b = 1.625 : 1$

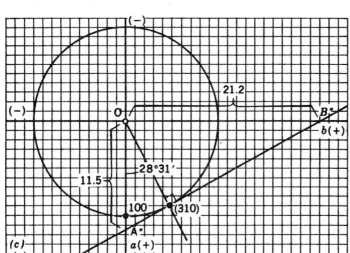

$(100) \wedge (310) = 28°31'$

$a : b = h(OA'') : k(OB'')$
$a : b = 3(11.5) : 1(21.2)$
$a : b = 34.5 : 21.2$
Setting $b = 1$, by dividing by 21.2, gives
$a : b = 1.627 : 1$

FIGURE 16.2 *(continued)*

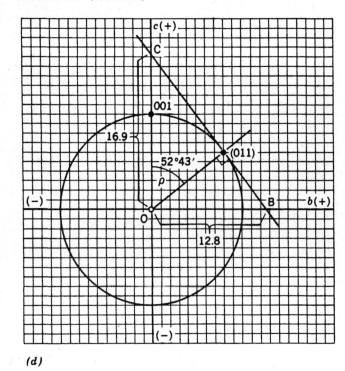

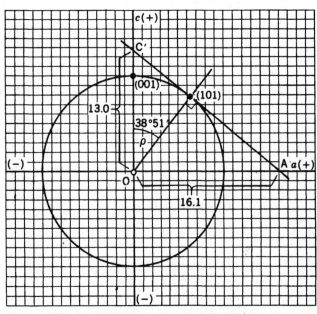

(d)

Scale

0   10

$(001) \wedge (011) = 52°43'$

$b : c = OB : OC$

$b : c = 12.8 : 16.9$

Setting $b = 1$, by dividing by 16.9, gives

$b : c = 1 : 1.32$

We have a prior statement for
$a : b = 1.63 : 1$

From these ratios follows
$a : b : c = 1.63 : 1 : 1.32$

The above, in turn, leads to
$a : c = 1.63 : 1.32$, or
setting $a = 1$, by dividing by 1.63, gives

$a : c = 1 : 0.81$

(e)

$(001) \wedge (101) = 38°51'$

$a : c = OA : OC'$

$a : c = 16.1 : 13.0$

Setting $a = 1$, by dividing by 16.1, gives

$a : c = 1 : 0.81$

A confirmation of the above (in Fig 16.2d).

FIGURE 16.3 Graphical determination of axial ratios. See text for discussion.

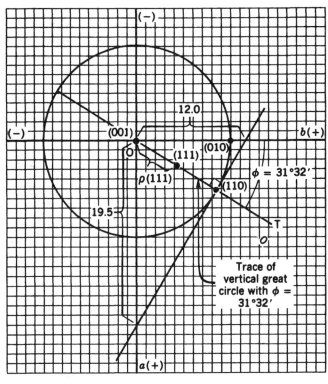

(a)

Scale

0 _____ 10

Position of (110) resulted in
a : b = 19.5 : 12.0
(see Fig. 16.2a).

Pole of (111) is located,
on the one vertical
great circle shown, at
ρ = 57°01′.

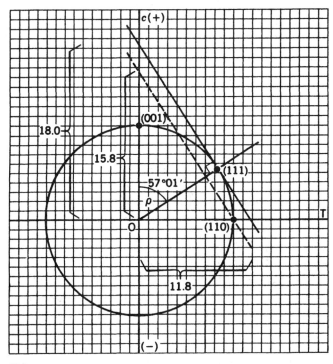

(b)

Pole of (111) is plotted in
a vertical section on the
basis of ρ = 57°01′.

The tangent to the
circle is the trace of (111).
However, in order to
keep the same scale
as in part *a*, draw
a plane parallel to the
tangent plane, through
the pole of 110.

This results in the unit measurements of
a : b : c = 19.5 : 12.0 : 15.8
which in turn results in
a : b : c = 1.62 : 1 : 1.32

allows for the location of the (111) face pole, and the trace of the (111) face as a tangent to the vertical great circle. This tangential plane intersects the vertical (+) end of c at 18.0 units and the horizontal OT trace at 11.8 units. This gives a relative $c:OT$ ratio of 1.52. However, to be able to compare this ratio directly with the measurements already obtained in Fig. 16.3a, one must keep the horizontal distance from the origin (O) to the face pole of (110) the same in both illustrations (Figs. 16.3a and b). Therefore, after having located the face pole of (110) on the trace of OT in Fig. 16.3b, a plane parallel to the tangent face is drawn through the face pole of (110). Now the distances from the origin to (110) are identical in both drawings. The only question remaining is, "Where does the properly oriented (111) face intersect c?" That distance turns out to be 15.8 units (see Fig. 16.3b).

We can now state the relative lengths of the axes, on the basis of our graphical plot of (111), as $a:b:c = 19.5:12.0:15.8$, which, when we set $b = 1$ (by dividing all numbers by 12), results in $a:b:c = 1.62:1:1.32$. This is the same as we had earlier, within graphical limits of error.

Any other (hkl) could have been chosen. For example, if ϕ and ρ measurements are available for a face such as (241), one would proceed as follows. Locate the (240) position, whether it is present on the crystal or not, on the basis of ϕ. Subsequently, locate the (241) pole in a vertical section as based on the ρ angle, as in Fig. 16.3b. Measure all relative lengths in graph units. This would result in $a:b:c = x$ units (OA):y units (OB):z units (OC). But because of the Miller index of 241, the correct $a:b:c = h$ (OA units):k (OB units):l (OC units). This then would result in $a:b:c = 2$ (OA units):4 (OB units):1 (OC units), which yields the correct axial ratio statement for an orthorhombic crystal on the basis of a (241) face.

b. *A worked example for determining axial ratios by trigonometry, on the basis of ϕ and ρ angles.* This example will also apply to an orthorhombic crystal, and for ease of comparison with the results obtained by graphical means, it will again be based on measurements of faces on barite.

Forms	ϕ	ρ
{001}	. . .	0°
{010}	0°	90°
{110}	31°32′	90°
{210}	50°49′	90°
{310}	61°28′	90°
{011}	0°	52°43′
{101}	90°	38°51′
{111}	31°32′	57°01′

Figure 16.4a shows a stereogram of the pole positions of faces belonging to the foregoing forms, but only in the positive octant. Subsequent illustrations show the angles

and trigonometric functions that apply, as well as the resulting axial ratio statements. Each of the faces given and their plotted positions were discussed earlier and are presented graphically in Figs. 16.4 and 16.5. The illustrated derivations in Figs. 16.4 and 16.5 are very similar to the earlier graphical methods, except that trigonometric functions are used. The captions to each of the illustrations in Figs. 16.4 and 16.5 provide step-by-step derivations of the axial ratio statements using trigonometric functions of ϕ and ρ. If the student is familiar with trigonometry, the desired axial ratio calculations are much quicker to perform than the graphical methods. As should be expected, the results for both methods are very similar, and in turn comparable to the ratios based on direct X-ray diffraction measurements.

c. *Plotting of face poles on the basis of known axial ratios.* If the axial ratio is known, faces can be plotted by reversing the foregoing procedures. For convenience a circle (that is, the equatorial plane of a stereonet) is drawn with radius = 10 divisions on square graph paper in Fig. 16.6a. The radius = 10 is taken as unity and is set equal to the b axis. The following steps apply to the plotting of (130) and (131) poles for an orthorhombic mineral with an axial ratio of $a:b:c = 0.41:1:0.26$. The (130) face has the following intercepts:

$$\frac{0.41}{1} : \frac{1}{3} : \frac{0.26}{\infty} = 1.23:1:0$$

With b = unity (10 divisions), the length of the a axis can be measured off as 12.3 divisions (at A). The trace of (130) is the join between A and the unit distance (= 10) along b. The pole of (130), therefore, is the point where the perpendicular from the origin to this trace intercepts the unit circle (the primitive).

The pole of (131) must lie somewhere along the radial line (great circle) from O to the (130) pole. The best way to illustrate the locating of the (131) pole is by a vertical section (Fig. 16.6b) in which the radial direction OT is horizontal. The intercepts of (131) are

$$\frac{0.41}{1} : \frac{1}{3} : \frac{0.26}{1} = 1.23:1:0.78$$

In Fig. 16.6b first measure off the distance OT′ (from Fig. 16.6a) along direction OT. Then measure 7.8 divisions along the (+) end of the c axis (as derived from the ratio value of 0.78). Joining points C and T′ gives the location of the trace of (131) in this vertical section. A line perpendicular to this trace gives the intersection point (the pole) on the vertical unit circle. By drawing a line that connects this pole with the (−) end of the c axis, we obtain an intersection point with the horizontal trace (along OT). The distance from O (the origin) to this intersection point is the distance that must be laid off in Fig. 16.6a to locate the pole of (131) on a stereogram. This turns out to be 4.2 divisions; this is measured out with a compass from O in Fig. 16.6a.

FIGURE 16.4 Determination of axial ratios by trigonometry. (a) Location of the poles of several faces of barite, in the positive octant (on a stereogram); the dashed lines represent great circles on which poles of faces in the same zone plot; (b) Some trigonometric functions. (c) Derivation of the axial ratio $a:b$ using cot ϕ (110). (d) Derivation of the axial ratio $a:b$ using cot ϕ (210). (e) Derivation of the axial ratio $a:b$ using cot ϕ (310). (f) Derivation of the axial ratio $b:c$ using tan ρ (011). (g) Derivation of the axial ratio $a:c$ using tanρ(101).

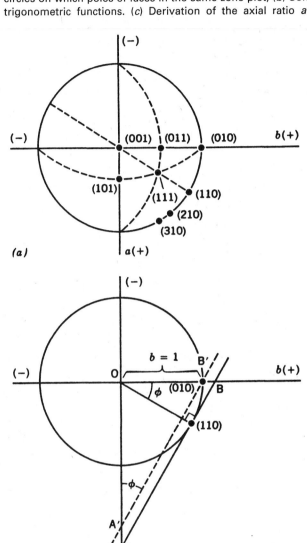

(a)

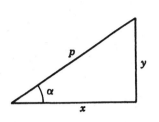

$$\sin \alpha = \frac{y}{p}; \cos \alpha = \frac{x}{p}$$

(b) $\quad \tan \alpha = \frac{y}{x}; \cot \alpha = \frac{x}{y}$

(c)

ϕ (110) = 31°32′

$$\cot \phi = \frac{OA}{OB} = \frac{OA'}{OB'}; OB' = b, \text{ set at } 1$$

$$\cot \phi = \frac{OA'}{1}$$

$$\cot \phi = OA'; \cot 31°32' = 1.6297$$
$$a:b = 1.6297:1$$

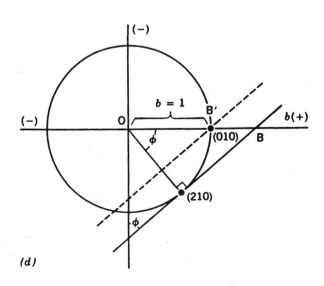

(d)

ϕ (210) = 50°49′

$$a:b = h(OA):k(OB) =$$
$$= h(OA'):k(OB')$$

$$OB' = b, \text{ set at } 1$$

$$\cot \phi = \frac{OA'}{1}; \cot 50°49' = 0.8151$$
$$a:b = h(0.8151):k(1)$$
$$= 2(0.8151):1(1)$$

$$a:b = 1.630:1$$

FIGURE 16.4 *(continued)*

ϕ (310) = 31°32'; cot 31°32' = 0.54371

$a : b = h(\cot \phi) : k(1)$
$\quad\quad = 3(0.54371) : 1$

(e) $a : b = 1.6311 : 1$

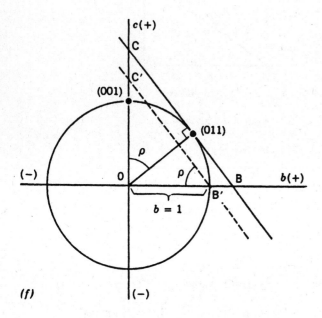

(f)

ρ (011) = **52°43'**

$b : c = OB : OC = OB' : OC'$

$OB' = b$, set at 1

$\tan \rho = \dfrac{OC'}{OB'} = \dfrac{OC'}{1}$

$\tan 52°43' = 1.3135$

$b : c = 1 : 1.3135$

Combining $a : b$ (from *c*, *d*, and *e*)
and $b : c$ results in an
axial ratio statement:
$a : b : c = 1.63 : 1 : 1.31$

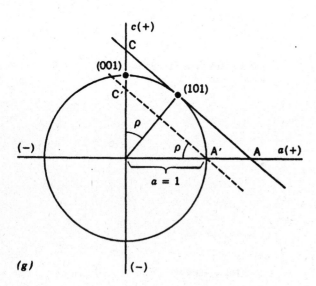

(g)

ρ (101) = **38°51**

$a : c = OA : OC = OA' : OC'$

$\tan \rho = \dfrac{OC'}{OA'}$; $OA' = a$, set at 1

$\tan 38°51' = 0.80546$

$a : c = 1 : 0.805$, which
confirms the result in part *f*

$a : c = 1.63 : 1.31$, which becomes
$a : c = 1 : 0.804$

FIGURE 16.5 Determination of axial ratios using trigonometry and the φ and ρ angles of face (111).

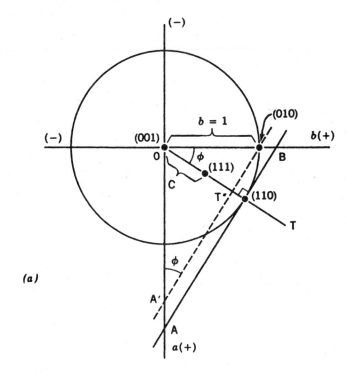

(a)

$\phi (111) = 31°32'; \rho (111) = 57° 01'$

The φ angle of 31°32'
is the same for (110) and (111)

In Fig.16.4 *c*

$\cot \phi = \dfrac{OA'}{1}$ gave the ratio

$a : b = 1.6297 : 1$

In order to refer any further
c : *b* calculations to *b* = 1,
we need the length of OT″ :

$\cos \phi = \dfrac{OT''}{1}$; $\cos 31°32' = 0.85239$

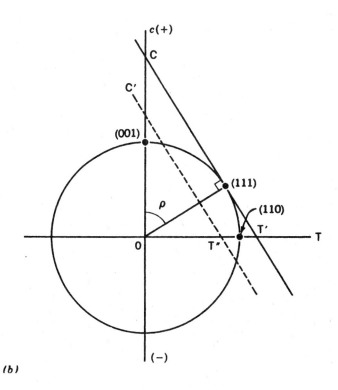

(b)

tan ρ(111) will provide
the *c* : *b* ratio.

$\tan \rho = \dfrac{OC}{OT'} = \dfrac{OC'}{OT''}$

$\tan 57°01' = 1.54085$

$OC' = OT'' \tan \rho$
$= 0.85234 \times 1.54085$
$= 1.3133$

This leads to
$a : b : c = 1.6297 : 1 : 1.3133$

FIGURE 16.6 Plotting of face poles on the basis of a given axial ratio.

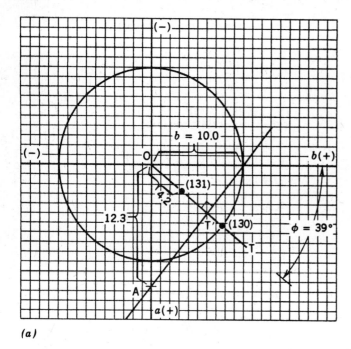

(a)

Scale

0   10

$a : b : c = 0.41 : 1 : 0.26$
For (130) intercepts are:

$$\frac{0.41}{1} : \frac{1}{3} : \frac{0.26}{\infty} =$$
$$1.23 : 1 : 0$$

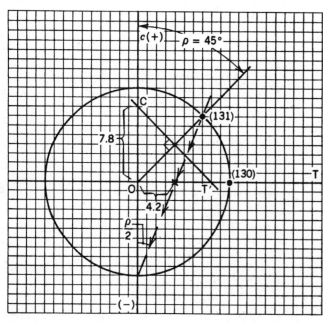

(b)

For (131) intercepts are:

$$\frac{0.41}{1} : \frac{1}{3} : \frac{0.26}{1} =$$
$$1.23 : 1 : 0.78$$

At this stage the ϕ angles of (130) and (131) can be measured from Fig. 16.6a with a protractor. This ϕ turns out to be 39°. The ρ angle for (131) can be measured in Fig. 16.6b and this is 45°. It is instructive to know which mineral constants were plotted and what the expected ϕ and ρ angles for (130) and (131) are. The mineral is chalcostibite; and the angles for its two forms are as follows:

	ϕ	ρ
{130}	38°45′	90°
{131}	38°45′	45°04′

MATERIALS

Square graph paper (provided in Fig. 16.7), a compass, protractor, and a calculator with trigonometric functions.

Also needed are ϕ and ρ angles for indexed faces, or interfacial angles for such faces. For plotting of face poles on the basis of axial ratios, an axial ratio statement is needed. All such data are given as part of the assignment. (See Tables 16.1 to 16.3.)

ASSIGNMENT

Graphical and trigonometric determination of axial ratios on the basis of interfacial angles and ϕ and ρ angles. Data are given in Tables 16.1 and 16.2. Furthermore, Table 16.3 provides data for the plotting of face poles for a mineral for which the axial ratio and the Miller indices of faces (to be plotted) are given.

TABLE 16.1   Interfacial Angles for Two Minerals

Point Group	
4/m2/m2/m	**2/m2/m2/m**
(100) Λ (101) = 47°50′	(010) Λ (110) = 45°1′
(101) Λ ($\overline{1}$01) = 84°20′	(001) Λ (011) = 41°45′
(101) Λ (011) = 56°41′	(001) Λ (111) = 51°38′
	(001) Λ (211) = 63°24′

TABLE 16.2 ϕ and ρ Angles for Two Minerals

Point Group					
4/m			**2/m2/m2/m**		
Form	ϕ	ρ	Form	ϕ	ρ
{001}	–	0°	{001}	–	0°
{013}	0°	35°54′	{010}	0°	90°
{011}	0°	65°16′	{100}	90°	90°
{114}	45°	37°31′	{130}	25°53′	90°
{112}	45°	56°55′	{120}	36°03′	90°
{123}	18°26′	59°47′	{011}	0°	26°41′
			{201}	90°	55°40′
			{111}	55°31′	41°36′
			{121}	36°03′	51°12′

TABLE 16.3   Axial Ratio for an Orthorhombic Mineral (2/m2/m2/m) and Miller Indices of Face Poles To Be Plotted on a Stereogram

$a:b:c = 0.608:1:0.721$

(110)

(021)

(111)

(130)

(031)

EXERCISE 16

FIGURE 16.7 Graph paper with 10 × 10 divisions per inch.

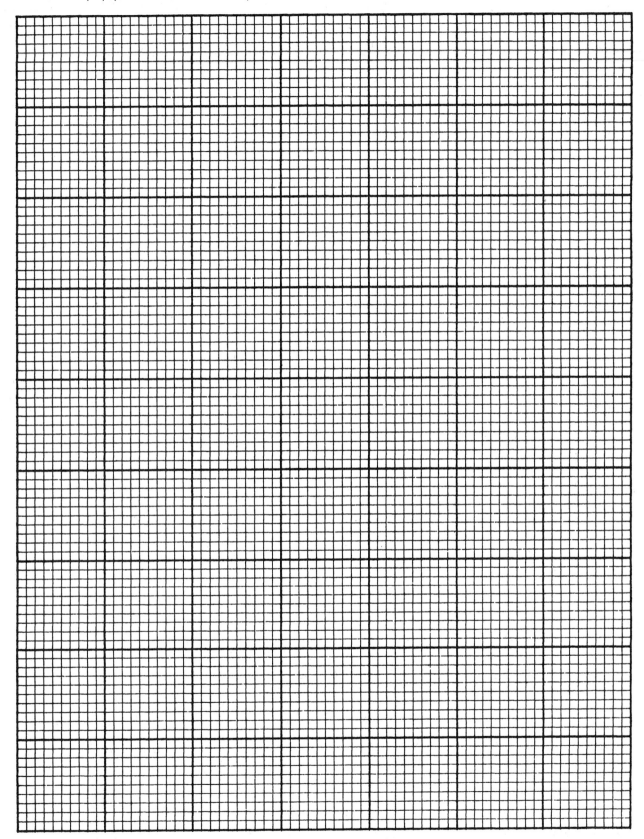

Please make photocopies of this form if more copies are needed.

Symmetry and Translation in One- and Two-Dimensional Patterns

PURPOSE OF EXERCISE

Identification of symmetry and translational elements in printed patterns. This process serves as an introduction to the internal structures of crystalline solids, which consist of atoms (or ions) in regular arrays in three dimensions. Symmetry and translations are fundamental properties of such arrays. As such, it is instructive to gain familiarity with these elements in printed one- and two-dimensional patterns.

FURTHER READING AND CD-ROM INSTRUCTIONS

Klein, C. and Dutrow, B., 2008, *Manual of Mineral Science*, 23rd ed., Wiley, Hoboken, New Jersey, pp. 143–158.

CD-ROM, entitled *Mineralogy Tutorials*, version 3.0, that accompanies *Manual of Mineral Science*, 23rd ed., click on "Module III", and subsequently on each of the two buttons "1-dimensional order" and "2-dimensional order".

Nesse, W. D., 2000, *Introduction to Mineralogy*, Oxford University Press, New York, pp. 6–8.

Perkins, D., 2002, *Mineralogy*, 2nd ed., Prentice Hall, Upper Saddle River, New Jersey, pp. 208–218.

Wenk, H. R. and Bulakh, A., 2004, *Minerals: Their Constitution and Origin*, Cambridge University Press, New York, New York, pp. 37–44.

Bloss, F. D., 1994, 2nd printing with minor revisions, *Crystallography and Crystal Chemistry*, Mineralogical Society of America, Chantilly, Virginia, pp. 140–159.

Stevens, P. S., 1991, *Handbook of Regular Patterns, An Introduction to Symmetry in Two Dimensions*, MIT Press, Cambridge, Massachusetts.

International Tables For Crystallography, vol. A, Theo Hahn, ed., 1983, D. Reidel Publishing Co., Dordrecht, The Netherlands, pp. 81–99.

Background Information: In exercise 8 you had a brief introduction to the symmetry content of letters, symbols, and common objects. In the subsequent exercises you concerned yourself with various aspects of the translation-free symmetry of crystals, that is, point group symmetry as expressed by the external form (= morphology) of crystals. All such point group symmetries are by definition translation-free because we can recognize only translation-free symmetry elements such as mirrors, rotation and rotoinversion axes, and a center of symmetry.

However, throughout the internal structure of crystals there are translations so extremely small (atomic scale) that they cannot be seen by the naked eye. Such translations are commonly expressed in nanometers or angstroms, in which 1 nanometer $= 10^{-7}$ cm, and 1 angstrom $= 10^{-8}$ cm. Indeed, the only way in which such distances can be seen directly is by transmission electron microscopy (see, for example, Fig. 7.18 in *Manual of Mineral Science*, 23rd ed., p. 163).

We must remind ourselves that the crystal form (of a mineral, or some other crystalline substance) is the external expression of the regularity of its internal atomic structure. The internal ordered arrangement of atoms (or of ions and of clusters of atoms and ions) in a crystalline structure may be thought of as motifs repeated by regular translations in three dimensions. In other words, *an atomic structure* of a crystalline material is the result of regular (= periodic) *translations of a motif in three dimensions*. In crystalline materials such motifs may be, for example, Cu atoms, SiO_4 tetrahedra, groups of SiO_4 tetrahedra, or triangular CO_3 groups. Whereas in a two-dimensional wallpaper pattern the motif may be the drawing of a flower, in a mineral the motif may consist of a grouping of SiO_4 tetrahedra about some large cation. In wallpaper patterns the motif is repeated infinitely by two translations to produce rolls of printed pattern. In the internal structure of crystals the motif (some chemical element or grouping) is translated, essentially infinitely, in three dimensions to produce the periodic array of the structure.

Illustrations of three-dimensional structures of minerals are found throughout Chapters 15 through 19 in the *Manual of Mineral Science*, 23rd ed. (see also other references listed). However, mineralogists and crystallographers commonly evaluate the symmetry elements and translations in such atomic structures by a more abstract representation of the structure known as the *lattice. A lattice is an imaginary pattern of nodes in which every node has an environment that is identical to that of any other node in the pattern. A lattice has no specific origin as it can be shifted parallel to itself.* A node is an abstract representation of a motif (be it in wallpaper patterns or in three-dimensional crystal structures). Using nodes, instead of the original motifs in evaluating a pattern, is very helpful in the assessment of the types of translations and symmetry elements inherent in a two-dimensional pattern or a three-dimensional structure. Once the abstract array of nodes has been derived from the original pattern, one can decide on the *unit cell* of the pattern. *A unit cell is the smallest unit of pattern that can be infinitely repeated to print the whole pattern.* A unit cell is outlined by connecting lattice nodes in a pattern. The translation directions in a pattern define the directions along which the unit cell must be repeated to complete the whole pattern. All the foregoing concepts are illustrated in Fig. 17.1 in two-dimensional patterns. Figure 17.1*a* is a two-dimensional pattern of flowers; Fig. 17.1*b* is the planar lattice deduced from this pattern by substituting a (lattice) node for each of the motifs (the flowers). A unit cell is outlined in both figures by dashed lines. The unit cell is the result of connecting four motifs in the two-dimensional pattern, which is equivalent to connecting four nodes in the plane lattice. It is useful at this stage to ask the question "How many motifs are contained within the unit cell?" This is most easily answered by inspection of Fig. 17.1*a*. Each corner of the unit cell contains the halves of two petals, which means that there is a total of one petal per corner. The unit cell has four corners, and it therefore follows that the unit cell contains only one flower pattern within it.

It should be clear from this discussion that when one prints a motif at each lattice node in Fig. 17.1*b*, one regenerates the original pattern from which the planar lattice was abstracted.

Figures 17.1*a* and 17.1*b* also allow for two sets of axes (in this case perpendiclar to each other) to be selected in the pattern. It is conventional to use the same axis notation as was used for the six crystal systems (such as *a*, *b*, and *c* or a_1, a_2 and *c*). The locations of the *a* and *b* axes in these two-dimensional patterns are noted as well as the units of translation along these axes (t_1 and t_2).

It may now be helpful to inspect a planar pattern with a more complicated motif and motif distribution, but still with orthogonal translation directions. Figure 17.1*c* shows such a pattern in which the total motif consists of parts of two other symbols in addition to the flower. The unit cell outlined in Fig. 17.1*c*, between the four diamond-shaped symbols, contains the following motif units: one whole flower in the center of the unit cell, four quarters of the diamond-shaped symbol at the corners of the unit cell, and four halves of the small circles along the edges of the unit cell. The motif therefore consists of one flower, one diamond-shaped symbol, and two small spheres. The planar lattice abstracted from this pattern is shown in Fig. 17.1*d*. Note that the plane lattices (in Figs. 17.1*b* and *d*) are of the same type (orthogonal axes, and a rectangular outline), even though the motifs are quite different in the two patterns. There are only five different types of planar lattices (also referred to as *nets*), and these are shown in Fig. 17.2. The planar lattices in Fig. 17.1 are equivalent to the *b* drawing in Fig. 17.2 (a rectangular lattice).

These five planar lattices have inherent symmetries. You are already familiar with most of the symmetry elements that they may contain such as mirror planes, rotations axes, rotoinversion axes, and centers of symmetry (*i*). There is

an additional translational symmetry element in patterns, namely a *glide line* which combines mirror reflection and translation. Motifs in one-dimensional patterns related by glide lines or mirrors (perpendicular to the page) are shown in Figs. 17.3*a* and *b* (in three-dimensional arrays the combinations of mirror reflections and translation are known as *glide planes*). Figure 17.3*c* shows a pattern based on a rectangular plane lattice (or net) that contains mirrors as well as glide lines. The mirrors (*m*) and glide lines (*g*) are interleaved parallel to the two axial directions of the pattern. Notice that the pattern in Fig. 17.1 contains only mirrors and no glide lines.

Several other aspects of the pattern in Fig. 17.3*c* are worth noting. In the lower left corner of the pattern a rectangular unit cell is outlined by shading. This unit cell contains two motifs per unit cell and is referred to as a *centered* (*c*) unit cell (with a centering motif or node in the center of the unit cell outline) or a *nonprimitive* unit cell. A *primitive cell* (*p*) is one that contains only one node per lattice (as, for example, in Figs. 17.1*b* and *d*). One could have chosen a primitive unit cell in the pattern of Fig. 17.3*c* as well, and such a cell choice is outlined by shading at the right-hand side of the pattern. These two different unit cell choices are equivalent to the illustrations of the centered and the noncentered lattice choices in Fig. 17.2*c*. Clearly the axes (that is, the translation directions) and the size of the translations in the two unit cell choices in Fig. 17.3*c* are different. The *a* and *b* axes of the nonprimitive unit cell are orthogonal, but the *a'* and *b'* axes of the primitive unit cell are not at 90° to each other.

Yet another feature of Fig. 17.3*c* is worth noting. If we were to replace each of the motifs (flowers) with a node, as we did in Fig. 17.1, the resulting planar lattice of Fig. 17.3*c* (shown in Fig. 17.4*a*) will contain all the mirrors and glide lines already shown in Fig. 17.3*c*. However, there are additional symmetry elements inherent in this planar lattice (or net). If you inspect the lattice carefully, you will note that 2-fold rotation axes, perpendicular to the plane of the page, occur at the position of every lattice node. There are additional 2-fold rotation axes (also perpendicular to the page) halfway along the lines that connect the nodes, as well as halfway along diagonals between the nodes. An evaluation of the *total symmetry content* of the lattice based on the pattern of Fig. 17.3*c* is shown in the upper left corner of Fig. 17.4*a*. One of the centered, orthogonal unit cells of this pattern has been enlarged in Fig. 17.4*b* and shows within it the total symmetry content (consisting of mirrors, glide lines, and 2-fold rotation axes). This centered orthogonal net can be classified as one of the 17 possible plane patterns (or plane groups) as illustrated in Fig. 17.5. If one compares Fig. 17.4*b* with the drawings in Fig. 17.5, one finds it to be identical to the drawing with the symbols *c2mm*. This notation should be reminiscent of the point group notation used in earlier exercises (see exercise 10). The small *c* means that the lattice is *centered* (nonprimitive), and the remaining

FIGURE 17.1 (*a*) Two-dimensional orthogonal pattern of flower motifs. (*b*) The lattice deduced from the pattern in *a*. (*c*) A similar two-dimensional pattern with a more complex motif. (*d*) The lattice deduced from the pattern in *c*.

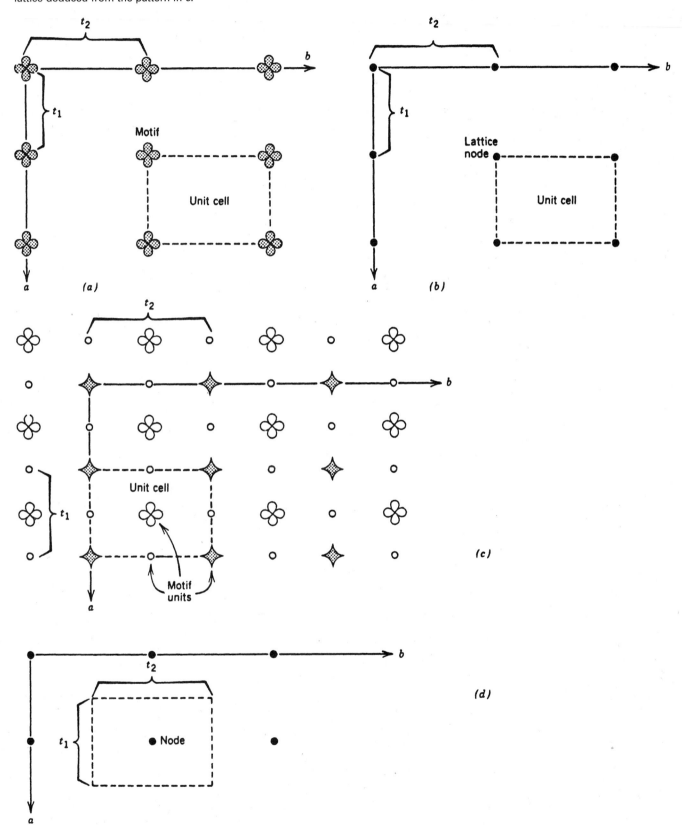

FIGURE 17.2 Five types of plane lattice or nets. The translation distances are represented by t_1 and t_2 and t_1' and t_2' as in (c). The angle between the translation vectors is γ. The symbol = means "is identically equal to" under all circumstances. The symbol $\neq$ implies nonequivalence. The unit cells of the lattices are outlined by a combination of arrows and dashed lines. In part c there are two choices of unit cell: one rectangular, the other with a diamond shape. (See *Manual of Mineral Science,* 23rd ed., p. 146.)

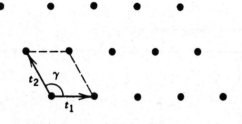

(a) Oblique lattice
$t_1 \neq t_2$
$\gamma \neq 90°$

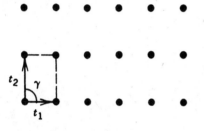

(b) Rectangular lattice
$t_1 \neq t_2$
$\gamma = 90°$

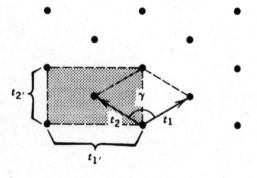

(c) Centered rectangular lattice
$t_1 = t_2$
$\gamma \neq 90°, 60°,$ or $120°$

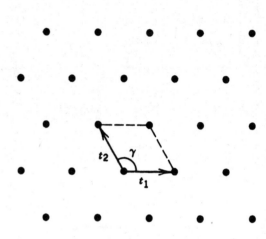

(d) Hexagonal lattice
$t_1 = t_2$
$\gamma = 60°,$ or $120°$

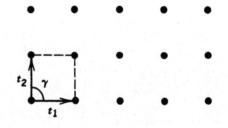

(e) Square lattice
$t_1 = t_2$
$\gamma = 90°$

FIGURE 17.3 (*a*) A glide line with translation *t*/2 and its effect on the distribution of flower motifs. (*b*) A mirror and its effect on the distribution of flower motifs. (*c*) A rectangular pattern of flower motifs in which the locations of mirrors (*m*) and glide lines (*g*) are shown. The two different choices of unit cell are indicated by shading (see text for discussion).

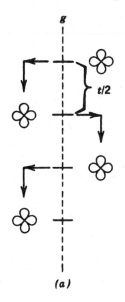

(a)

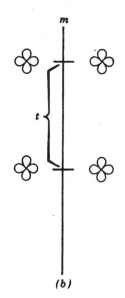

(b)

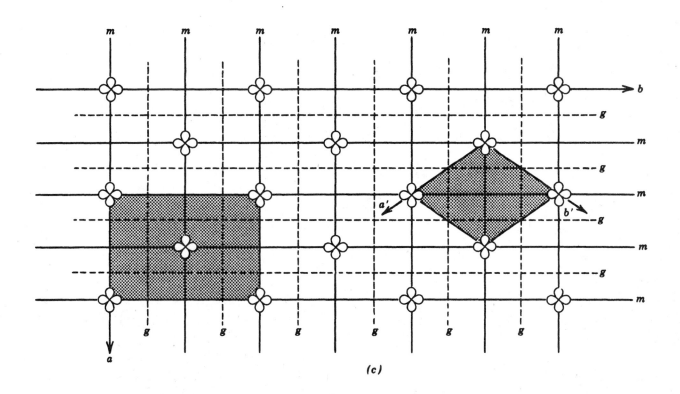

(c)

FIGURE 17.4 (*a*) The plane lattice derived from the pattern in Fig. 17.3. Here mirrors (*m*), glide lines (*g*), and 2-fold rotation axes are located within the lattice. (*b*) An enlarged view of one rectangular (centered) unit cell choice and its symmetry content. (*c*) The 4*mm* symmetry of the flower motif.

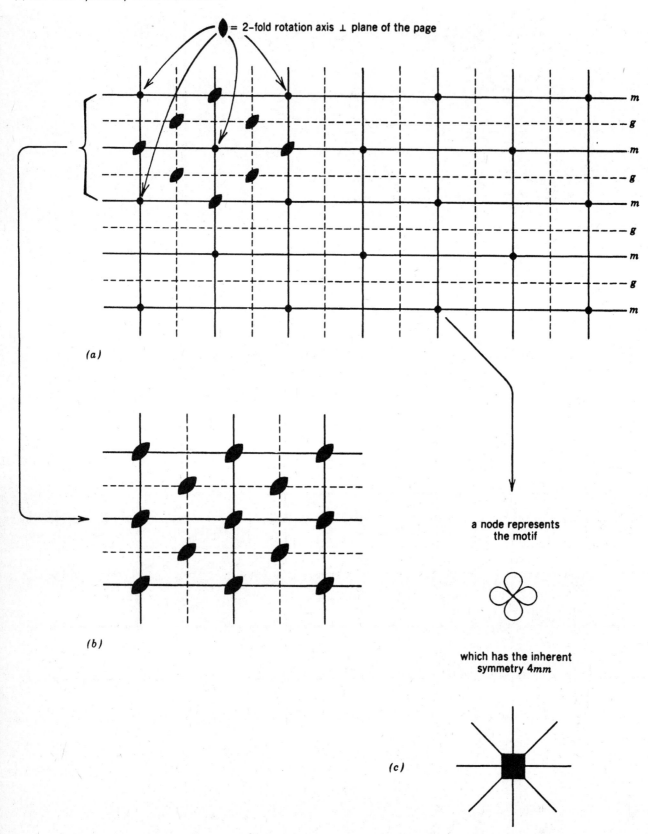

⬥ = 2–fold rotation axis ⊥ plane of the page

(a)

(b)

a node represents the motif

which has the inherent symmetry 4*mm*

(c)

FIGURE 17.5 Representation of the 17 space groups of plane patterns (= plane groups). Heavy solid lines and dashed lines represent mirrors and glides, respectively, perpendicular to the page.

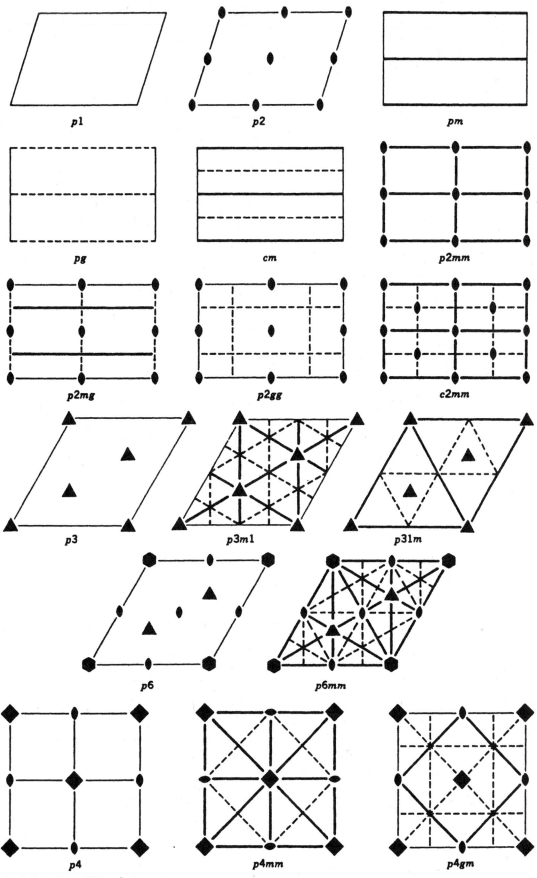

symbolism states that it contains 2-fold rotation axes perpendicular to the page and mirrors along the two axial directions (the presence of the glide lines is not identified in the symbolic statement).

The 17 plane groups represent the 17 unique arrangements of symmetry elements in two dimensions. They result from the interaction of the symmetry (the point group) of the motif with the symmetry of the various planar lattices. There are only 10 possible two-dimensional point group symmetries. These are illustrated in Fig. 17.6 and listed in the first column of Table 17.1; the 17 plane groups are illustrated in Fig. 17.5 and listed in the second column of Table 17.1. With reference to Figs. 17.3 and 17.4, one can illustrate how the point group symmetry of the motif interacts with the lattice symmetry. Notice that the motif has symmetry 4mm (the flower; see Fig. 17.4c), but the overall lattice symmetry is only 2mm. This means that the final pattern displays the symmetry of the lattice when the symmetry elements of the motif are aligned with the corresponding symmetry elements of the lattice. If the motif has less symmetry than the lattice, the pattern will express the motif's lesser degree of symmetry, with the symmetry elements of the motif aligned with the corresponding symmetry elements of the lattice.

Before addressing specific aspects of the assignment ahead, we should look at some more general aspects of printed patterns. A sheet of paper has two sides: one side may contain a drawing and the backside may be blank. As such the two sides are different. This means that there are no symmetry elements that lie in the plane of the paper. That is, there is no mirror plane parallel to the paper, nor are there axes of rotational symmetry parallel to the paper. There may, however, be a number of symmetry elements that are perpendicular to the plane of the drawing. These are mirrors (m), glide lines (g), and rotation axes (1, 2, 3, 4, and 6).

Any repetitive, ordered drawing also contains translational elements. In a one-dimensional row there is only one translation direction and only one translational spacing. In a two-dimensional pattern there are two translational directions, with either a single spacing or two different ones. There are also various choices of angles between the translation directions (see Fig. 17.2). In a three-dimensional ordered array there are three translation directions, and the possibility of only one spacing, two different spacings, three different spacings, as well as various possible angles between the translation directions.

MATERIALS

Figures 17.7 and 17.8 with printed patterns, the last three of which are redrawn from M. C. Escher, the late Dutch graphic artist; sheets of transparent (tracing) paper; millimeter ruler, protractor, soft pencil, and eraser. Sheets of various patterned wallpapers can be added, if so desired.

ASSIGNMENT

Figures 17.7 and 17.8 form the basis of this assignment.

1. Inspect the drawings carefully and locate the motif or motifs. The motif of a pattern is the smallest unit of the pattern that, when repeated by symmetry and translational elements, will generate the whole pattern. Notice that the motif may contain symmetry within it (see Fig. 17.6).

2. A one-dimensional, ordered pattern is the result of a constant translation along the direction of the row. A two-dimensional pattern is the result of translation of the motif along two noncollinear directions. Two such noncollinear directions are chosen as the outline of the unit cell. The unit cell of a two-dimensional pattern is the quadrilateral whose sides are these unit translations. These noncollinear unit translations may or may not be the shortest noncollinear translations in the pattern. This unit cell outline is most easily seen by replacing the actual printed pattern with the imaginary pattern of the lattice, that is, by substituting nodes for the motifs on your transparent overlay. The choice of unit cell in a two-dimensional pattern is restricted to one of the following five (see Fig. 17.2):

a parallelogram (an oblique choice)

a rectangle (which may or may not contain an internal motif)

a diamond (with equal translations)

a square

a rhombus or triangle (also known as hexagonal)

The best choice of unit cell in a pattern is not arbitrary. Crystallographers have adopted the following conventions concerning the choice.

a. The edges of the unit cell should, if possible, coincide with symmetry elements of the pattern.
b. The edges should be related to each other by the symmetry of the pattern.
c. The smallest possible cell in accordance with conventions **a** and **b**, above, should be chosen.

If the unit cell of your choice contains nodes only at its corners, it is described as *primitive* (*p*). A unit cell choice in which an additional node is located at the center of the cell is known as *centered* (*c*).

3. After having located the translation directions of the pattern, and having recognized the repeated motif, put a transparent overlay (tracing paper) over the pattern. On this overlay substitute large round dots (known as "nodes") for each motif. Do this throughout the pattern. The point you choose is arbitrary, but once you have chosen a specific location in the pattern for a node location, you must be sure that you continue to use the exact same node location throughout the pattern. The pattern of nodes you have generated (in a two-dimensional ordered array) is the *net*

FIGURE 17.6 The point group symmetry of two-dimensional motifs.

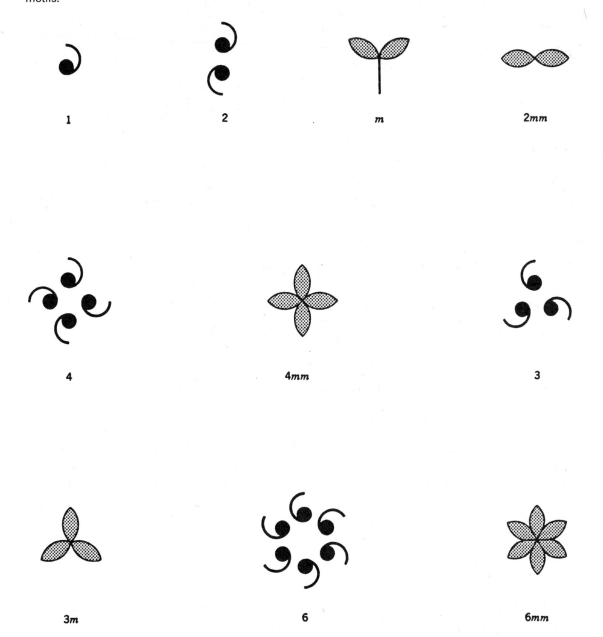

TABLE 17.1 Two-Dimensional Point Groups and Space Groups

Lattice	Point Group	Space Group
Oblique *p*	1	p1
	2	p2
Rectangular *p* and *c*		pm
	m	pg
		cm
	2mm	p2mm
		p2mg
		p2gg
		c2mm
Square *p*	4	p4
	4mm	p4mm
		p4gm
Hexagonal *p*	3	p3
	3m	p3m1[a]
		p31m[a]
	6	p6
	6mm	p6mm

There are two distinct groups for *3m*—*p3m1* and *p31m*. They have the same total symmetry content and shape. However, the conventional location of cell edges (as defined by three axes, a_1, a_2, a_3) differs by 30° in the two groups. In *p3m1* the mirror lines bisect the 60° angle between cell edges; in *p31m* the reflection lines coincide with the cell edges (see drawings on the right hand side).

SOURCE: From International Tables for Crystallography, vol. A, 1983 D. Reidel Publishing Co., Dordrecht, The Netherlands.

(an abstract rendition of the original pattern). On this net (on transparent paper) outline the most appropriate choice or choices of unit cell [in some patterns there may be more than one choice, e.g., primitive (*p*) versus centered (*c*)]. If there are choices, mark them all and carefully note how they differ. With a pencil and ruler draw the translation directions, parallel to the unit cell choice, throughout the net.

4. Locate all symmetry elements with the appropriate symbols on the transparent overlay. Use the symbols given in Fig. 14.1 and use a dashed line symbol for a glide line. Do this throughout a large part of the net. Note where these symmetry elements are located with respect to your choice of unit cell (see the last section of point 2 above). Generally it is best to choose corners (of the unit cell) coincident with major symmetry elements (e.g., rotations). Furthermore, edges of your unit cell choice may coincide with mirror planes or be perpendicular to such mirrors.

5. Measure and record the length of the translations in the one-dimensional patterns. Note the presence of any symmetry elements (e.g., *m*, *g*, and/or rotations). Record your findings in Table 17.2.

6. Measure and record the lengths of the two translations in the two-dimensional patterns. Also note the angle between your two chosen directions. Record your findings in Table 17.2.

7. Decide on the shape of your choice of unit cell. See point 2 as well as Fig. 17.2. Record this in Table 17.2.

8. Table 17.2 allows for additional evaluations. For the one-dimensional rows only a few data are requested. However, for the two-dimensional patterns a formal assessment of the total symmetry content should be made. There are only 17 possible options for regularly ordered two-dimensional patterns. These 17 possible space groups are listed in Table 17.1, and their symmetry elements and lattices are illustrated in Fig. 17.5. In looking these over you see the numeral 1 in several places (e.g., *p1*, *p3m1*, and *p31m*). The 1 implies 1-fold rotation, which is equivalent to 0° or 360° rotation. (This means that after 0° rotation, or after 360°, the 1-fold rotational operation has generated the identical object, or motif; because of this the 1-fold rotation is also known as "identity.")

9. Locate all symmetry elements on the lattice tracing of the two-dimensional patterns, and assign each pattern to one of the 17 plane groups. Your tracing paper pattern will be equivalent to one of the plane groups in Fig. 17.5 only if you have indeed plotted *all* of the symmetry elements. Please note that the origins of the unit cells in Fig. 17.5 are really arbitrary, but they are agreed upon by convention. You may have to shift the origin of your pattern to conform with the conventions of Fig. 17.5. If your pattern still does not fit one of the plane groups, you have probably overlooked a symmetry element.

FIGURE 17.7 One-dimensional (linear) patterns.

One-dimensional patterns

(a)

(b)

(c)

(d)

(e)

(f)

Student Name

FIGURE 17.8 Two-dimensional (planar) patterns. Patterns *i, j,* and *k* are redrawn from illustrations by M. C. Escher, as published by Caroline H. MacGillavry, 1976, *Fantasy and Symmetry;*

The Periodic Drawings of M.C. Escher. Harry N. Abrams, New York; copyright © 2006 The M. C. Escher Company–Holland. All rights reserved.

Two - dimensional patterns

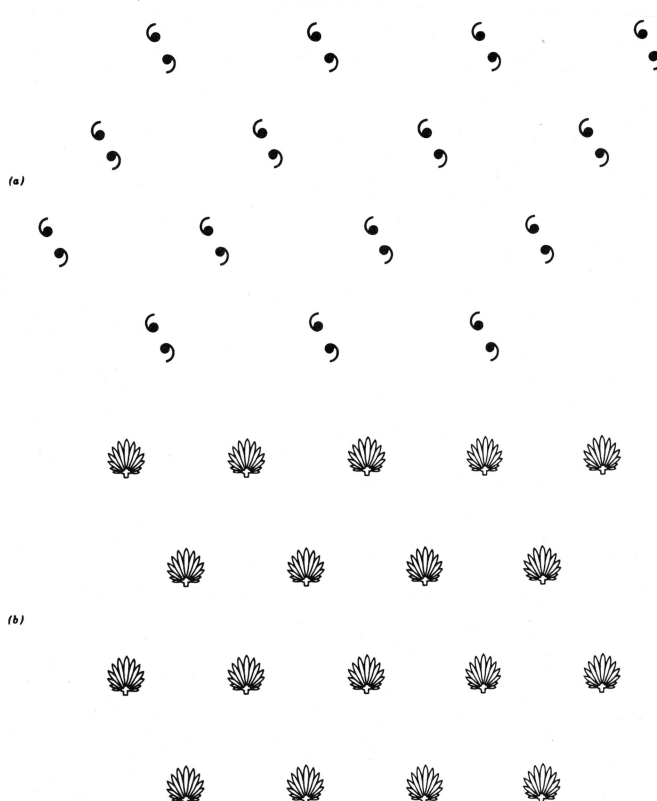

Student Name

Two - dimensional patterns

(c)

(d)

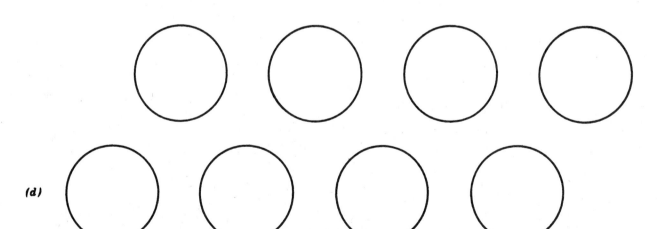

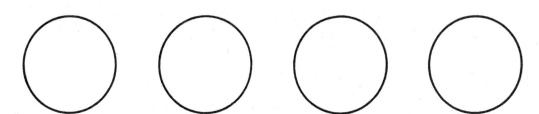

FIGURE 17.8 *(continued)*

Two - dimensional patterns

(e)

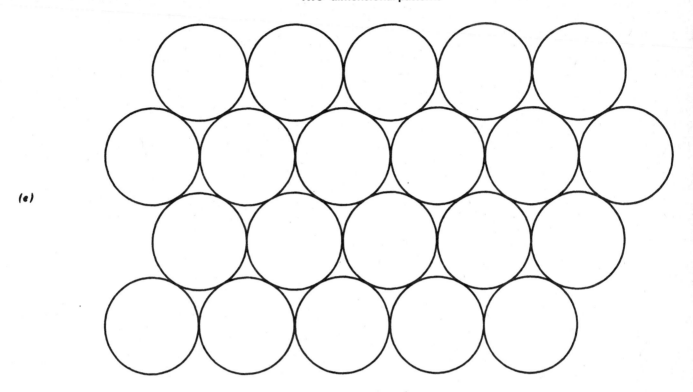

(f)

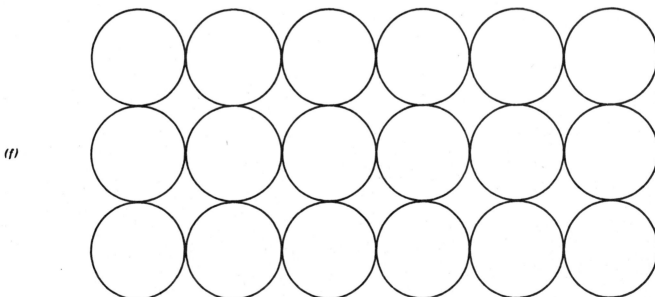

Student Name

FIGURE 17.8 *(continued)*

Two-dimensional patterns

(g)

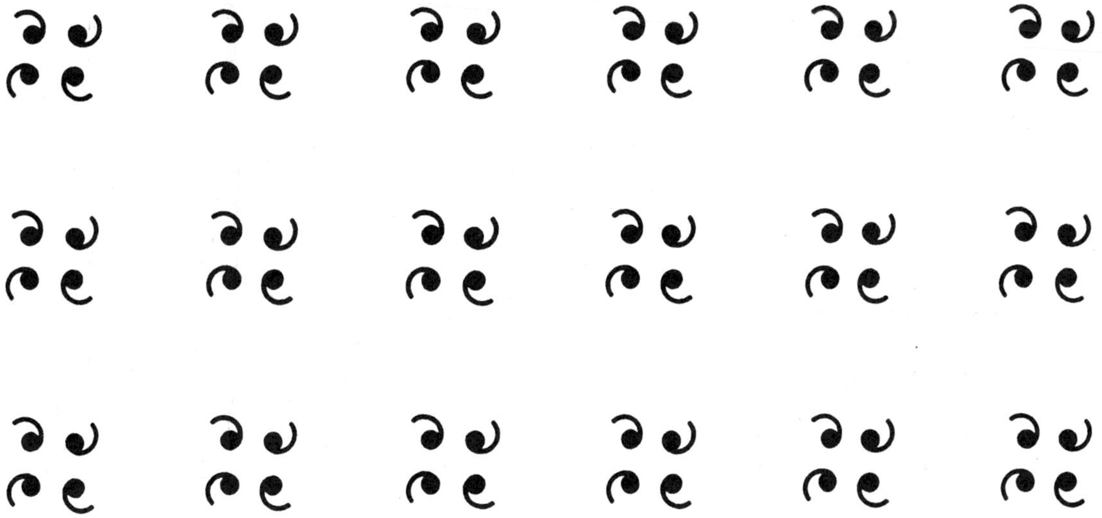

(h)

(i)

Student Name

FIGURE 17.8 *(continued)*

(j)

(k)

Student Name

TABLE 17.2 Data Obtained from the One- and Two-Dimensional Patterns in Figs. 17.7 and 17.8

One-Dimensional Patterns (Fig. 17.7)	a	b	c	d	e	f
Translation distance along row						
Symmetry content of motif (if any)						
Additional symmetry element and its orientation						

Two-Dimensional Patterns (Fig. 17.8)	a	b	c	d

Translation distance along:
    direction 1 = t_1
 	
    direction 2 = t_2				

Angle between t_1 and t_2				
Unit cell choice (*p* or *c*)				
Shape of unit cell				
What is the motif? (give a sketch)				
Notation for plane group				

TABLE 17.2 *(continued)*

Two-Dimensional Patterns (Fig. 17.8)	e^a	f^b	g	h	i^c	j^c	k^c
Translation distance along: direction 1 = t_1							
direction 2 = t_2							
Angle between t_1 and t_2							
Unit cell choice (p or c)							
Shape of unit cell							
What plane net is this?							
What is the motif? (give a sketch)							
What is its point group symmetry?							
Plane group symbol for pattern							

[a]This arrangement of spheres represents the densest possible packing of spheres in a regular array in two dimensions (known as simple hexagonal packing).

[b]This arrangement of spheres is less space-conserving than the one in *e* (as such less dense).

[c]Patterns *i*, *j*, and *k* are redrawn from illustrations by M. C. Escher, as published by Caroline H. MacGillavry, 1976, *Fantasy and Symmetry; The Periodic Drawings of M. C. Escher*. Harry N. Abrams, New York. © 2006 The M. C. Escher Company–Holland. All rights reserved. *www.mcescher.com*

EXERCISE 18

Space Lattices

PURPOSE OF EXERCISE

Understanding space lattice choices and their projections as an introduction to the notation and graphical representation of space groups.

FURTHER READING AND CD-ROM INSTRUCTIONS

Klein, C. and Dutrow, B., 2008, *Manual of Mineral Science*, 23rd ed., Wiley, Hoboken, New Jersey, pp. 156–168.

CD-ROM, entitled *Mineralogy Tutorials*, version 3.0, that accompanies *Manual of Mineral Science*, 23rd ed., click on "Module III", and subsequently the button "3-dimensional order". This will give a new set of listings. Click the button "Generation of 10 Bravais Lattices". Also click "Screw Axes" and "Glide Planes".

Nesse, W. D., 2000, *Introduction to Mineralogy*, Oxford University Press, New York, pp. 8–12.

Perkins, D., 2002, *Mineralogy*, 2nd ed., Prentice Hall, Upper Saddle River, New Jersey, pp. 215–225.

Wenk, H. R. and Bulakh, A., 2004, *Minerals: Their Constitution and Origin*, Cambridge University Press, New York, New York, pp. 37–45.

Bloss, F. D., 1994, 2nd printing with minor revisions, *Crystallography and Crystal Chemistry*, Mineralogical Society of America, Chantilly, Virginia, pp. 140–159.

International Tables For Crystallography, vol. A, Theo Hahn, ed., 1983, D. Reidel Publishing Co., Dordrecht, The Netherlands, p. 734–744, and space group illustrations, pp. 101–709.

International Tables For Crystallography, *Brief Teaching Edition of vol. A*, Theo Hahn, ed., 1985, D. Reidel Publishing Co., Dordrecht, The Netherlands, p. 119.

Background Information

a. *On the choice of unit cell within a space lattice.* In exercises 9, 10, 14, and 15 various aspects of the 32 point groups (= crystal classes) were introduced. Such point groups are translation-free. In exercise 17 the combinations of symmetry elements and translations in one- and two-dimensional patterns were considered. This led to the recognition of the 17 plane groups, as a result of combining the inherent point group symmetry of a motif with the five plane lattices (or nets). In three-dimensional structures the combination of the 32 point groups and all possible lattice types (there are 14 of these, referred to as the *Bravais lattices*) leads to 230 possible ways in which symmetry and translation can interact. These 230 possible combinations of symmetry and translation are known as the 230 *space groups*, which are

230 possible three-dimensional architectures of symmetry elements. In addition to the symmetry elements already introduced, that is, a center of symmetry (i), mirrors (m), and rotational axes $(1, 2, 3, 4, 6$ and $\bar{1} = i, \bar{2} = m, \bar{3} = 3.i, \bar{4},$ and $\bar{6} = 3/m)$, space groups contain various translational elements. These are (1) pure translations between equivalent positions in an ordered structure (see exercise 17 for examples of such translations in one and two dimensions), (2) translations that are combined with rotational axes (giving rise to *screw axes*), and (3) translations combined with mirrors (giving rise to *glide planes*). In the first part of this exercise we will concentrate on the purely translational aspects of space groups and their notation.

The translational aspect of an ordered, three-dimensional crystal structure is exemplified by the lattice (or lattice type) of that structure. A lattice is defined as follows (*Manual of Mineral Science*, 23rd ed., p. 156). *A lattice is an imaginary pattern of points (or nodes) in which every point (or node) has an environment that is identical to that of any other point in the pattern. A lattice has no specific origin as it can be shifted parallel to itself* (see also exercise 17). The three-dimensional lattices are formed through the addition of a third translation direction (a vector) to the five plane lattices of Fig. 17.2. Just as we saw that there are only five ways in which points (= nodes) can be arranged periodically in two dimensions, there are only 14 ways to arrange points periodically in three dimensions. These are the 14 *Bravais lattices*.

The 230 space groups incorporate in their notation a statement of lattice type. In each of the six crystal systems there is an ordered repeat pattern that can be based on a *primitive lattice* (a primitive lattice is one that contains only corner nodes, and as such is noncentered). This means that out of the 14 possible Bravais lattice choices, six are primitive (referred to as *P*). There is an additional primitive lattice in the hexagonal system, known as rhombohedral (*R*), with the shape of a rhombohedron. The additional seven lattices, distributed among four of the crystal systems, are nonprimitive and have specialized centering nodes. Examples of all such centerings (*A, B, C, F,* and *I*) are given in Fig. 18.1. The distribution of the 14 space lattice types among the six crystal systems is shown in Fig. 18.2 and Table 18.1.

In exercise 17 translational repeats were studied in one-dimensional patterns, and in two-dimensional patterns unit cell were selected. In two-dimensional patterns the unit cell choices are restricted to five types. In three-dimensional structures the choice of space lattice is restricted to one of the 14 Bravais lattice types.

As in two-dimensional patterns, in three-dimensional structures the choice of lattice type is not arbitrary. As

FIGURE 18.1 (*a*) A primitive unit cell with orthogonal axes *x, y*, and *z* and with the lengths along these axes (*a, b*, and *c*) unequal. (*b*) Possible centerings in the unit cell outlined in part *a*. (*c*) A primitive unit cell outlined by two equal axes, a_1 and a_2, and a third, *c*, perpendicular to them. This type of lattice occurs in crystals known as hexagonal. (*d*) A rhombohedral unit cell outlined by translation direction a_R (the edge of the rhombohedron) and α_R (the angle between two edges). The edges are symmetrical with respect to a $\bar{3}$ axis along the *c* direction.

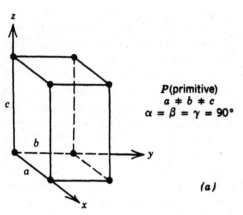

P(primitive)
$a \neq b \neq c$
$\alpha = \beta = \gamma = 90°$

(*a*)

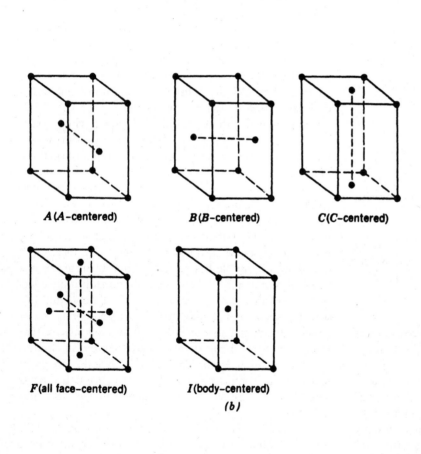

A (*A*-centered)   *B* (*B*-centered)   *C* (*C*-centered)

F (all face–centered)   *I* (body–centered)

(*b*)

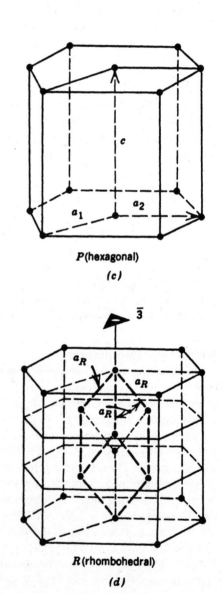

P (hexagonal)

(*c*)

R (rhombohedral)

(*d*)

FIGURE 18.2 The 14 unique types of space lattices, known as the Bravais lattices. Axial lengths are indicated by *a, b,* and *c* and axial angles by α, β, and γ. Each lattice type has its own symmetry constraints on lengths of edges *a, b,* and *c* and angles between edges, α, β, and γ. In the notations the nonequivalences of angles or edges that usually exist but are not mandatory are set off by parentheses.

*In the monoclinic system the unit cell can be described by a body-centered cell or a *C*-face-centered cell through a change in choice of the length of the *a* axis and the angle β. Vectorially these relations are: $a_I = c_c + a_c$; $b_I = b_c$; $c_I = -c_c$; and $a_I \sin \beta_I = a_c \sin \beta_c$. Subscripts *I* and *c* refer to the unit cell types. (From *Manual of Mineral Science*, 23rd ed., p. 162.)

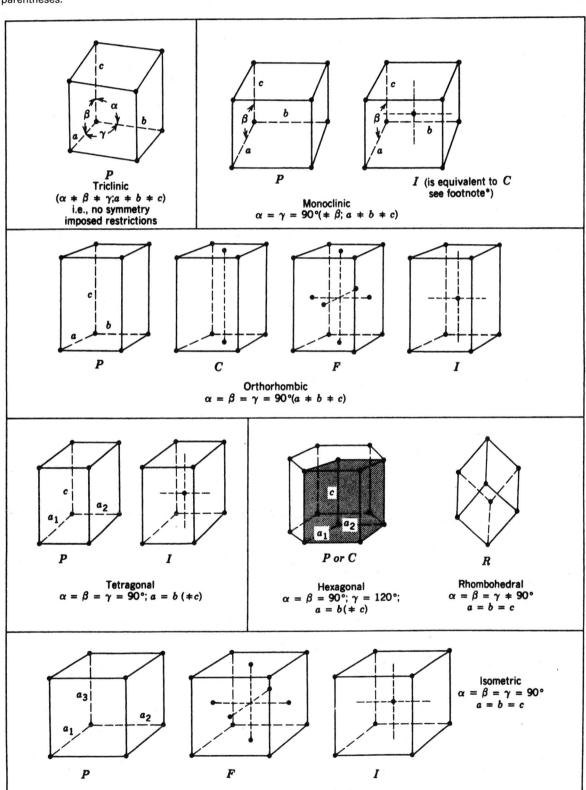

P
Triclinic
(α ≠ β ≠ γ; *a* ≠ *b* ≠ *c*)
i.e., no symmetry
imposed restrictions

P   **I** (is equivalent to **C**
                  see footnote*)
Monoclinic
α = γ = 90°(≠ β; *a* ≠ *b* ≠ *c*)

P   **C**   **F**   **I**
Orthorhombic
α = β = γ = 90°(*a* ≠ *b* ≠ *c*)

P   **I**
Tetragonal
α = β = γ = 90°; *a* = *b* (≠ *c*)

P or C
Hexagonal
α = β = 90°; γ = 120°;
a = *b*(≠ *c*)

R
Rhombohedral
α = β = γ ≠ 90°
a = *b* = *c*

P   **F**   **I**
Isometric
α = β = γ = 90°
a = *b* = *c*

TABLE 18.1 Description of Space Lattice Types and Distribution of the 14 Bravais Lattices Among the Six Crystal Systems

Name and Symbol		Location of Nonorigin Nodes	Multiplicity of Cell
Primitive (*P*)		. . . .	1
Side-centered	(*A*)	Centered on *A* face (100)	2
	(*B*)	Centered on *B* face (010)	2
	(*C*)	Centered on *C* face (001)	2
Face-centered	(*F*)	Centered on all faces	4
Body-centered	(*I*)	An extra lattice point at center of cell	2
Rhombohedral	(*R*)	A primitive rhombohedral cell	1

Primitive (*P*) in each of the six crystal systems	= 6
Body-centered (*I*) in monoclinic, orthorhombic, tetragonal, and isometric	= 4
Side-centered (*A* = *B* = *C*) in orthorhombic	= 1
Face-centered (*F*) in orthorhombic and isometric	= 2
Rhombohedral (*R*) in hexagonal	= 1
	Total = 14

SOURCE: From *Manual of Mineral Science*, 23rd ed., p. 163.

noted in exercise 17, crystallographers have drawn up the following restrictions concerning the choice.

1. The edges of the unit cell* (of the lattice) should, if possible, coincide with symmetry axes of the lattice.

2. The edges should be related to each other by the symmetry of the lattice.

3. The smallest possible cell in accordance with restrictions **1** and **2**, above, should be chosen.

In any regular (= ordered) three-dimensional array of (lattice) nodes, a primitive (*P*) lattice type can always be chosen (the *P* choices are the six primitive lattices, each belonging to one of the six crystal systems, among the 14 Bravais lattice types). However, it is frequently desirable and appropriate to choose a nonprimitive unit cell (or lattice type). In practice, in the study of crystal structures the investigator will determine the three-dimensional translation (lattice) pattern of the structure being investigated by, for instance, X-ray diffraction techniques, on the basis of the distribution of X-ray diffraction maxima (e.g., exposed spots on a film). He or she will in accordance with point **3** try to describe the lattice in terms of the smallest unit cell. Such a choice might turn out to be a primitive (*P*) lattice that has an awkward relationship to the symmetry elements (mirrors, glide planes, rotation and screw axes) of the structure. If such is the case, the investigator will subsequently decide on a nonprimitive lattice in accordance with points **1** and **2**.

At this stage, it will be useful to illustrate not only lattice-type choices but also the various ways in which three-dimensional (space) lattices are illustrated, or projected, onto a two-dimensional page. In Fig. 18.3*a* unit cells of two lattice types are shown in three-dimensional (perspective) drawings. In these two drawings the nodes are connected by lines in order to outline the unit cells of the lattice; normally a lattice, as stated in its definition, consists only of nodes (without connecting lines or bars). The left-hand unit cell is compatible with a crystal structure in which the structural elements are arranged in an isometric array. Therefore, the unit cell has a cubic outline. It is primitive because it contains nodes only at the corners of the unit cell. The "node content" (or multiplicity) of this unit cell = 1, because each corner node (there are eight of them) contributes only one-eighth of a node to the cell outlined: $8 \times \frac{1}{8} = 1$. This primitive lattice is abbreviated as *P*.

The unit cell on the right-hand side has a parallelepiped shape with three edges of different lengths, but all at 90° to each other. Such a unit cell choice is compatible with a crystal structure in the orthorhombic system. This is clearly not a primitive lattice (or unit cell) choice, because in addition to eight corner nodes there are six nodes centered on six faces. The centerings are on top and bottom, front and back, and right- and left-hand sides. This unit cell contains $8 \times \frac{1}{8} = 1$ node for the corner nodes and an additional $6 \times \frac{1}{2} = 3$ nodes for the nodes on the six faces (each node on a face contributes one-half of that node to the unit cell), resulting in a total of four nodes per unit cell. By definition this is a nonprimitive unit cell, with a multiplicity of 4, and is known as all-face-centered, abbreviated as *F*.

The representations of unit cells or lattice types in a perspective view as in Fig. 18.3*a* are not common. Normally, three-dimensional crystal structures, their

*Unit cell definition: A unit cell is the smallest volume or parallelepiped within the three-dimensional repetitive pattern of a crystal that contains a complete sample of the atomic or ionic groups making up the pattern.

FIGURE 18.3 Unit cells and their projections. (*a*) Perspective views of an isometric *P* cell and an orthorhombic *F* cell. (*b*) Projection of these same cells onto the plane of the page (= 0 level) in extended lattices.

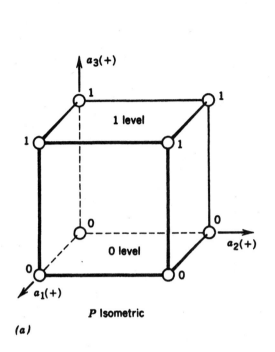
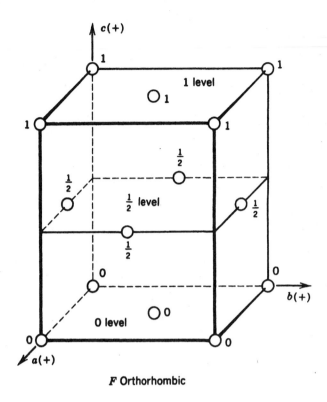

P Isometric

(a)

F Orthorhombic

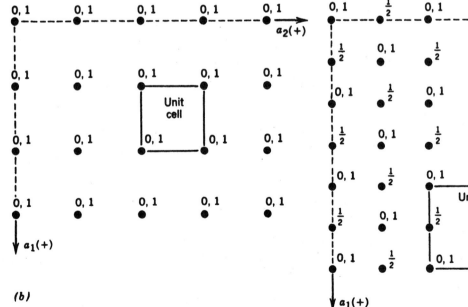
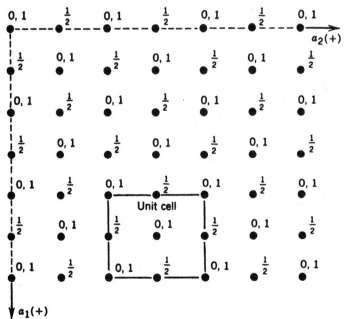

(b)

lattice types, and intrinsic symmetry operations are shown as projections onto a (two-dimensional) page. In order to make such projections onto a page understandable, the original heights of nodes above the page are shown as whole numbers or fractions of the vertical unit cell dimension. In the isometric P lattice the "ground floor" is the plane defined by the a_1 and a_2 axes. Conventionally this is the 0 level; accordingly the nodes in this level have 0s next to them. The upper level in this lattice is equivalent to the lower level, except for one upward translation along the z direction (here identified as the a_3 axis).* The nodes in this upper level ("one floor up") are all accompanied by a 1. The orthorhombic F-type unit cell has a "lower floor," identified by 0s along the nodes; it also has an "upper level" (equivalent to the 0 level) identified by 1s alongside the nodes. However, it also has a "mezzanine," halfway between the 0 and 1 levels, on which four of the six centering nodes are located. This "mezzanine" is half way up, between the 0 and 1 levels, and as such the nodes in this in-between level are accompanied by the fractional statement $\frac{1}{2}$. All the translations are shown in the positive octant defined by three coordinate axes. As such, each number or fraction could have been preceded by a $+$ sign: $+1$, $+\frac{1}{2}$, and so on. This would be necessary in order to distinguish any node positions in planes below the 0 level, which would be in the negative direction of the vertical coordinate axis (e.g., -1, $-\frac{1}{2}$, etc.). Translations of $+1$ and -1 are implicit for all lattices, which means that 0, 1 implies 2, 3, 4, -1, -2 . . . and $\frac{1}{2}$ implies $1\frac{1}{2}$, $2\frac{1}{2}$, $-\frac{1}{2}$, $-1\frac{1}{2}$ . . . .

In Fig. 18.3b the two unit cells in Fig. 18.3a are shown as projections onto the 0 level, the plane of the page. The scale of the illustration in Fig. 18.3b is smaller than that of Fig. 18.3a in order to show several unit cells, that is, the projection of an extended space lattice onto a two-dimensional page. Nodes that superimpose in such a projection scheme are accompanied by two numbers, namely 0 and 1, to remind the reader that two nodes have been collapsed onto each other in the projection. It is not uncommon to see illustrations of lattices in publications in which the 0 and 1 are left off for corner nodes and only those nodes with fractional locations along the z-coordinate axis are identified by fractions (e.g., $\frac{1}{2}$, $\frac{1}{3}$, or $\frac{1}{4}$). The reader then concludes that the unmarked nodes occur at 0 and at 1.

It was earlier stated that a primitive unit cell or lattice type can always be selected in any structure, even though a multiple lattice (or unit cell) would be more appropriate. Figure 18.4a shows the perspective drawings of a centered unit cell, as well as the outline of a possible primitive cell in the same array of nodes. Figure 18.4a shows an all-face-centered (F) unit cell

with cubic outline, as well as a noncentered (primitive, P) cell with a parallelepiped shape. Both unit cell choices will account for all the nodes in an extended lattice, but clearly the F choice, although larger (in volume) than the P choice, is preferable because symmetry axes (in this case 4s) and mirror planes coincide with the edges of the F cell, not the P cell. Figure 18.4b shows the cubic outline for a cell with one internal centering node (I), as well as a primitive (P) cubic unit cell at an angle to the I cell. Please note that only one centering node is contained within the volume of the I cell (node 5); the other three centering nodes shown in the drawing (nodes 1, 3, 8) are the centers of adjoining unit cells and are part of an extended lattice. In Figs. 18.4c and d the perspective unit cell drawings (of parts a and b) are projected onto the 0 level [crystallographically this would be referred to as the (001) plane] in extended isometric lattices. Remembering that these are isometric lattices, then clearly the P choice, in both cases, is one that is *not* in accordance with the need for coincidence of unit cell edges and symmetry elements of the lattice. Please study these projections carefully so that you understand how they were obtained. The assignments in this exercise will draw upon your ability to visualize three-dimensional unit cell choices from two-dimensional projections.

b. *On the symmetry content of unit cells and space lattices.* The definition of a lattice states ". . . an imaginary pattern of nodes in which every node has an environment that is identical to that of any other node. . . ." This implies that the symmetry about each node (in a specific lattice) is identical. In the following discussion three examples of combinations of lattice type and symmetry will be given. You will note that in these examples all symmetry elements (as given in point group notation) are perpendicular to the projection page (see Figs. 18.5, 18.6 and 18.7). These symmetry types were selected as an introduction to the subject. More general symmetry is considered in exercise 19.

Figure 18.5a gives a perspective drawing of a primitive tetragonal unit cell that incorporates the translational symmetry of a structure with internal symmetry 422. The space group symbol for such a structure would be $P422$ (for further details on space group notation see exercise 19). The primitive unit cell has only corner nodes and each of these nodes must reflect the 422 symmetry. The incorporation of this symmetry is shown in Fig. 18.5b. From our earlier exercises on point group symmetry you recall that the 4-fold rotation axis is the unique c axis, that the first 2 (in the symbol) refers to the a_1 and a_2 axes, and that the next 2 reflects the symmetry about the diagonals to the a_1 and a_2 axes. This knowledge allows for the construction in Fig. 18.5b. Figure 18.5c shows this same information in the standard projected form, a projection down the c axis. This diagram, however, shows only the symmetry elements specifically listed as part of 422. It does not show additional elements that are implied (or generated) at

---

*The notation for axial directions (such as a, b, and c, or a_1, a_2, and c) as used by crystallographers is equivalent to axes denoted as x, y, and z, or x_1, x_2, and z.

FIGURE 18.4 Alternate choices of unit cells in two cubic lattices. (*a*) Perspective view of an isometric unit cell, showing a *P* versus *F* choice. (*b*) Perspective view of an isometric unit cell, showing a *P* versus *I* choice. (*c*) Projection of the unit cells in part *a* onto the plane of the page (= 0 level) in an extended isometric lattice. (*d*) Projection of the unit cells in *b* onto the plane of the page (= 0 level) in an extended isometric lattice. For clarity the nodes for the *P* choice have been given corresponding numbers in the perspective and projected illustrations.

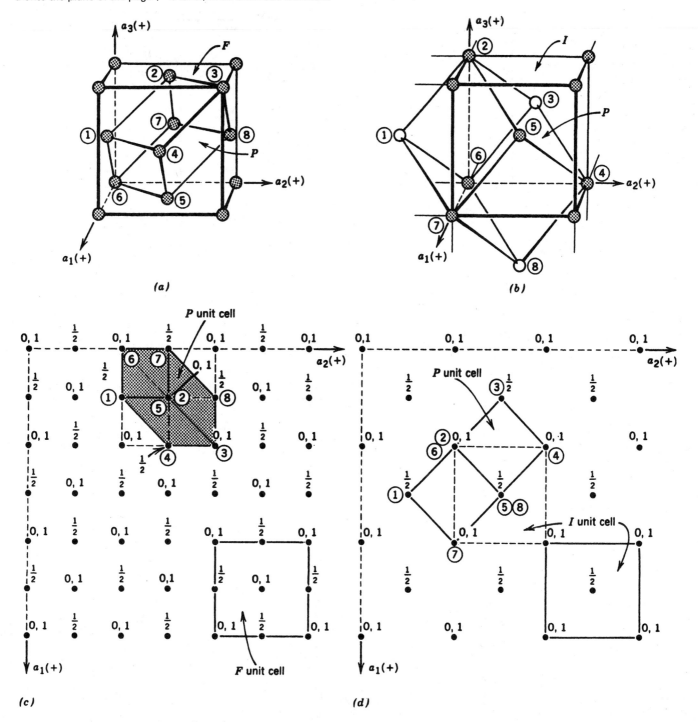

FIGURE 18.5 Graphical derivation of the total symmetry content in *P*422. See text for discussion.

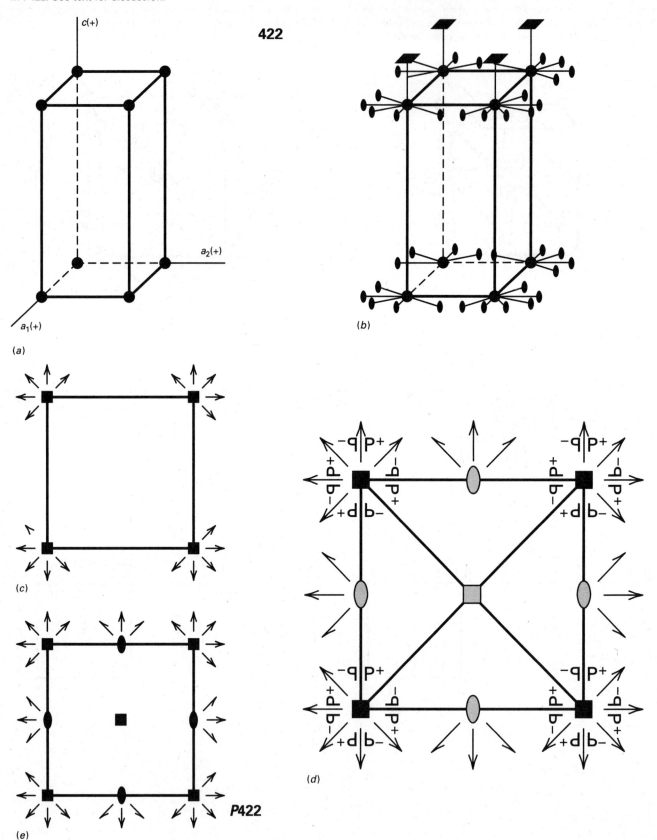

additional locations in the unit cell. The easiest way to evaluate the possible presence of such additional symmetry elements at locations other than the nodes (in the unit cell) is to draw a motif unit and let the 422 symmetry repeat it throughout the unit cell.

In Fig. 18.5d an asymmetric motif, that is, a motif without internal symmetry, was chosen, the capital letter P. In the *Manual of Mineral Science*, 23rd ed., another type of asymmetric motif is commonly used, namely a large comma. If all motif units lie in the same plane as, for example, the 0 level, there is no need to concern oneself with indications of height (for the motif P) above the page. If, however, a motif is slightly above the page, it would be shown as P$^+$, indicating some arbitrary distance upward in the + direction of the c axis; if the location of the motif is halfway up, this c-axis repeat would be shown as P$\frac{1}{2}$. A notation such as P$^{1/2+}$ would indicate a motif above the plane by a translation slightly more than one-half of the unit cell edge along c. The position that is equivalent to P$^+$, but in the negative direction of the c axis, would be shown as P$^-$. Such notation allows for the visualization of, for example, the presence of horizontal 2-fold axes as follows:

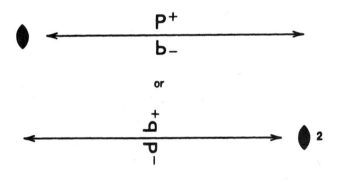

It is wise to locate the original P motif in a random (or general) position, with a + notation along the c axis rather than on a symmetry element. In Fig. 18.5d the motif unit P is repeated by the 422 symmetry elements as well as the translations in the primitive unit cell. Careful inspection of this diagram should reveal additional symmetry elements at the center of the cell (4-fold rotation axis) and at the centers of the cell edges (2-fold rotations perpendicular to the page). However, there are also horizontal symmetry axes halfway along the edges. These are horizontal 2-fold axes perpendicular to the edges. There are also horizontal 2-fold screw axes at 45° to the edges. In the horizontal plane of projection 2-fold axes are normally shown by $\longrightarrow$ and 2-fold screw axes by $\longrightarrow$. A 2-fold screw axis combines 2-fold rotation with a translation of one-half the identity period (along the axis). For definitions of various screw axis types, see exercise 19. Figure 18.5e shows the total symmetry content of a tetragonal P lattice, consistent with symmetry 422. In short, Fig. 18.5e

is the complete graphical representation of the lattice and symmetry shown in Fig. 18.5b.

The example in Fig. 18.5 was for a primitive lattice. In Fig. 18.6 the symmetry content of a centered unit cell is shown for the orthorhombic system, with internal symmetry 222. Figure 18.6a shows the C-centered unit cell. Figure 18.6b incorporates the 222 symmetry at each of the lattice nodes. Figure 18.6c shows this same information projected down the c axis. This figure does not illustrate the additional (implied) symmetry elements that are the result of a C-centered cell. This additional symmetry is best revealed by drawing motif units distributed about each of the nodes in the unit cell, in accordance with the 222 symmetry. This is shown in Fig. 18.6d. Upon careful inspection of the relationship of the clusters of motif units to each other—that is, the centering cluster versus the corner clusters—it becomes clear that 2-fold axes perpendicular to the page exist halfway between the centering node and the corners. In addition, horizontal 2-fold screw axes are interleaved between the already located 2-fold axes. The total symmetry content of this centered cell with symmetry 222 is shown in Fig. 18.6c.

The last two examples (Figs. 18.5 and 18.6) have involved only rotational axis symmetry. It is useful to briefly illustrate a centered unit cell with mirror symmetry. Figure 18.7 shows an I-centered tetragonal cell, with internal symmetry 4mm. Figure 18.7a shows a perspective drawing of the I cell. Figure 18.7b locates the symmetry elements 4mm about each of the nodes in an expanded drawing; this is shown in projection (down the c axis) in Fig. 18.7c. In an expanded view of this projection (Fig. 18.7d), all these same symmetry elements are plotted, as well as all the symmetry-required motif units (a capital P). Once these clusters of motif units have been located, inspection of the projection allows for the location of additional symmetry elements. All such implied symmetry elements are oriented perpendicular to the page. They are screw axes (shown by ▸ and ◗), which involve rotation as well as translation along the direction of the axes, and glide planes (shown by dot and dash–dot patterns), which combine reflection and translation. (See exercise 19 for the graphical significance of the symbols used.) Here it suffices to note that the space group notation I4mm allows for only partial representation of all the symmetry implied in the Hermann–Mauguin (or international) symbol. These examples illustrate that *some symmetry elements are not given explicitly in the space group symbol because they can be derived*. Symmetry elements that occur in space groups, on account of centering of the unit cell, are always omitted. In this respect it should be noted that all the 14 Bravais lattice types are parallelpipeds which require opposite and parallel faces. Such opposite and parallel face pairs are related by a center of symmetry (= inversion, i). As such all space lattice types are said to be *centric*.

FIGURE 18.6 Graphical derivation of the total symmetry content in C222. See text for discussion.

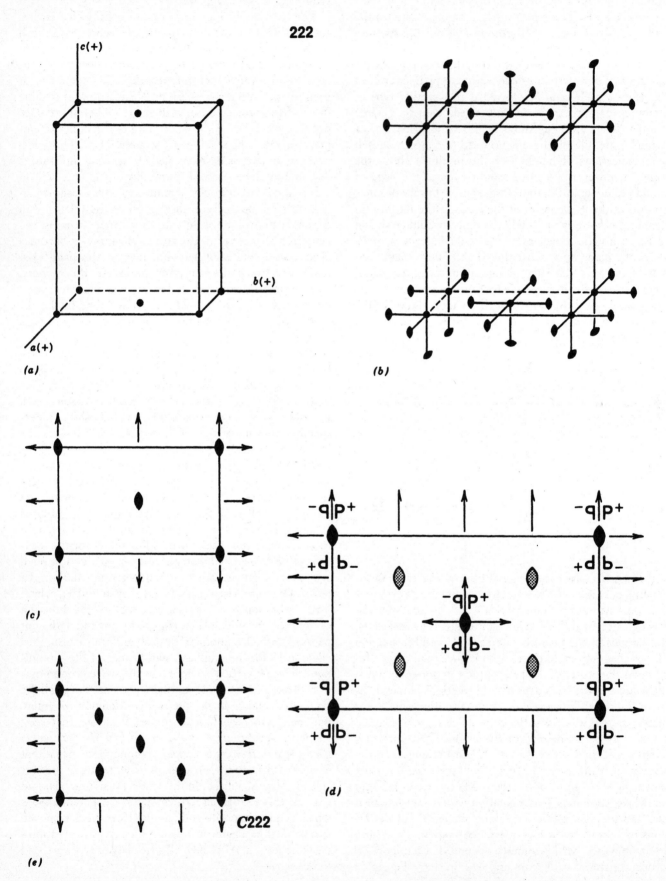

FIGURE 18.7 Graphical derivation of the total symmetry content in *I4mm*. See text for discussion.

4mm

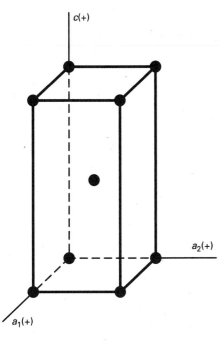

(a)

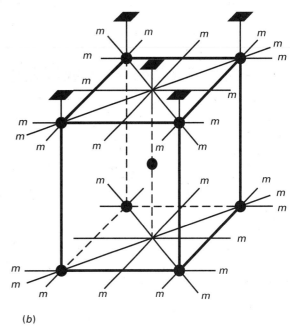

(b)

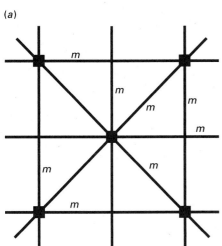

(c)

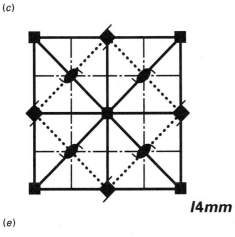

I4mm

(e)

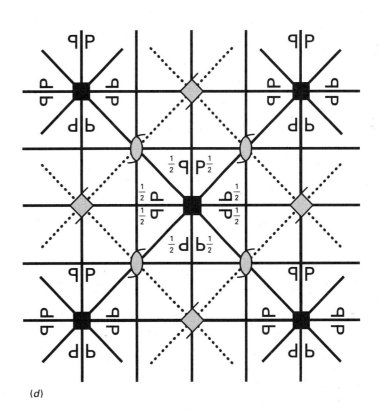

(d)

MATERIALS

Illustrations in Figs. 18.8, 18.9, and 18.10, ruler, soft pencil, and transparent (tracing) paper for Figs. 18.9 and 18.10.

ASSIGNMENTS

A. *Space lattice choices.* Figure 18.8 shows projections of extended lattices on the plane of the page. In tetragonal, hexagonal, and orthorhombic lattices such projections are down the c axis, in a [001] or [0001] direction. For the monoclinic case, the 2-fold axis perpendicular to the page is the b axis; as such the projection is down b in a [010] direction. This monoclinic orientation is referred to as the second setting (with b the unique = 2-fold axis). In these lattice projections the corner nodes, that is, equivalent nodes that occur in the 0 as well as 1 level of the lattice, are not identified by (0,1). Instead they are left without labels for reasons of clarity. Only nodes that occur in fractional levels are identified.

1. In each of the projections of Fig. 18.8; draw as many choices of unit cell as you can interpret.

2. For each unit cell choice give the type name, abbreviation, and number of nodes per unit cell.

3. From the various unit cell choices drawn, decide on which is most appropriate; show this in the drawing.

4. Locate and label coordinate axes in accordance with your decision in item **3**.

B. *Symmetry content of lattices.* Figure 18.9 gives six illustrations chosen from the 17 possible plane groups. These are the result of combinations of the five possible plane lattices or nets (see exercise 17) and various symmetry elements. These represent two-dimensional patterns, and all motif units lie, by definition, in the plane of the page. The shapes of the unit cells were considered in exercise 17, but not the combination of unit cell (or lattice) type and symmetry. This part of the exercise serves as an introduction to space lattice symmetry, as shown in Fig. 18.10.

1. In Fig. 18.9, on the basis of the distribution of motif units (the motif is an asymmetric capital P), determine the location and presence of various symmetry elements. Do this by placing a transparent overlay over the page. Trace the outline of the unit cell *but not the motif units*. Locate all symmetry elements in their appropriate positions using standard symbols. (Such symmetry elements are rotational axes, mirrors, and glide planes; label glide planes with g.) In your overlay you will have abstracted the total symmetry content of the given six plane groups. You can compare your results with illustrations of the 17 plane groups in Fig. 17.5.

Figure 18.10 gives projected illustrations of several space lattices and their associated motif units.

2. On the basis of the illustrations in this figure, determine for each the lattice type and the location and presence of various symmetry elements. Do this by placing a transparent overlay over the page. Trace the outline of the unit cell *but not the motif units*. Name the unit cell type. Locate all symmetry elements, using standard symbols in their appropriate positions. (Such symmetry elements are rotational axes; screw axes—indicate by s; mirrors; and glide planes—indicate by g.) In your overlay you will have abstracted the total symmetry content of six of the 230 space groups. At this stage it is unnecessary to reveal the exact translation directions involved in screws and glides; we will do this more thoroughly in exercise 19.

EXERCISE 18

Student Name

FIGURE 18.8 (*a*) Tetragonal lattice. (*b*) Hexagonal lattice and reference unit cells. (*c*) Orthorhombic lattice. (*d*) Monoclinic lattice with the *b* axis perpendicular to the page.

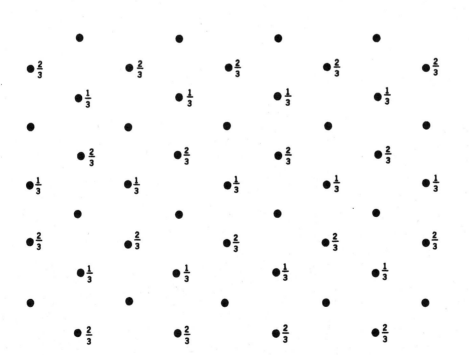

(a)   Tetragonal lattices

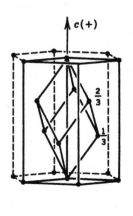

(b)   Hexagonal lattice and reference unit cells

EXERCISE 18

Student Name

FIGURE 18.8 *(continued)*

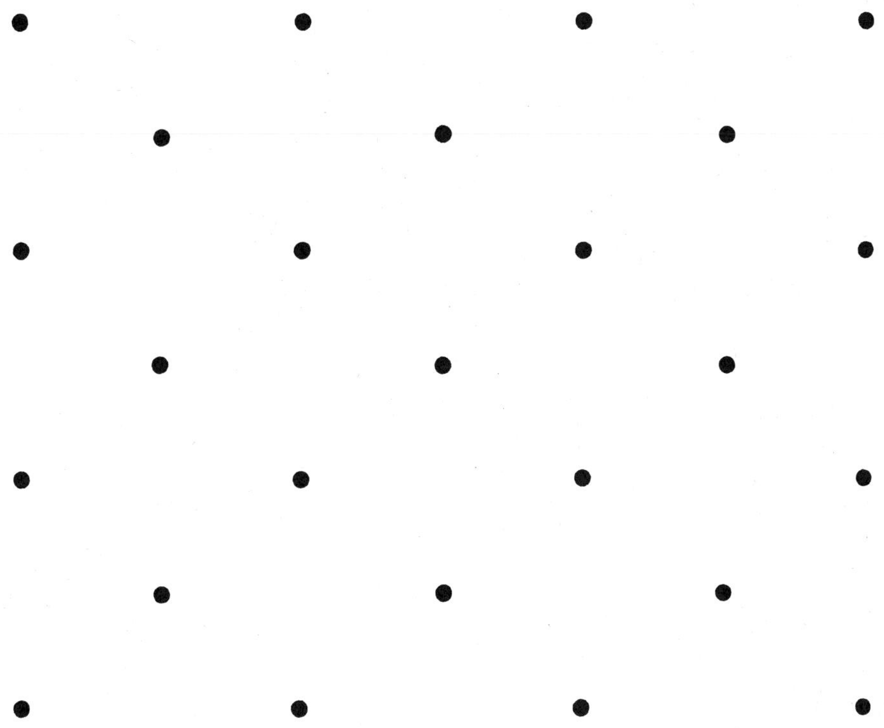

(c) **Orthorhombic lattice**

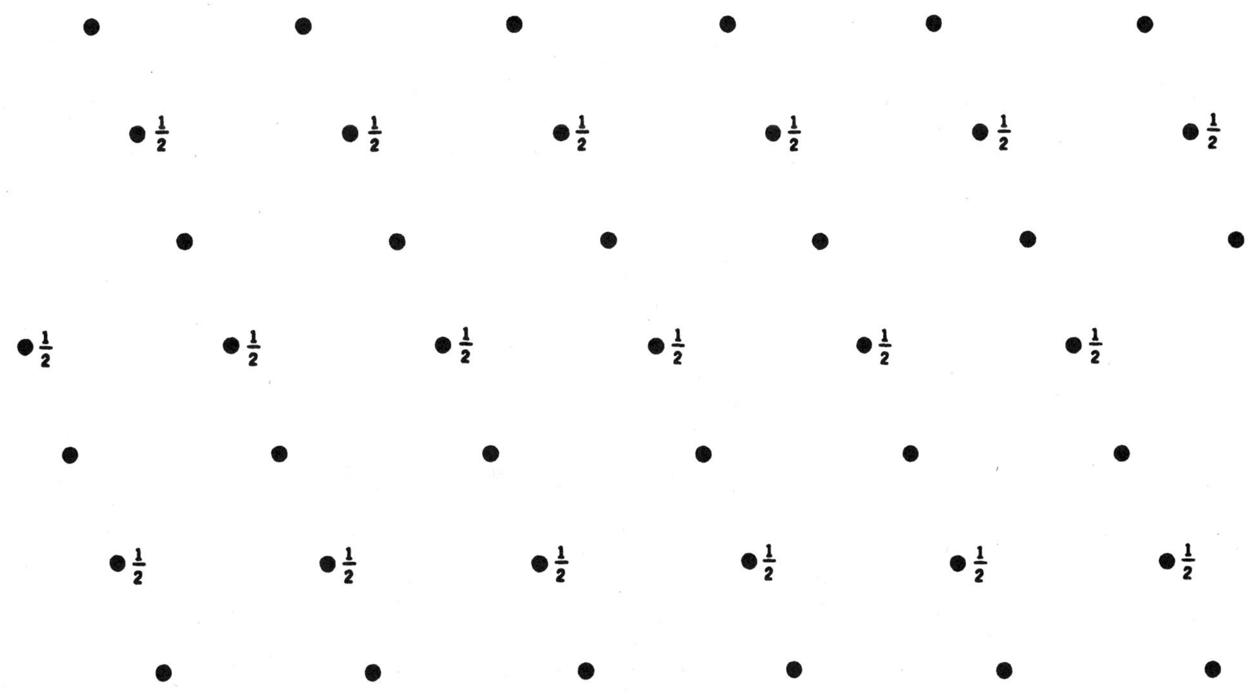

(d) **Monoclinic lattice with the *b* axis perpendicular to the page**

EXERCISE 18

FIGURE 18.9 Motif distributions according to six of the 17 possible plane groups.

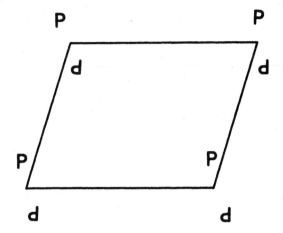

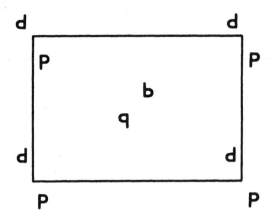

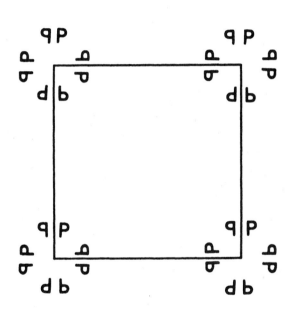

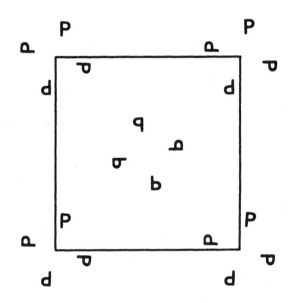

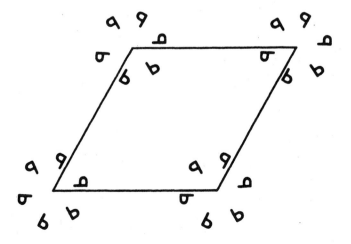

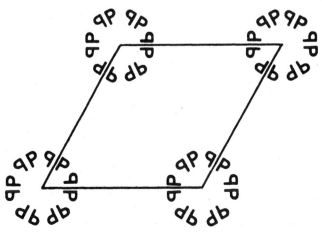

FIGURE 18.10 Six projected illustrations of space lattices and their associated motifs.

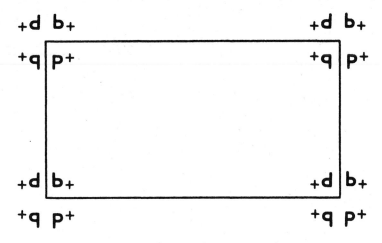

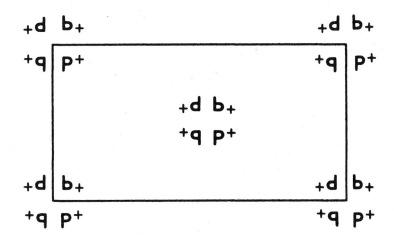

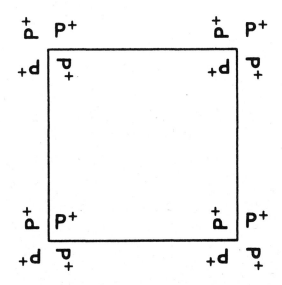

FIGURE 18.10 *(continued)*

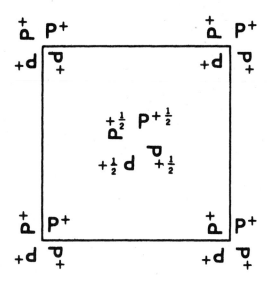

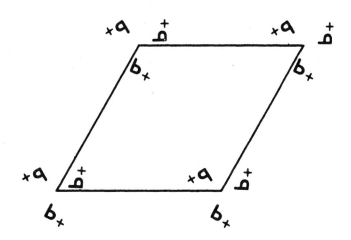

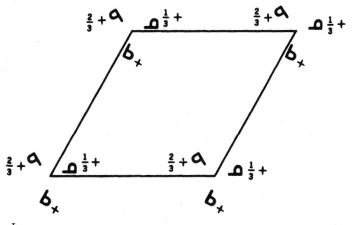

Space Groups—An Introduction to their Graphical Representation

PURPOSE OF EXERCISE

To gain familiarity with and understanding of the graphical representations of the 230 space groups as in volume A of the *International Tables for Crystallography*.

FURTHER READING AND CD-ROM INSTRUCTIONS

Klein, C. and Dutrow, B., 2008, *Manual of Mineral Science*, 23rd ed., Wiley, Hoboken, New Jersey, pp. 164–168, and 208–216.

CD-ROM, entitled *Mineralogy Tutorials*, version 3.0, that accompanies *Manual of Mineral Science*, 23rd ed., click on "Module III", and subsequently the button "3-dimensional order". This to be followed, consecutively, by clicking on the buttons on the next screen entitled: "Screw Axes," "Glide Planes," and "Space Group Elements in Structures".

Nesse, W. D., 2000, *Introduction to Mineralogy*, Oxford University Press, New York, pp. 17–19.

Perkins, D., 2002, *Mineralogy*, 2nd ed., Prentice Hall, Upper Saddle River, New Jersey, pp. 229–232 and p. 234.

Wenk, H. R. and Bulakh, A., 2004, *Minerals: Their Constitution and Origin*, Cambridge University Press, New York, New York, pp. 79–83.

Bloss, F. D., 1994, 2nd printing with minor revisions, *Crystallography and Crystal Chemistry*, Mineralogical Society of America, Chantilly, Virginia, pp. 160–179.

Buerger, M. H., 1956, *Elementary Crystallography*, Wiley, New Jersey, pp. 199–475.

Boisen, M. B., Jr. and Gibbs, G. V., 1985, *Mathematical Crystallography*, Reviews in Mineralogy, v. 15, Mineralogical Society of America, Chantilly, Virginia, 406p.

International Tables For Crystallography, vol. A, Theo Hahn, ed., 1983, D. Reidel Publishing Co., Dordrecht, The Netherlands, Space Group illustrations, pp. 101–709.

International Tables For Crystallography, Brief Teaching Edition of vol. A, Theo Hahn, ed., 1985, D. Reidel Publishing Co., Dordrecht, The Netherlands, pp. 59–119.

Background Information: In exercises 9, 10, 14 and 15 we dealt with various aspects of the 32 point groups (= crystal classes), which represent nontranslational symmetry. In exercise 17 we introduced translation and symmetry in one- and two-dimensional patterns, and in exercise 18 we introduced a purely translational three-dimensional concept, the space lattice. We noted that there are 14 lattice types (known as the Bravais lattices) distributed among the six crystal systems. In exercise 17 additional translational symmetry elements were mentioned as well: (1) screw axes,

and (2) glide planes, both of which contain translation as part of their symmetry operation. A *screw axis* is defined as *a rotational operation with a translation (t) parallel to the axis of rotation*. A *glide plane* is defined as *a mirror reflection with a translation component parallel to the mirror*. The various possible screw axis types, their graphical symbols, as well as glide planes and their graphical symbols are given in Fig. 19.1 and Tables 19.1 and 19.2. When all translational elements—that is, lattice types, screw axes, and glide planes—are taken into account, in conjunction with the nontranslational symmetries of the 32 point groups, we arrive at 230 different ways in which all these combinations can be arranged in space. These are known as the 230 *space groups*. The geometrical and mathematical derivation of these 230 possibilities is formidable and beyond the scope of an introductory mineralogy course. However, you may wish to look at an elegant mathematical treatment of the subject by Boisen and Gibbs (1985), or the more geometrical treatment by Buerger (1956).

In this exercise we will restrict ourselves to an outline of space group nomenclature and the graphical representation thereof. A complete tabulation and associated illustrations of the 230 space groups are given in the *International Tables for Crystallography*, volume A, 1983. In the following remarks the conventions laid down in the *International Tables* will be followed.

Every space group is expressed by an international (or Hermann-Mauguin) symbol. This consists first of a capital letter that indicates the lattice type, followed by a statement of the essential symmetry elements. The second part of the space group symbol is analogous to that of the 32 point groups; however, translational elements may be present in the symmetry statement of the space group, which by definition are absent in the analogous point group notations. This relationship of translational to translation-free symbolism is extremely useful, for if the space group notation is given, one can immediately derive therefrom the analogous point group symmetry and crystal system. For example, the space group $I4_1/a$ is related to the $4/m$ point group in the tetragonal system; the lattice type I is body-centered. The 4-fold screw axis (4_1) is equivalent to a 4-fold rotation axis (parallel to c) as observed morphologically. The glide plane a, with a glide component of $a/2$, is equivalent to the mirror (m). A listing of the 230 space groups and their analogous (translation-free), *isogonal*[*] point groups is given in Table 19.3. In this

---

[*]Isogonal (from the Greek roots *iso*, equal, and *gonio*, meaning angle) implies that rotation and screw axes with the same rotational repeat have the same rotational angle (e.g., 60° in a 6-fold rotation and 6-fold screw axis).

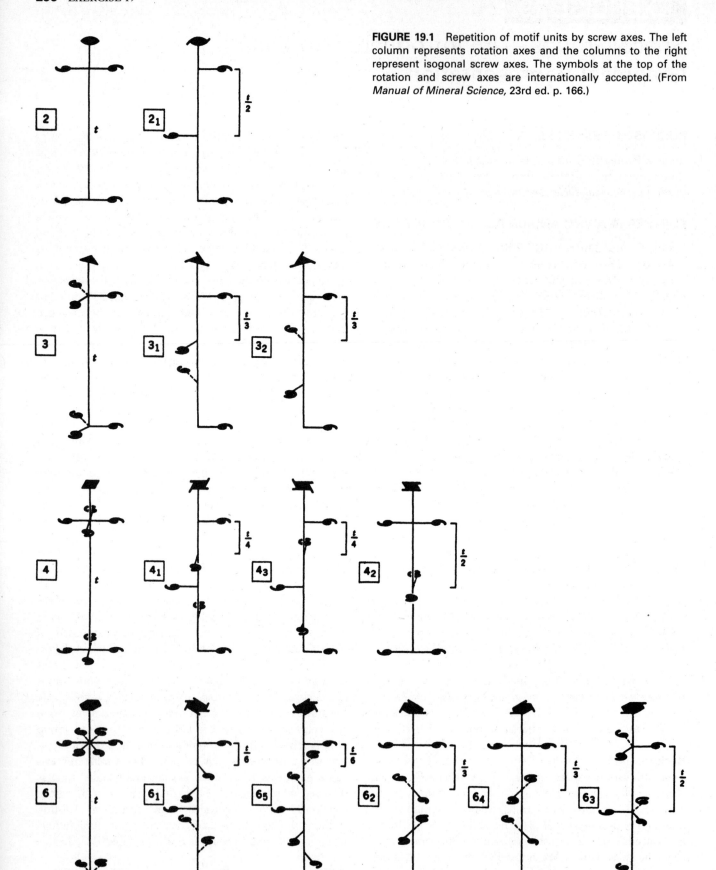

FIGURE 19.1 Repetition of motif units by screw axes. The left column represents rotation axes and the columns to the right represent isogonal screw axes. The symbols at the top of the rotation and screw axes are internationally accepted. (From *Manual of Mineral Science,* 23rd ed. p. 166.)

TABLE 19.1 Symbols for Symmetry Axes (All Graphical Symbols Are for Axes Normal to the Page, Unless Otherwise Noted)

Symbol	Symmetry Axis	Graphical Symbol	Type of Translation (If Present)	Symbol	Symmetry Axis	Graphical Symbol	Type of Translation (If Present)
1	1-fold rotation	None	None	4	4-fold rotation	■	None
$\bar{1}$	1-fold rotoinversion	○	None	4_1	4-fold screw (right-handed)		$\frac{1}{4}c$
2	2-fold rotation	(lens)	None	4_2	4-fold screw (neutral)		$\frac{2}{4}c = \frac{1}{2}$
		→ (parallel to paper)		4_3	4-fold screw (left-handed)		$\frac{3}{4}c$
2_1	2-fold screw	(lens)	$\frac{1}{2}c$	$\bar{4}$	4-fold rotoinversion	◪	None
		→ (parallel to paper)	$\frac{1}{2}a$ or $\frac{1}{2}b$	6	6-fold rotation	⬡	None
3	3-fold rotation	▲	None	6_1	6-fold screw (right-handed)		$\frac{1}{6}c$
3_1	3-fold screw (right-handed)	▲	$\frac{1}{3}c$	6_2	6-fold screw (right-handed)		$\frac{2}{6}c$
3_2	3-fold screw (left-handed)	▲	$\frac{2}{3}c$	6_3	6-fold screw (neutral)		$\frac{3}{6}c = \frac{1}{2}$
$\bar{3}$	3-fold rotoinversion	▲	None	6_4	6-fold screw (left-handed)		$\frac{4}{6}c$
				6_5	6-fold screw (left-handed)		$\frac{5}{6}c$
				$\bar{6}$	6-fold rotoinversion	⬡	None

TABLE 19.2 Symbols for Mirror and Glide Planes

Symbol	Symmetry Plane	Graphical Symbol		Nature of Glide Translation
		Normal to Plane of Projection	Parallel to Plane of Projection*	
m	Mirror	——————	⌐ ⟍120°	None
a, b	Axial glide plane	– – – – – –	⌐↓←	*a*/2 along [100] or *b*/2 along [010]
c		··················	None	*c*/2 along the *c* axis
n	Diagonal glide plane	—·—·—·—·—	⌐↙	*a*/2 + *b*/2; *a*/2 + *c*/2; *b*/2 + *c*/2; or *a*/2 + *b*/2 + *c*/2 (tetragonal and isometric)
d	Diamond glide plane	—·←·—·— —·—·→·—	⌐↙	*a*/4 + *b*/4; *b*/4 + *c*/4; *a*/4 + *c*/4; or *a*/4 + *b*/4 + *c*/4 (tetragonal and isometric)

*When planes are parallel to the paper, heights other than zero are indicated by writing the *z* coordinate next to the symbol (e.g., $\frac{1}{4}$ or $\frac{3}{4}$). The arrows indicate the direction of the glide component.

SOURCE: From *International Tables for Crystallography*, vol. A, 1983, D. Reidel Publishing Co., Dordrecht, The Netherlands.

TABLE 19.3 The 230 Space Groups, and the 32 Crystal Classes (Point Groups). The Space Group Symbols Are, in General, Unabbreviated.

Crystal Class	Space Group
1	$P1$
$\bar{1}$	$P\bar{1}$
2	$P2$, $P2_1$, $C2$
m	Pm, Pc, Cm, Cc
$2/m$	$P2/m$, $P2_1/m$, $C2/m$, $P2/c$, $P2_1/c$, $C2/c$
222	$P222$, $P222_1$, $P2_12_12$, $P2_12_12_1$, $C222_1$, $C222$, $F222$, $I222$, $I2_12_12_1$
$mm2$	$Pmm2$, $Pmc2_1$, $Pcc2$, $Pma2$, $Pca2_1$, $Pnc2$, $Pmn2_1$, $Pba2$, $Pna2_1$, $Pnn2$, $Cmm2$, $Cmc2_1$, $Ccc2$, $Amm2$, $Abm2$, $Ama2$, $Aba2$, $Fmm2$, $Fdd2$, $Imm2$, $Iba2$, $Ima2$
$2/m2/m2/m$	$P2/m2/m2/m$, $P2/n2/n2/n$, $P2/c2/c2/m$, $P2/b2/a2/n$, $P2_1/m2/m2/a$, $P2/n2_1/n2/a$, $P2/m2/n2_1/a$, $P2_1/c2/c2/a$, $P2_1/b2_1/a2/m$, $P2_1/c2_1/c2/n$, $P2/b2_1/c2/m$, $P2_1/n2_1/n2/m$, $P2_1/m2_1/m2/n$, $P2_1/b2/c2_1/n$, $P2_1/b2_1/c2_1/a$, $P2_1/n2_1/m2_1/a$, $C2/m2/c2/m$, $C2/m2/c2_1/a$, $C2/m2/m2/m$, $C2/c2/c2/m$, $C2/m2/m2/a$, $C2/c2/c2/a$, $F2/m2/m2/m$, $F2/d2/d2/d$, $I2/m2/m2/m$, $I2/b2/a2/m$, $I2/b2/c2/a$, $I2/m2/m2/a$,
4	$P4$, $P4_1$, $P4_2$, $P4_3$, $I4$, $I4_1$
$\bar{4}$	$P\bar{4}$, $I\bar{4}$
$4/m$	$P4/m$, $P4_2/m$, $P4/n$, $P4_2/n$, $I4/m$, $I4_1/a$
422	$P422$, $P42_12$, $P4_122$, $P4_12_12$, $P4_222$, $P4_22_12$, $P4_322$, $P4_32_12$, $I422$, $I4_122$
$4mm$	$P4mm$, $P4bm$, $P4_2cm$, $P4_2nm$, $P4cc$, $P4nc$, $P4_2mc$, $P4_2bc$, $I4mm$, $I4cm$, $I4_1md$, $I4_1cd$
$\bar{4}2m$	$P\bar{4}2m$, $P\bar{4}2c$, $P\bar{4}2_1m$, $P\bar{4}2_1c$, $P\bar{4}m2$, $P\bar{4}c2$, $P\bar{4}b2$, $P\bar{4}n2$, $I\bar{4}m2$, $I\bar{4}c2$, $I\bar{4}2m$, $I\bar{4}2d$
$4/m2/m2/m$	$P4/m2/m2/m$, $P4/m2/c2/c$, $P4/n2/b2/m$, $P4/n2/n2/c$, $P4/m2_1/b2/m$, $P4/m2_1/n2/c$, $P4/n2_1/m2/m$, $P4/n2_1/c2/c$, $P4_1/m2/m2/c$, $P4_2/m2/c2/m$, $P4_2/n2/b2/c$, $P4_2/n2/n2/m$, $P4_2/m2_1/b2/c$, $P4_2/m2_1/n2/m$, $P4_1/n2_1/m2/c$, $P4_2/n2_1/c2/m$, $I4/m2/m2/m$, $I4/m2/c2/m$, $I4_1/a2/m2/d$, $I4_1/a2/c2/d$
3	$P3$, $P3_1$, $P3_2$, $R3$
$\bar{3}$	$P\bar{3}$, $R\bar{3}$
32	$P312$, $P321$, $P3_112$, $P3_121$, $P3_212$, $P3_221$, $R32$
$3m$	$P3m1$, $P31m$, $P3c1$, $P31c$, $R3m$, $P3c$
$\bar{3}2/m$	$P\bar{3}1m$, $P\bar{3}1c$, $P\bar{3}m1$, $P\bar{3}c1$, $R\bar{3}m$, $R\bar{3}c$
6	$P6$, $P6_1$, $P6_5$, $P6_2$, $P6_4$, $P6_3$
$\bar{6}$	$P\bar{6}$
$6/m$	$P6/m$, $P6_3/m$
622	$P622$, $P6_122$, $P6_522$, $P6_222$, $P6_422$, $P6_322$
$6mm$	$P6mm$, $P6cc$, $P6_3cm$, $P6_3mc$
$\bar{6}m2$	$P\bar{6}m2$, $P\bar{6}c2$, $P\bar{6}2m$, $P\bar{6}2c$
$6/m2/m2/m$	$P6/m2/m2/m$, $P6/m2/c2/c$, $P6_3/m2/c2/m$, $P6_3/m2/m2/c$
23	$P23$, $F23$, $I23$, $P2_13$, $I2_13$
$2/m\bar{3}$	$P2/m\bar{3}$, $P2/n\bar{3}$, $F2/m\bar{3}$, $F2/d\bar{3}$, $I2/m\bar{3}$, $P2_1/a\bar{3}$, $I2_1/a\bar{3}$
432	$P432$, $P4_232$, $F432$, $F4_132$, $I432$, $P4_332$, $P4_132$, $I4_132$
$\bar{4}3m$	$P\bar{4}3m$, $F\bar{4}3m$, $I\bar{4}3m$, $P\bar{4}3n$, $F\bar{4}3c$, $I\bar{4}3d$
$4/m\bar{3}2/m$	$P4/m\bar{3}2/m$, $P4/n\bar{3}2/n$, $P4_2/m\bar{3}2/n$, $P4_2/n\bar{3}2/m$, $F4/m\bar{3}2/m$, $F4/m\bar{3}2/m$, $F4/m\bar{3}2/c$, $F4_1/d\bar{3}2/m$, $F4_1/d\bar{3}2/c$, $I4/m\bar{3}2/m$, $I4_1/a\bar{3}2/d$

SOURCE: From *International Tables for Crystallography*, vol. A, 1983, D. Reidel Publishing Co., Dordrecht, The Netherlands.

table all 230 space groups are identified by the "long" or full Hermann–Mauguin symbol. Examples are *P2/m2/m2/m* and *P4/m2/m2/m*. The commonly used short symbols for these two space groups would be *Pmmm* and *P4/mmm*, respectively. In such "short" space group (and analogous point group) notations, the 2-fold axes are implied as the result of the intersection of two mutually perpendicular mirror planes in a line of 2-fold symmetry. In the *International Tables for Crystallography*, full as well as short symbols are used. Here we will use only the more informative full symbols.

Understanding the "cork screw" motion (right-handed, left-handed, or neutral) of screw axes, be it from perspective drawings of screw axes (Fig. 19.1) or in two-dimensional projections of screw axis symmetry (e.g., *Manual of Mineral Science*, 23rd ed., Fig. 9.40, p. 211), is relatively straightforward. However, it may be useful to stress here that when we have determined that the *c* axis of a tetragonal crystal relates prism faces by a 4-fold rotation axis, this exact same vertical prism may well be the result of a 4-fold screw axis in the internal structure (see Fig. 19.2*b*). Morphologically we would have concluded 4-fold rotational symmetry, but the basic symmetry element may well have involved atomic (angstrom size) translations as implied by 4_1 (see Fig. 19.2*a*).

Although evaluation of the presence of screw axes or their direction of motion is relatively straightforward in the graphical representation of space groups, the visualization of the various translations and translation directions of glide planes may be somewhat more difficult. Table 19.2 gives the translational elements and graphical symbols for all possible glide planes. Additional representations of the effects of glide planes on motif units are shown in *Manual of Mineral Science*, 23rd ed., Fig. 9.42, p. 213. In order to interpret the types of standard representations used in the *International Tables for Crystallography*, you will find it useful to study Fig. 19.3, which portrays glide translations and glide directions in a three-dimensional sketch as well as in projections.

The goal in this exercise is an understanding of the types of illustrations (for space groups) as used in the *International Tables for Crystallography*, volume A. For each of the 230 space groups the *International Tables* show two diagrams, both of which are projections on the (001) or (0001) plane (down the *c* axis) unless otherwise noted. One such diagram shows the distribution of symmetry elements in a plan view of the unit cell, and the other diagram shows the location of all the motif units (in the *general position*) that result from the operation of the various symmetry and translational elements (see Fig. 19.6 as an example). In this and earlier exercises we have commonly spoken about a motif unit and a symmetrically equivalent motif unit as generated by a specific symmetry element. For example, the left hand is reflected into an equivalent right hand by a mirror. It is useful to generalize this concept with reference to points, instead of motif units. This leads to the

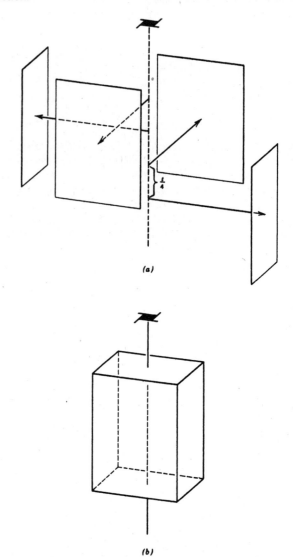

FIGURE 19.2 (*a*) Repetition and translation by a 4-fold screw axis of atomic domains (in a crystal structure). (*b*) External faces on the crystal whose internal (vertical) axis is one of screw motion.

definition of an *equivalent point* (or *equipoint*) as a point in a unit cell that is symmetrically related to any or all other such points through the operation of all symmetry and translational elements inherent in the unit cell. An equivalent point can be located in a *special* or a *general position* in a unit cell.* If the chosen point lies on a symmetry element such as a rotation axis or mirror plane, it is said to be in a *special* position because the point cannot be repeated by these symmetry operators. If the point, however, lies on a screw axis or glide plane, it is repeated by the operators even though in a special (or specialized) position. A *general* position is a locale in

---

*See exercise 11 for an analogous discussion of special and general forms and exercise 18 for the definition of a motif in a special versus a general position.

FIGURE 19.3 (*a*) Sketch of the various glide planes and their translation components with reference to orthorhombic coordinate axes. (*b*) Symbols for glide and mirror planes when such planes are parallel to the plane of projection (001) or (0001). (*c*) Symbols for glide and mirror planes when such planes are perpendicular to the standard plane of projection. (Adapted from Figs. 7.9 and 7.11 in *Crystallography and Crystal Chemistry: An Introduction*, by F. Donald Bloss, copyright © 1994, Mineralogical Society of America, reprinted by permission of the author.).

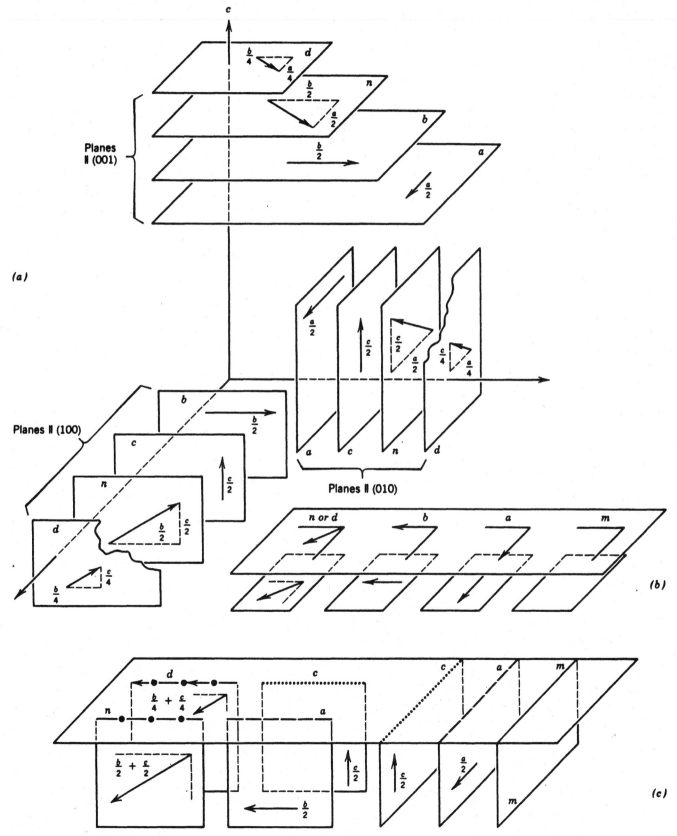

the unit cell that is removed from symmetry, translational, and corrdinate axial elements. As such, a general position is generally defined as one with coordinates x, y, and z, in the unit cell, where x, y, and z represent random fractions of the unit cell edge along the coordinate axes a, b, and c, respectively. The diagrams in the *International Tables* that show (001) or (0001) projections of atoms in the general position, therefore, depict the distribution of these atoms (or motif units) in a random position, one that is unconstrained by symmetry or translational elements. Before we evaluate the conventional symmetry and general equipoint diagrams as presented in the *International Tables*, it is instructive to look at some interactions of symmetry elements and equipoints, as shown graphically. Until now we have used a nonsymmetric motif unit (e.g, a capital P) to show the interrelation of symmetry and motif. The motif unit used in the *International Tables* is a small open circle, representing an atom in the general position. In the standard (001) projection plane it is customary not to label the axial direction. Unless otherwise indicated, the origin is assumed to be the top left corner of the illustration with the positive direction of the b axis pointing toward the right (see top of Fig. 19.4). In Fig. 19.4 examples of a 2-fold rotation and a 2-fold screw axis are shown as well as a mirror and various glide planes and a center of symmetry operating on an equipoint in the general position. The open circles represent equivalent positions (with general coordinates x, y, and z). An equipoint that has a "handed" (right-versus left-handed) relationship to another equipoint as through a mirror or glide reflection, or an inversion, is identified with a comma. The original general position is identified by the open circle, and an equipoint that is related to it by a reversal of hand is shown with the same small circle but with a comma inside. Figure 19.5 shows standard projections of symmetry elements and equipoints on the (001) plane. The same convention for the right- versus left-handed relationship is shown by sequences of atoms alternatingly with and without commas. When two equipoints superimpose through the presence of a mirror parallel to (001), the projection surface, the equivalent points are split in half with a comma in one half (the heights along z are shown by + and −, or $\frac{1}{2}$+, and $\frac{1}{2}$−). When a mirror plane parallel to (001) crosses the unit cell at $\frac{1}{4}z$ (and the equivalent $\frac{3}{4}z$), the symbol for the mirror is accompanied by a fractional notation ($\frac{1}{4}$ in this case; see the last illustration in Fig. 19.5). Because of the location of such a mirror plane, general equipoints with coordinates x, y, and $z+$ become upon reflection x, y, $\frac{1}{2} - z$, as shown in the illustration.

With the foregoing background information, we should be ready to understand what the graphical representations of some relatively simple space groups mean. Keep in mind that the examples as well as assignments in this exercise are restricted to understanding some of the basic aspects of space group representation. If you wish to be challenged by some very complex and intricate space group symmetry and representations, look through pages 592 to 707 of volume A of the *International Tables for Crystallography*, where space groups in the isometric system are depicted. In Figs. 19.6 and 19.7 the top diagrams (labeled *a*) show the distribution of symmetry elements in a plan view of the unit cell, and the lower diagrams (labeled *b*) show the location of equipoints in the general position, commensurate with the information given in *a*. Figure 19.6 represents the orthorhombic space group $P2/b2/a2/n$, for which the short symbol is *Pban*, and the isogonal point group symbols are $2/m2/m2/m$, and *mmm*, respectively. Figure 19.6*a* shows the presence of 2-fold rotation axes along the *a*, *b*, and *c* crystallographic axes. (Note that 2-fold axes perpendicular to the page are shown by the standard symbol used in point group notation, but by full arrows when parallel to the plane of projection.) No mirrors, only glide planes, are present as shown in the space group notation. Two glide planes (with axial glides parallel to *b* and *a*) are shown intersecting the *a* and *b* axes at $\frac{1}{4}a$ and $\frac{3}{4}a$, and $\frac{1}{4}b$ and $\frac{3}{4}b$, respectively. The diagonal glide plane is perpendicular to the *c* axis and its presence is shown by the symbol at the top left. Centers of symmetry are present in four locations in the unit cell, located at $\frac{1}{4}, \frac{1}{4}, 0; \frac{3}{4}, \frac{3}{4}, 0; \frac{3}{4}, \frac{1}{4}, 0$, and $\frac{1}{4}, \frac{3}{4}, 0$, in terms of x, y, and z coordinates. Figure 19.6*b* shows the location and distribution of the general equivalent positions for this space group. This diagram reveals clearly why there is a central 2-fold axis parallel to the *c* axis; it is shown in the central cluster of equipoints, a cluster that is equivalent to any of the four corner clusters because of the *n* glide. Study Fig. 19.6*b* carefully and convince yourself that you can locate all the symmetry elements shown in Fig. 19.6*a* by the distribution of general equipoints shown here. It is noteworthy that the number of general equivalent positions in a space group with a *P* lattice is equal to the number of planes in the general form {*hkl*} of the corresponding point group. The general form in $2/m2/m2/m$ is a rhombic dipyramid with eight faces; in Fig. 19.6*b* there are eight general positions within the unit cell. Such a simple relationship does not hold for centered unit cells.

Figure 19.7 shows the orthorhombic space group *Imm*2, which is isogonal with point group *mm*2. Figure 19.7*a* locates the symmetry elements (mirrors and 2-fold axes) in the unit cell. Because the unit cell is body-centered, additional symmetry elements occur as well (see also exercise 17), namely diagonal glide planes parallel to *a* and *b*, and 2-fold screw axes parallel to *c* that are interleaved with the 2-fold rotation axes. Figure 19.7*b* shows the location and distribution of the general equivalent positions for this space group. Again, as in Fig. 19.6*b*, study the location of these equivalent points with care so that you can account for each of their locations and relations in terms of the symmetry and translational elements shown in Fig. 19.7*a*.

FIGURE 19.4 Graphical representation of equipoints, 2-fold and 2-fold screw axes, mirror and glide planes, as well as a center of symmetry on the standard (001) projection.

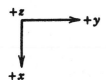

General coordinate
axis orientation

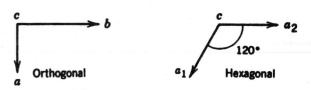

Orientation of crystallographic
axes

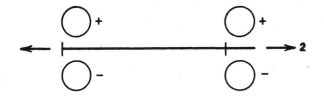

2–fold rotation parallel to *b*

2–fold screw axis parallel to *b*

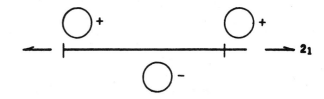

Mirror plane parallel to
(100)

Glide plane with glide component
parallel to *b*; plane parallel
to (100)

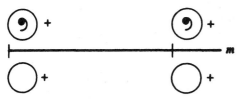

Glide plane with glide
component parallel to *c*; plane
parallel to (100)

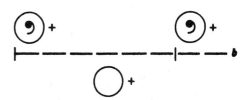

Diagonal glide plane
parallel to (100)

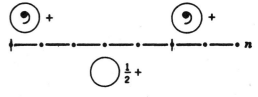

Diamond glide plane
parallel to (100)

Center of symmetry (=
inversion, *i*)

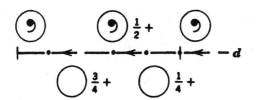

FIGURE 19.5 Graphical representation of mirror and glide planes onto the (001) projection when such mirrors or glide planes are parallel to the plane of projection.

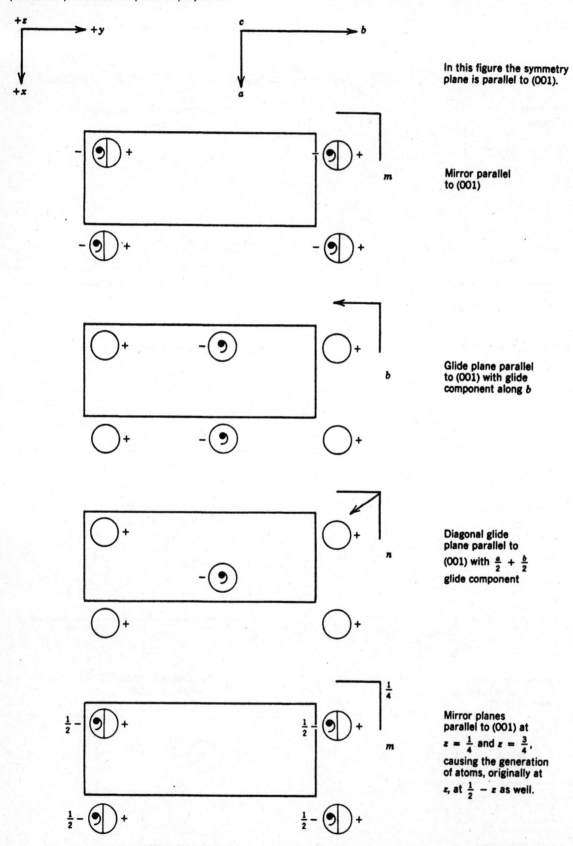

In this figure the symmetry plane is parallel to (001).

Mirror parallel to (001)

Glide plane parallel to (001) with glide component along b

Diagonal glide plane parallel to (001) with $\frac{a}{2} + \frac{b}{2}$ glide component

Mirror planes parallel to (001) at $z = \frac{1}{4}$ and $z = \frac{3}{4}$, causing the generation of atoms, originally at z, at $\frac{1}{2} - z$ as well.

FIGURE 19.6 Conventional diagrams for space group *P2/b2/a2/n*.
(From: *The International Tables for Crystallography,* vol. A, 1983,
D. Reidel Publishing Co., Dordrecht, The Netherlands, p. 260.)

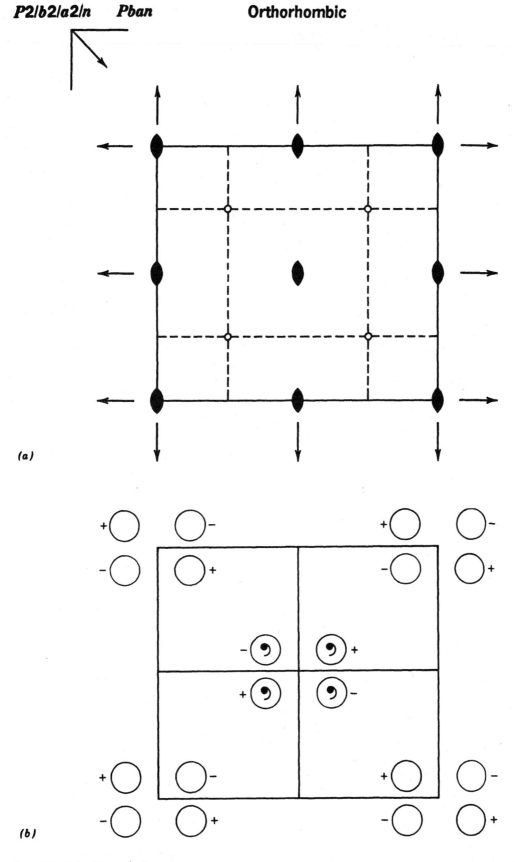

P2/b2/a2/n   *Pban*   **Orthorhombic**

(a)

(b)

FIGURE 19.7 Conventional diagrams for space group *Imm*2. (From *The International Tables for Crystallography,* vol. A, 1983, D. Reidel Publishing Co., Dordrecht, The Netherlands, p. 246).

*Imm*2   **Orthorhombic**

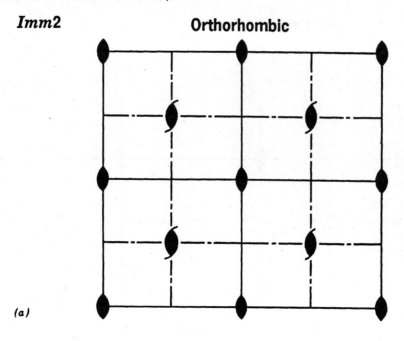

(a)

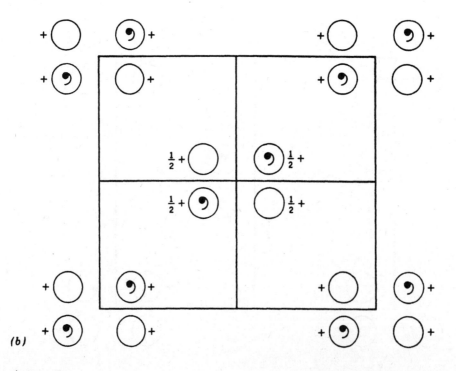

(b)

MATERIALS

Illustrations in Figs. 19.8, 19.9, 19.10 and 19.11; some translucent overlays, ruler, soft pencil, and eraser. The illustrations on these four pages have been taken, with some omissions, from the *International Tables for Crystallography*. As you will wish your final, completed assignment to be of high quality and neatness, you may decide to photocopy each of these figures several times for some rough sketch work.

ASSIGNMENT

The illustrations in Figs. 19.8 to 19.11 are direct reproductions from the *International Tables for Crystallography*, except for the omission of some information. This exercise consists of inserting the omitted information in accordance with the data provided. You will be asked to fill in (1) the symmetry elements on the basis of the distribution of equipoints and (2) the location of equipoints on the basis of symmetry elements provided. Your final diagrams should be carefully constructed, and you must use the internationally accepted symbols for symmetry elements as given in Figs. 19.1 and 19.3 and Tables 19.1 and 19.2. As noted earlier, you may wish to do some preliminary sketch work on photocopies of the assignment illustrations. Your final interpretation can be done directly on the figures, or even better on a transparent overlay.

1. In Fig. 19.8 the space group is given as *Pmm2*. The symmetry is illustrated in the top diagram. You are asked to complete, in the unit cell outline in the lower half of the page, the distribution of the general equivalent points on the basis of this symmetry content. The location of a general equipoint is given in the lower diagram.

2. In the upper part of Fig. 19.9 you are provided with the general equivalent point distribution for space group $P2_1/m2_1/m2/n$. In the lower half you are asked to locate all the symmetry elements compatible with this space group notation and equipoint distribution. Make sure you use the standard symbols.

3. In Fig. 19.10 you are provided only with the distribution of general equipoints in a tetragonal unit cell. You are asked to provide, in the lower half of the figure, all the symmetry elements compatible with the diagram at the top. You should also give the space group notation.

4. Figure 19.11 shows the symmetry content for monoclinic space group *P2/m*. Here the unique 2-fold axis is chosen as perpendicular to the page. The *a* and *c* axes are identified. In the lower half of the diagram you are asked to locate all the general equivalent points compatible with this space group notation. Because a mirror is present parallel to the plane of projection, there must be superposition of equivalent points—see Fig. 19.5 for illustrations of how such superimposed points must be represented.

When all the illustrations are completed, hand your most careful and neatest effort to the instructor. He or she may wish to show you the published results as given in the *International Tables for Crystallography*.

Student Name

FIGURE 19.8 Assignment on equivalent points in a space group representation.

Pmm2

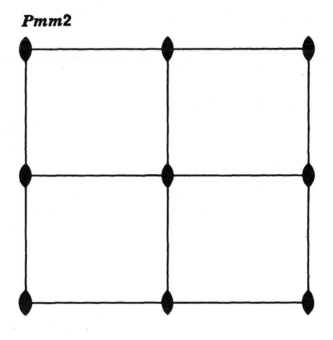

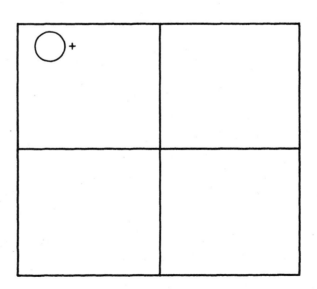

FIGURE 19.9 Assignment on symmetry content in a space group representation.

P2₁/m2₁/m2/n

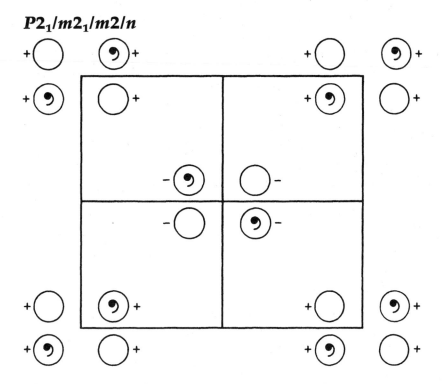

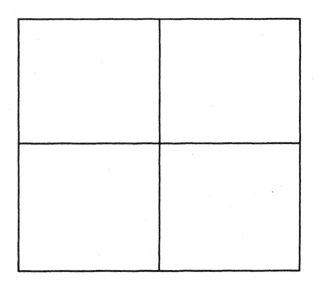

FIGURE 19.10 Assignment on symmetry content in a space group representation.

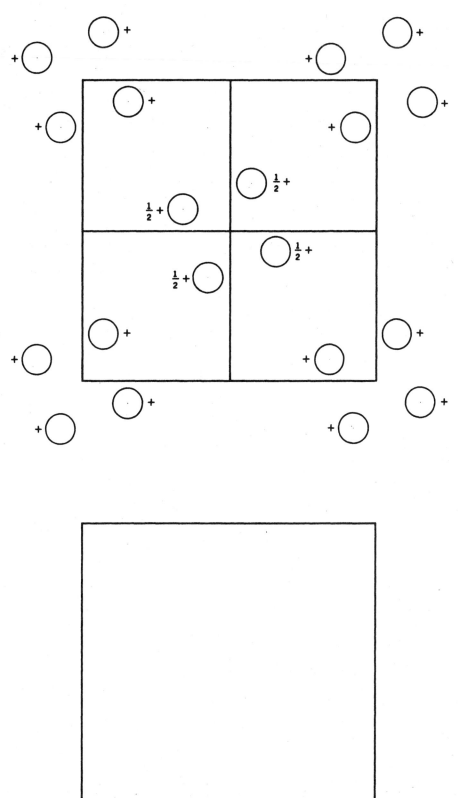

FIGURE 19.11 Assignment on equivalent points in a space group representation.

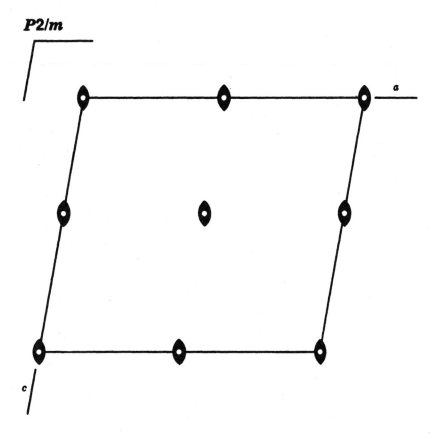

P2/m

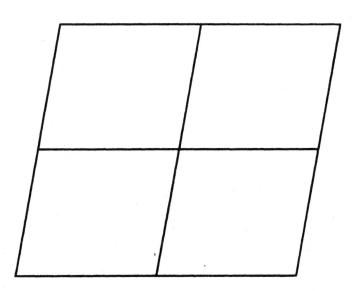

Recognition of Space Group Elements in Selected Crystal Structures

PURPOSE OF EXERCISE

To gain familiarity with space group elements (lattice type, rotation and screw axes, mirrors, and glide planes) inherent in crystal structures. Such elements are most easily observed and located in models of crystal structures (ball and stick models, or polyhedral models), but in this exercise we are restricted to the study of two-dimensional projections of crystal structures. The availability of some of the appropriate crystal structure models will greatly enhance this exercise.

FURTHER READING AND CD-ROM INSTRUCTIONS

Klein, C. and Dutrow, B., 2008, *Manual of Mineral Science*, 23rd ed., Wiley, Hoboken, New Jersey, pp. 648–652.
CD-ROM, entitled *Mineralogy Tutorials*, version 3.0, that accompanies *Manual of Mineral Science*, 23rd ed., click on "Module III", and subsequently on the button "3-dimensional order" that reveals the next screen. On that click on "Space Group Elements in Structures".
Nesse, W. D., 2000, *Introduction to Mineralogy*, Oxford University Press, New York, pp. 17–19.
Perkins, D., 2002, *Mineralogy*, 2nd ed., Prentice Hall, Upper Saddle River, New Jersey, pp. 229–232 and p. 234.
Wenk, H. R. and Bulakh, A., 2004, *Minerals: Their Constitution and Origin*, Cambridge University Press, New York, New York, pp. 79–83.
Bloss, F. D., 1994, 2nd printing with minor revisions, *Crystallography and Crystal Chemistry*, Mineralogical Society of America, Chantilly, Virginia, pp. 160–179.
Bragg, L. and Claringbull, G. F., 1965, *Crystal Structures of Minerals*, vol. 4, of *The Crystalline State*, G. Bell and Sons, London, 409p.

Background Information: In exercises 18 and 19 you were introduced to the various operations of space group elements, as revealed by the distribution and orientation of motif units. In the present exercise no new concepts will be introduced, except that motif units in real crystal structures are atoms, ions, ionic groups, and clusters of ions, instead of the abstract commas, circles, or capital Ps used earlier. For example, lattice translation (that is, lattice type) is expressed by the location of identical clusters of atoms or ions or ionic groups. The only aspect of this exercise that is different from what we did in exercises 18 and 19 is the notation for the location of atoms (or ions) in a two-dimensional projection. When a crystal structure is projected onto a two-dimensional page, the plane of projection is identified [e.g., (001), or (0001), or (010), etc.], and the actual locations in such a projection of the centers of the atoms (or ions), that is, the x and y coordinates of these centers, are shown graphically. The third dimension, that of the z coordinate of the atom's center, is commonly shown by a number that represents the height of that atom above (or below) the plane of projection, in terms of the percent of the unit cell length along the z direction. Figure 20.1a shows a unit cell of low quartz projected down the c axis onto the (0001) plane. This is an open "ball and stick" representation of the low quartz structure. Balls represent the ions and sticks connect nearest neighbors. The numbers next to the Si and O ions specify their z coordinates. Figure 20.1b shows the location of the space group elements that are compatible with this structural arrangement of SiO_4 tetrahedra. The space group notation for this arrangement is $P3_121$, or $P3_221$. The vertical 3-fold screw axes are shown as well as the locations of the horizontal 2-fold screw and 2-fold rotation axes. The numbers next to the axes define the z coordinates of the axes. For example $\frac{1}{6}, \frac{2}{3}$ next to a 2-fold screw axis means that two such horizontal screw axes are present in the unit cell, at heights $z = \frac{1}{6}$, and $z = \frac{2}{3}$. Figure 20.1c shows the standard space group representation for symmetry elements and lattice type of $P3_221$. Figure 20.2 shows a polyhedral representation of the structure of low quartz, this time with the enantiomorphous space group $P3_121$. In this figure only a few space group elements are identified, and only the z coordinates of the Si ions are noted as fractions.

The space group elements that are probably most difficult to discern in a two-dimensional projection of a crystal structure are the elements (mirrors, glide planes, rotational and screw axes) that are parallel to the plane of projection. The location and the position of the 2-fold rotation and 2_1 screw axes parallel to (0001) in the low quartz structure (see Fig. 20.1) are examples. Locating such elements is much less of a problem when you deal with an actual three-dimensional atomic model of a crystal structure.

In this exercise we will introduce you stepwise into the evaluation of space group elements in two-dimensional projections, beginning with relatively simple structures and leading you into more complex structures at the end.

FIGURE 20.1 (*a*) The unit cell of the structure of low quartz projected on (0001). (*b*) Space group elements in the low quartz structure. (*c*) Standard representation of space group elements for $P3_221$.

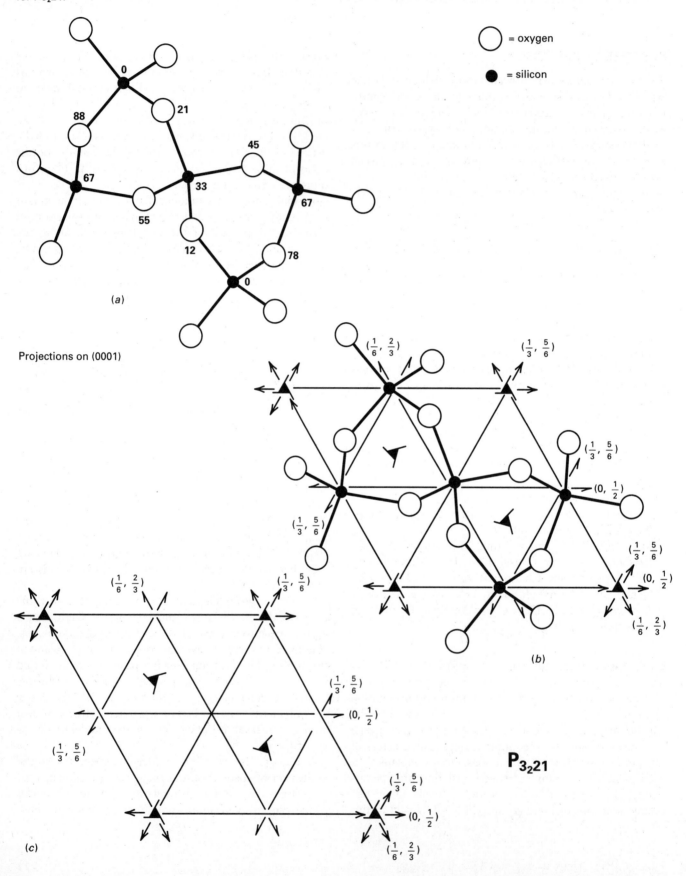

○ = oxygen

● = silicon

(*a*)

Projections on (0001)

(*b*)

(*c*)

$P3_221$

FIGURE 20.2 Projection of the low quartz structure, in polyhedral representation, onto (0001). This structure has space group $P3_121$. The z coordinates are given only for the centers of the Si atoms. (After J. J. Papike and M. Cameron, 1976, *Reviews of Geophysics and Space Physics,* vol. 14, p. 63.)

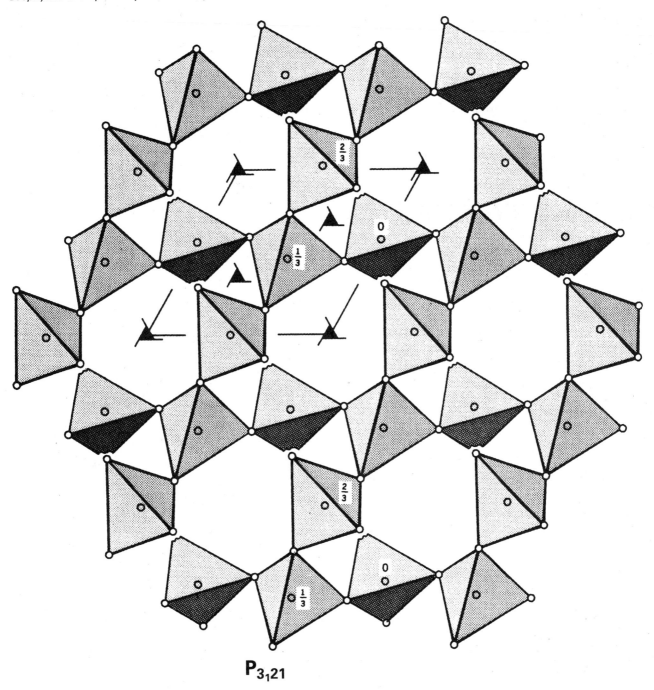

P3₁21

MATERIALS

Figures 20.3 through 20.9, ruler, protractor, soft pencil, and eraser. Sheets of transparent paper are helpful if you wish to trace neat space group representations on the basis of the structures given. You may find that locating and interpreting space group operators are the most difficult when these are parallel to the plane of projection; you will find this work easier in crystal structure models than on projections. Therefore, if crystal structure models are available, use them to aid your understanding of space group elements as derived from structures. The most important crystal structure models for this exercise are hexagonal closest packing, epidote, sanidine, and clinopyroxene (diopside).

ASSIGNMENTS

The assignments are based on the structure projections in Figs. 20.3 to 20.9. For each of the illustrations you will be asked to determine and to locate a lattice (type), and all other space group elements. The best way to show space group elements and their notation is by standard graphical representations as discussed in exercise 19. Such illustrations are best derived by putting a transparent overlay over the structure projection and outlining lattice type and other space group elements with the standard symbols given in exercises 18 and 19.

The following comments, specific to each of the structure illustrations, should aid you in the process of deriving space group elements from structure projections.

1. Figure 20.3a. This represents a vertical stacking of equal-sized spheres. You would obtain such a structure if you very carefully glued a cubic arrangement of eight Ping-Pong balls. Figure 20.3b is the projection of this two-layer arrangement onto the base. More spheres are shown than in Fig. 20.3a to provide you with additional space for the exercise.

Assignment: (a) Locate the coordinate axes and label them. (b) Locate all symmetry elements that you note perpendicular as well as parallel to the page of projection. (c) Give the space group notation. (d) Trace a graphical illustration of the space group onto a transparent overlay.

2. Figure 20.3c. This represents a vertical stacking of equal-sized spheres. You would obtain such a structure if you very carefully glued together two hexagonal arrays of Ping-Pong balls. Figure 20.3d is the projection of this two-layer arrangement onto the base. More spheres are shown than in Fig. 20.3c to provide you with additional space for the exercise.

Assignment: Do the same as requested for assignment 1, points a through d.

3. Figure 20.4. Hexagonal closest packing of equal-sized spheres in an *ABABABA* . . . . sequence. See item 1 for the various steps in the assignment. This assignment is best visualized with a model, or with several layers of glued Ping-Pong balls.

4. Figure 20.5. Structure of orthorhombic hemimorphite. (a) Outline a unit cell; give cell type. (b) Locate all space group elements in this unit cell. (c) Give a space group notation. (d) Trace a graphical illustration of the space group on transparent paper.

5. Figure 20.6. The low-temperature structure of acanthite. (a) Outline a unit cell; give its type; measure the nonorthogonal angle. (b) Locate all space group elements in the unit cell (those perpendicular as well as parallel to the page of projection). (c) Give a space group notation. (d) Trace a graphical illustration of the space group on transparent paper.

6. Figure 20.7. The structure of epidote. (a) Outline a unit cell; give its type; measure the nonorthogonal angle. (b) Locate all space group elements in the unit cell. (c) Give a space group notation. (d) Trace a graphical illustration of the space group on transparent paper.

7. Figure 20.8a. The structure of high-temperature $KAlSi_3O_8$, sanidine. (a) Outline a unit cell; the cell is c-centered; measure the β angle. (b) Locate all space group elements in the unit cell. (c) Using Fig. 20.8b, identify and locate any additional space group elements. (d) Give a space group notation. (e) Trace a graphical illustration of the space group using Fig. 20.8a.

8. Figure 20.9a. The structure of diopside. (a) In the outlined unit cell measure the β angle; this is a c-centered cell. (b) Locate all space group elements in the unit cell. (c) Using Fig. 20.9b identify and locate any additional space elements. (d) Give a space group notation. (e) Trace a graphical illustration of the space group using Fig. 20.9a.

Student Name

FIGURE 20.3 (a) Three-dimensional arrangement of equal-sized spheres known as simple cubic stacking. The bottom layer of spheres is identified by 0 and the top layer by 1 (these numbers represent the distances along the z coordinate).

(b) Projection of the arrangement in part a along the z (= c) axis. (c) Three-dimensional arrangement of equal-sized spheres known as simple hexagonal stacking. (d) Projection of the arrangement in part c along the z (= c) axis.

(a)

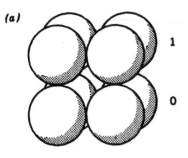

Simple cubic stacking

Space group: – – – – – – – –

(b)

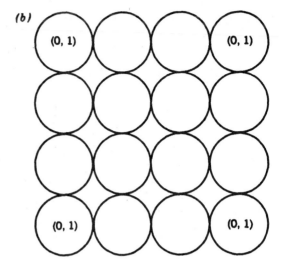

(c)

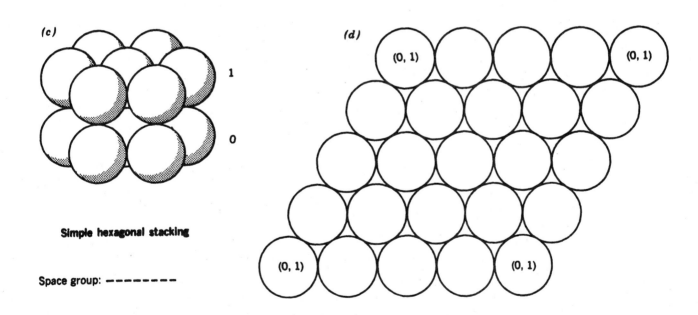

Simple hexagonal stacking

Space group: – – – – – – – –

FIGURE 20.4 (a) Three-dimensional stacking of equal-sized spheres in "hexagonal closest packing." Spheres in the 0 and 1 levels (along the z coordinate) are located in the A position; spheres in the interlayer (at $\frac{1}{2}$ along z) are located in the dimples marked B. (b) Projection of the arrangement in part a along the z (= c) axis.

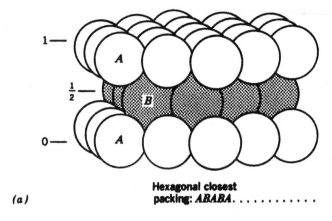

(a)

Hexagonal closest
packing: *ABABA*. . . . .

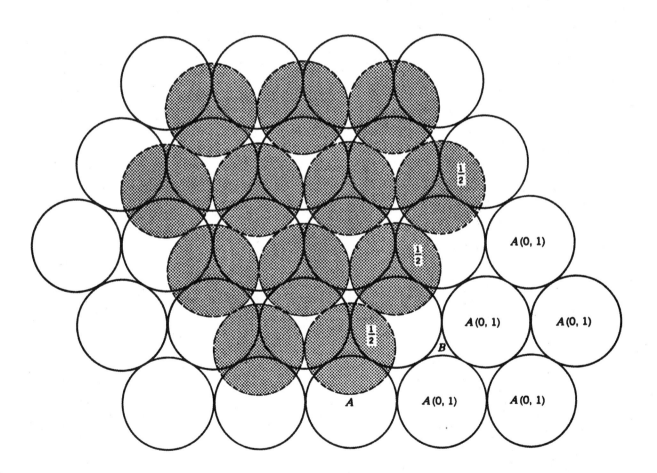

(b)

Student Name

FIGURE 20.5 The orthorhombic structure of hemimorphite–$Zn_4 (Si_2O_7)(OH)_2 \cdot H_2O$–projected on (001). Notice the isolated water molecules. (Modified after L. Bragg and G. F. Claringbull, 1965, *Crystal Structures of Minerals,* p. 202, copyright © 1965 by G. Bell and Sons Ltd., used by permission of Cornell University Press, Ithaca, N.Y.)

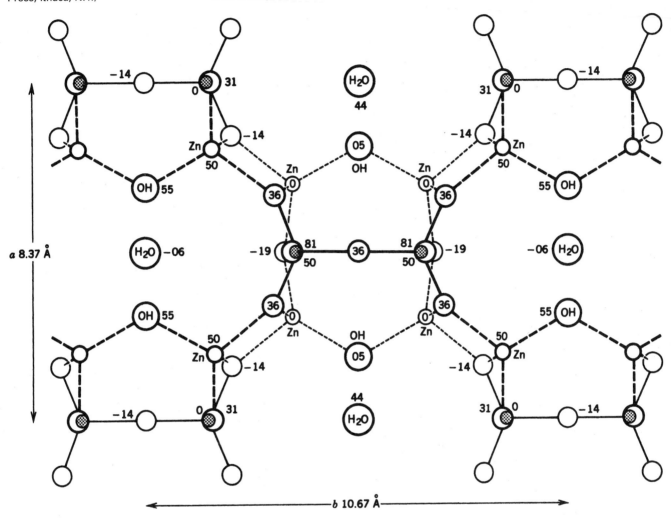

Space group: ----------

Student Name

FIGURE 20.6 The structure of acanthite, Ag_2S, projected on (010). (Modified after L. Bragg and G. F. Claringbull, 1965, *Crystal Structures of Minerals,* p. 47, copyright © 1965, by G. Bell and Sons Ltd., used by permission of Cornell University Press, Ithaca, N.Y.)

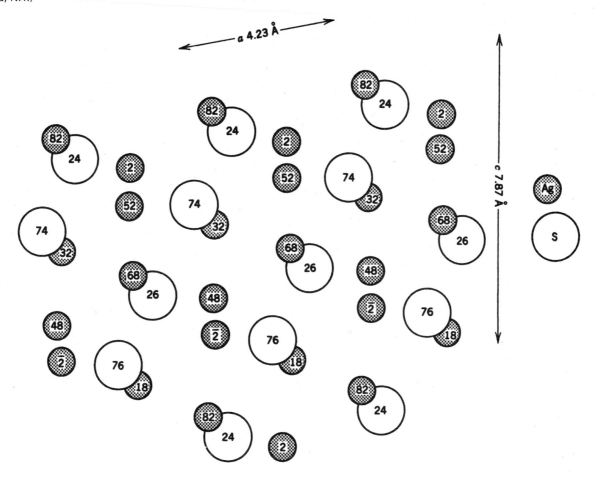

Space group: ─ ─ ─ ─ ─ ─ ─

EXERCISE 20

FIGURE 20.7 The structure of epidote—Ca$_2$(Al,Fe)Al$_2$O(SiO$_4$) (Si$_2$O$_7$)(OH)—projected on (010). Si–O bonds are solid. Superimposed oxygens are symmetrically displaced. (Modified after L. Bragg and G. F. Claringbull, 1965, *Crystal Structures of Minerals,* p. 210, copyright © 1965, by G. Bell and Sons Ltd., used by permission of Cornell University Press, Ithaca, N.Y.).

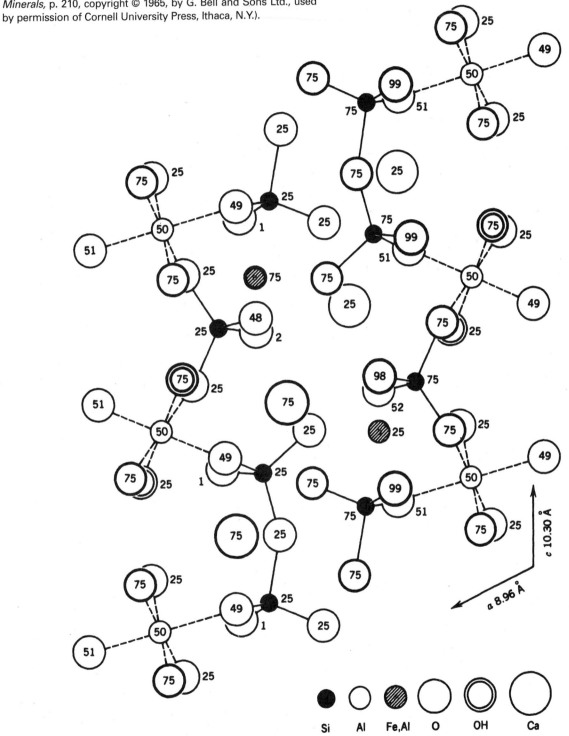

Si   Al   Fe,Al   O   OH   Ca

Space group: ————————

FIGURE 20.8 (a) The structure of sanidine, KAlSi₃O₈, projected on (010). Only atoms in the lower half of the unit cell are shown; those in the upper half are their reflections by a mirror at height 50. (Modified after L. Bragg and G. F. Claringbull, 1965, *Crystal Structures of Minerals*, p. 297, copyright © 1965, by G. Bell and Sons Ltd., used by permission of Cornell University Press, Ithaca, N.Y.) (b) Polyhedral representation of the structure of sanidine projected on ($\bar{2}$01). (From *Manual of Mineral Science*, 23rd ed., Fig. 18.48, p. 472 after J. J. Papike and M. Cameron, 1976, *Reviews of Geophysics and Space Physics*, vol. 14, p. 66.)

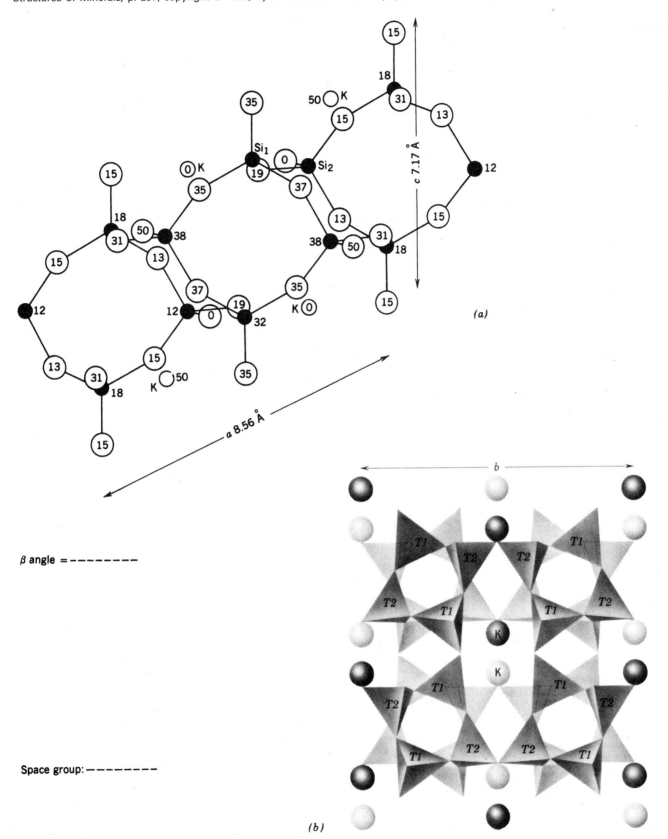

(a)

(b)

β angle = — — — — — — —

Space group: — — — — — — —

Student Name

FIGURE 20.9 (*a*) The structure of diopside, CaMgSi$_2$O$_6$, projected on (010). (After L. Bragg and G. F. Claringbull, 1965, *Crystal Structures of Minerals,* p. 231, copyright © 1965, by G. Bell and Sons Ltd., used by permission of Cornell University Press, Ithaca, N.Y.) (*b*) Polyhedral representation of the structure of diopside projected on (100). (From *Manual of Mineral Science,* 23rd ed., p. 449; after J. J. Papike and M. Cameron, 1976, *Reviews of Geophysics and Space Physics,* vol. 14, p. 66.)

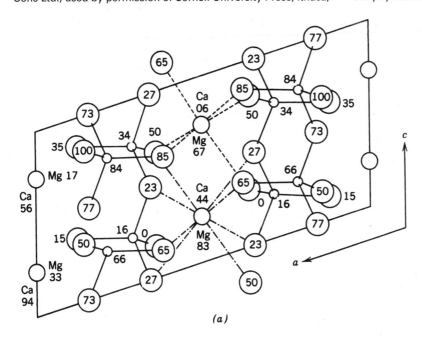

(*a*)

β angle = _____

Space group: _____

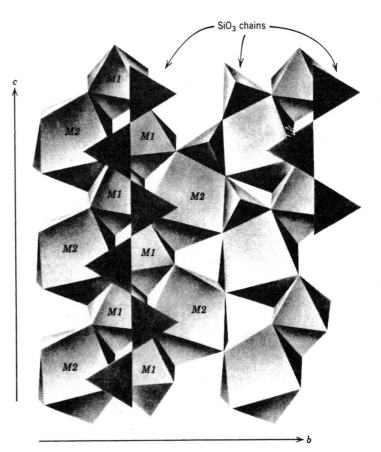

(*b*)

Identification of an Unknown by X-Ray Powder Diffraction Film

PURPOSE OF EXERCISE

To gain an understanding of the various steps involved in the identification of a crystalline substance using the photographic record (film) of its diffraction pattern from a powdered sample. You will need to refer to the Powder Diffraction File published by the International Center for Diffraction Data.

FURTHER READING

Klein, C. and Dutrow, B., 2008, *Manual of Mineral Science*, 23rd ed., Wiley, Hoboken, New Jersey, pp. 307–321.

Nesse, W. D., 2000, *Introduction to Mineralogy*, Oxford University Press, New York, pp. 160–168.

Perkins, D., 2002, *Mineralogy*, 2nd ed., Prentice Hall, Upper Saddle River, New Jersey, pp. 251–270.

Wenk, H. R. and Bulakh, A., 2004, *Minerals: Their Constitution and Origin*, Cambridge University Press, New York, New York, pp. 117–130.

Bloss, F. D., 1994, 2nd printing with minor revisions, *Crystallography and Crystal Chemistry*, Mineralogical Society of America, Chantilly, Virginia, pp. 454–479.

Bish, D. L. and Post, J. E., eds., 1989, *Modern Powder Diffraction*, vol. 20, Reviews in Mineralogy, Mineralogical Society of America, Chantilly, Virginia, 369p.

Cullity, B. D. and Stock, S. R., 2001, *Elements of X-ray Diffraction*, 3rd edition, Prentice Hall, Upper Saddle River, New Jersey, 664p.

Fang, J. M. and Bloss, F. D., 1966, *X-ray Diffraction Tables*, Southern Illinois University Press, Carbondale, Illinois.

Jenkins, R. and Snyder, R. L., 1966, *Introduction to X-ray Powder Diffractometry*, John Wiley and Sons, Hoboken, New Jersey, 403p.

Nuffield, E. W., 1966, *X-ray Diffraction Methods*, John Wiley and Sons, Hoboken, New Jersey, 409p.

Powder Diffraction File (Inorganic) published by International Center for Diffraction Data, 12 Campus Blvd., Newton Square, Pennsylvania, USA.

Background Information: Irradiating powdered material with X rays and analyzing the results of this interaction on film is a relatively simple and very powerful technique in the unambiguous identification of minerals. In this exercise we will concern ourselves only with the handling and interpretation of the results of such an experiment, as provided by a record of diffraction arcs on a strip film. You are referred to any of the references given above for an introduction to the physics of X rays and the processes that occur when crystalline materials interact with a high-intensity X-ray beam. Some of you may have the opportunity to observe X-ray equipment in a research laboratory at your college or university; however, few of you will probably gain any hands-on experience in such a laboratory. If you do, so much the better. Here it is assumed that you have some basic understanding of the physical processes involved, but that you will not be doing any X-ray experiments yourself. This exercise will present you with a drawing of an X-ray powder film on which you can do various measurements to identify the substance.

In the most commonly used X-ray powder film technique a powdered sample is irradiated with X rays inside a metal camera that contains a strip film. Figure 21.1*a* shows such a camera with a strip film wrapped flush against its inner cylindrical wall, and a spindle of powdered material in its center. Such a camera is commonly called a *powder camera*, or a *Debye–Scherrer camera* after the two German scientists who discovered that there is a characteristic diffraction pattern from powdered crystalline materials. Figures 21.1*b* and *c* show the circular form of the strip film when it is inside the camera, as well as its appearance when it is stretched out horizontally. The general distribution of arcs (commonly referred to as X-ray diffraction lines) is also shown. Notice also that the strip film has two holes, one (hole 1) to let the X rays into the camera for interaction with the powdered sample, and the other (hole 2) to let the undiffracted part of the beam out (in a straight line from its position of entry). It is stopped by lead glass at the end of the beam catcher. A mechanical cutter is used to cut these holes; if the film were not eliminated in these two places, it would very quickly become opaque owing to the intense X-ray bombardment. The diffraction arcs on the film are concentric about the two holes as shown in Fig. 21.1*c*. X-ray powder cameras are commonly manufactured so that the distance between the center of the left arcs (located inside the left-hand hole of the film) and the center of the right arcs (inside the right-hand hole of the film) is 180 mm; such a camera has a radius of 57.3 mm. Referring back to Fig. 21.1*b*, it is obvious that the arc subtended between the center of the left entry port and the center of the right exit port of the camera is 180°. A camera with a radius of 57.3 mm therefore provides a 180-mm distance for 180 degrees of arc. In other words, each millimeter measured along the film equals 1°.

FIGURE 21.1 (a) Metal powder diffraction camera, with strip film flush against its cylindrical wall. The film is shown schematically by the pattern of diffraction arcs. Two concentric cones of diffraction are shown as well around film hole 2. (b) The strip film shown in the cylindrical shape in which it is fastened inside the powder camera. The continuous film strip between the two centers (inside holes 1 and 2) represents 180° of arc. (c) The conventional and flat orientation of the strip film after exposure. Film hole 2 is positioned at the left. Several dashes circles show that many diffraction arcs appear in pairs, as left- and right-hand parts of the same diffraction cone (see also part a).

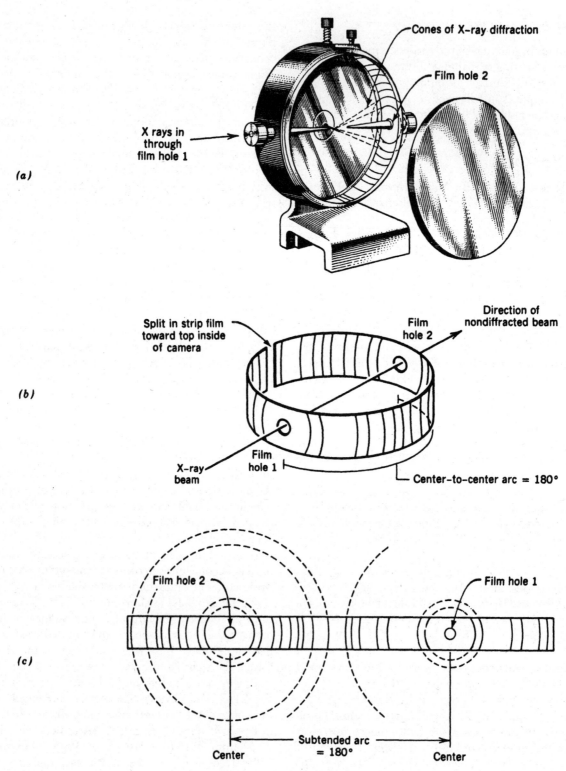

The irradiated powder at the center of the camera consists of a small amount of very finely ground mineral mixed with a binder, such as flexible collodion or Duco cement. A similar mount can also be obtained by filling a very fine capillary tube (made of lead-free glass) with a powdered sample.

The exposure time necessary to produce a well-developed X-ray powder film from a powdered sample may range from several hours up to 20 hours. The exposure time depends on many factors, among them (1) the amount of powdered material in the spindle, (2) the intensity of the X-ray tube used, and (3) the chemical composition of the sample being irradiated.

After exposure, the film is developed and fixed using standard darkroom procedures. When it has completely dried, it is ready for measurement and interpretation. A black-line drawing of such a film is shown in Fig. 21.2*a*. In an X-ray laboratory the original film (essentially transparent except for the regions of darkening where the arcs are) would be viewed on a light box and measured using a highly accurate metric scale.

In Fig. 21.2*a* several aspects of the appropriate measurement of such a film are illustrated and tabulated. All X-ray lines are numbered from left to right. Two arcs that are part of the same cone are given the same number. The locations of the lines are measured with a millimeter scale along the central line of the film. (The zero mark of such a scale is most conveniently located to the left of the last line on the left.) The millimeter measurements are indicated by x_1, x_2, x_3, etc. The locations (in millimeters) for pairs are shown as x_1 and x_1', etc. The method by which the left center and the right center of the diffraction cones are obtained is shown directly below the film in Fig. 21.2*a*. To obtain a properly increasing progression of millimeter values from the left center to the right center (this distance is 180 mm), we subtract the location of the left center from all readings to the right, until the right center is reached (see Fig. 21.2*a* and also Table 21.2, part *B*).

In Fig. 21.2*b* a metric scale has been superimposed on a drawing of an actual film in such a way that it is exactly centered along the horizontal centers of the arcs. The size of the illustrated film is identical to that of the original film with parts of the left and right ends cut off because the vertical dimension of the printed page is less than the actual film length. This type of film, with two holes and the ends of the strip located inside the top of the camera (see Fig. 21.1*b*), is known as a Straumanis mounting.

With the aid of the metric scale two reference locations can be obtained for the film (use Fig. 21.2*a* and part *A* of Table 21.2). The exact center of the left-hand arcs can be located by bisecting the distance between the corresponding diffraction lines (they are two arcs which, when connected, form one circle) on both sides of the hole in the film. This is illustrated schematically in Fig. 21.2*a*. The location of one of the left diffraction arcs is x_1' (in millimeters) and the

equivalent right arc is x_1 (again in millimeters). The center of these two arcs (that is, the center of the one and the same diffraction cone of which these two curves are the left- and right-hand parts) is found at $(x_1' + x_1)/2$. The center of the diffraction arcs at the right hand of the film can be located in a similar way (see Fig. 21.2*a*). If the film has not shrunk or expanded (which happens mainly during developing and fixing techniques and subsequent drying in the darkroom), the distance between the right and left centers is 180 mm. (The camera was specifically constructed to yield this distance.) In Fig. 21.2*b* the drawing was made with the centers exactly 180 mm apart; however, printing and paper stretching (or shrinking) may change this a little. If the distance is less than 180 mm, the film is said to have shrunk; if it is greater than 180 mm, it is said to have stretched. In either case a shrinkage (or stretching) correction can be applied, but we will not pursue this correction further.

Now that the locations of the left and right centers are known, you can proceed to measure the locations of all the diffraction lines (between the left and right centers) with respect to the location of the left center (in hole 2). That is, once you have obtained the millimeter location of each of the lines, you subtract from it the value for the left center (see Table 21.2, part *B*). In this way you have set the left center as the 0-mm position, and all your other measurements should increase, in an orderly fashion, from left to right toward larger millimeter values. (These values should be tabulated in Table 21.2, part *B*, as requested in the assignment.) Because of the direct correspondence between the distance of 1 mm and one degree of arc, the millimeter values are equivalent to degrees of arc (for diffraction cones). The position of a diffraction cone is expressed in degrees of angle of diffraction known as 2θ; see Fig. 21.3. This angular measurement can be related via the Bragg equation to the corresponding interplanar spacing (d) in the crystalline substance. The Bragg equation states that

$$n\lambda = 2d \sin \theta$$

where n is an integer such as 1, 2, or 3 for first-, second-, and third-order reflections. Here n is assumed to be 1 in the solving of the equation. The wavelength of the X-radiation used is λ. The film in Fig. 21.2*b* was obtained with CuK$\overline{\alpha}$ radiation with $\lambda = 1.5418$ Å. The interplanar spacings of the crystalline material under investigation are designated by d, and θ is the angle of diffraction.

Before we solve the Bragg equation, two of the parameters in the equation, namely λ and θ, need some further discussion. An X-ray beam that is generated inside an X-ray tube by the interaction of electron bombardment on a specific metal target (e.g., Cu or Fe or Mo, etc.) is said to be characteristic of the target metal used. In other words, a Cu target tube emits X radiation that is specifically related to Cu. The characteristic beam that results is

Student Name

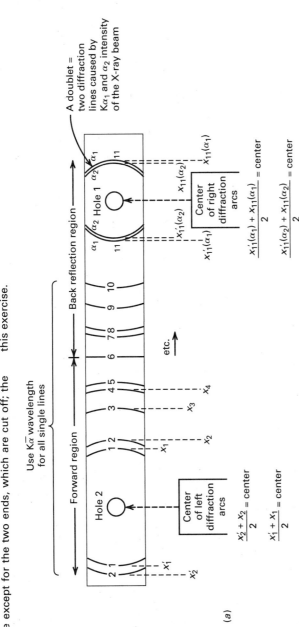

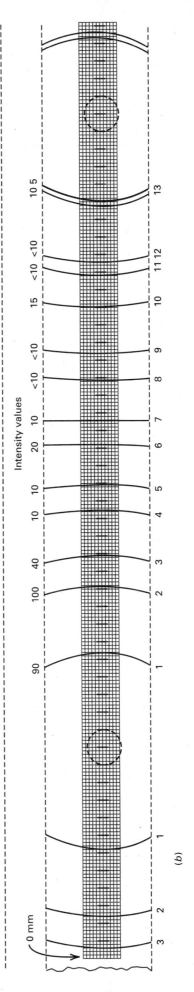

FIGURE 21.2 (a) Schematic illustration of a powder diffraction strip film and several aspects of its measurement. (b) Blackline illustration of a powder diffraction film of an unknown material, to original size except for the two ends, which are cut off; the film is longer than the height of the page. The original film was taken with CuK radiation. A millimeter scale has been superimposed. This illustration is the basis for actual measurements in this exercise.

TABLE 21.1 Table for the Determination of d for 0.1 Intervals of 2θ, Using Cuk$\overline{\alpha}$ Radiation ($\lambda = 1.5418$ Å) for the Solution to the Bragg Equation, $n\lambda = 2d \sin \theta$.

2θ	.0	.1	.2	.3	.4	.5	.6	.7	.8	.9
2	44.1716	42.068	40.1564	38.410	36.8105	35.338	33.979	32.721	31.552	30.464
3	29.446	28.499	37.509	26.773	25.986	25.243	24.542	23.879	23.251	22.655
4	22.089	21.552	21.037	20.548	20.082	19.636	19.209	18.800	18.409	18.034
5	17.673	17.327	16.994	18.673	16.365	16.068	15.781	15.504	15.237	14.979
6	14.730	14.488	14.2551	14.029	13.810	13.598	13.392	13.192	12.988	12.810
7	12.626	12.450	12.277	12.109	11.946	11.787	11.632	11.481	11.334	11.191
8	11.0513	10.915	10.782	10.652	10.526	10.402	10.281	10.163	10.048	9.9355
9	9.8255	9.7176	9.6124	9.5091	9.4082	9.3093	9.2126	9.1178	9.0250	8.9341
10	8.8450	8.7576	8.6721	8.5880	8.5057	8.4249	8.3456	8.2678	8.1915	8.1166
11	8.0432	7.9708	7.8998	7.8302	7.7617	7.6944	7.6283	7.5634	7.4995	7.4367
12	7.3730	7.3142	7.2543	7.1957	7.1379	7.0810	7.0251	6.9699	6.9157	6.8624
13	6.8098	6.7580	6.7071	6.6589	6.6074	6.5587	6.5107	6.4634	6.4168	6.3708
14	6.3256	6.2809	6.2368	6.1935	6.1507	6.1085	6.0669	6.0259	5.9854	5.9454
15	5.9060	5.8671	5.8288	5.7909	5.7535	5.7166	5.6802	5.6442	5.6088	5.5737
16	5.5391	5.5049	5.4711	5.4378	5.4049	5.3723	5.3402	5.3084	5.2771	5.2461
17	5.2154	5.1852	5.1552	5.1257	5.0964	5.0675	5.0390	5.0107	4.9828	4.9552
18	4.9279	4.9009	4.8742	4.8478	4.8216	4.7958	4.7702	4.7450	4.7199	4.6952
19	4.6707	4.6465	4.6225	4.5988	4.5753	4.5521	4.5291	4.5063	4.4838	4.4615
20	4.4394	4.4175	4.3959	4.3744	4.3532	4.3322	4.3114	4.2908	4.2704	4.2502
21	4.2302	4.2103	4.1907	4.1713	4.1520	4.1329	4.1140	4.0953	4.0767	4.0583
22	4.0401	4.0220	4.0042	3.9864	3.9689	3.9515	3.9342	3.9171	3.9001	3.8833
23	3.8667	3.8502	3.8338	3.8176	3.8015	3.7855	3.7697	3.7540	3.7385	3.7231
24	3.7078	3.6926	3.6776	3.6627	3.6479	3.6332	3.6187	3.6043	3.5900	3.5758
25	3.5617	3.5477	3.5339	3.5201	3.5065	3.4930	3.4796	3.4662	3.4530	3.4399
26	3.4269	3.4140	3.4012	3.3885	3.3759	3.3634	3.3510	3.3386	3.3264	3.3143
27	3.3022	3.2903	3.2784	3.2666	3.2549	3.2433	3.2318	3.2203	3.2090	3.1977
28	3.1865	3.1754	3.1644	3.1534	3.1426	3.1318	3.1210	3.1104	3.0998	3.0893
29	3.0789	3.0685	3.0582	3.0480	3.0379	3.0278	3.0178	3.0079	2.9980	2.9882
30	2.9785	2.9689	2.9592	2.9497	2.9402	2.9308	2.9214	2.9122	2.9029	2.8938
31	2.8847	2.8756	2.8666	2.8577	2.8488	2.8400	2.8312	2.8225	2.8139	2.8053
32	2.7968	2.7883	2.7798	2.7715	2.7631	2.7549	2.7466	2.7385	2.7303	2.7223
33	2.7143	2.7063	2.6984	2.6905	2.6827	2.6749	2.6671	2.6595	2.6518	2.6442
34	2.6367	2.6292	2.6217	2.6143	2.6069	2.5996	2.5923	2.5851	2.5779	2.5707
35	2.5636	2.5565	2.6495	2.5425	2.5355	2.5286	2.5218	2.5149	2.5081	2.5014
36	2.4947	2.4880	2.4813	2.4747	2.4682	2.4616	2.4551	2.4487	2.4422	2.4358
37	2.4295	2.4232	2.4169	2.4106	2.4044	2.3982	2.3921	2.3860	2.3799	2.3738
38	2.3678	2.3618	2.3559	2.3500	2.3441	2.3382	2.3324	2.3266	2.3208	2.3151
39	2.3094	2.3037	2.2981	2.2924	2.2869	2.2813	2.2758	2.2703	2.2648	2.2593
40	2.2539	2.2485	2.2432	2.2378	2.2325	2.2273	2.2220	2.2168	2.2116	2.2064
41	2.2012	2.1961	2.1910	2.1859	2.1809	2.1759	2.1709	2.1669	2.1609	2.1560
42	2.1511	2.1462	2.1414	2.1365	2.1317	2.1270	1.1222	2.1175	2.1127	2.1080
43	2.1034	2.0987	2.0941	2.0895	2.0849	2.0804	2.0758	2.0713	2.0668	2.0623
44	2.0579	2.0534	2.0490	2.0446	2.0402	2.0359	2.0316	2.0273	2.0230	2.0187
45	2.0144	3.0102	2.0060	2.0018	1.9976	1.9935	1.9893	1.9852	1.9811	1.9770
46	1.9729	1.9689	1.9649	1.9609	1.9569	1.9529	1.9489	1.9450	1.9411	1.9372
47	1.9333	1.9294	1.9255	1.9217	1.9179	1.9141	1.9103	1.9065	1.9028	1.8990
48	1.8953	1.8916	1.8879	1.8842	1.8806	1.8769	1.8733	1.8697	1.8661	1.8625
49	1.8589	1.8554	1.8519	1.8483	1.8448	1.8413	1.8378	1.8344	1.8309	1.8275
50	1.8241	1.8207	1.8173	1.8139	1.8105	1.8072	1.8039	1.8005	1.7972	1.7939

TABLE 21.1 *(continued)*

2θ	.0	.1	.2	.3	.4	.5	.6	.7	.8	.9
51	1.7906	1.7874	1.7841	1.7809	1.7776	1.7744	1.7712	1.7680	1.7648	1.7617
52	1.7585	1.7554	1.7523	1.7491	1.7460	1.7430	1.7399	1.7368	1.7338	1.7307
53	1.7277	1.7247	1.7217	1.7187	1.7157	1.7127	1.7098	1.7068	1.7039	1.7009
54	1.6980	1.6951	1.6922	1.6894	1.6865	1.6836	1.6808	1.6779	1.6751	1.6723
55	1.6695	1.6667	1.6639	1.6612	1.6584	1.6556	1.6529	1.6502	1.6474	1.6447
56	1.6420	1.6393	1.6367	1.6340	1.6313	1.6287	1.6260	1.6234	1.6208	1.6182
57	1.6156	1.6130	1.6104	1.6078	1.6053	1.6027	1.6002	1.5976	1.5951	1.5926
58	1.5901	1.5876	1.5851	1.5826	1.5801	1.5777	1.5752	1.5728	1.5703	1.5679
59	1.5655	1.5631	1.5607	1.5583	1.5559	1.5535	1.5512	1.5488	1.5465	1.5441
60	1.5418	1.5395	1.5371	1.5348	1.5325	1.5302	1.5279	1.5257	1.5234	1.5211
61	1.5189	1.5166	1.5144	1.5122	1.5099	1.5077	1.5055	1.5033	1.5011	1.4989
62	1.4968	1.4946	1.4924	1.4903	1.4881	1.4860	1.4839	1.4817	1.4796	1.4775
63	1.4754	1.4733	1.4712	1.4691	1.4671	1.4650	1.4629	1.4609	1.4588	1.4568
64	1.4547	1.4527	1.4507	1.4487	1.4467	1.4447	1.4427	1.4407	1.4387	1.4367
65	1.4347	1.4328	1.4308	1.4289	1.4269	1.4250	1.4231	1.4211	1.4192	1.4173
66	1.4154	1.4135	1.4116	1.4097	1.4079	1.4060	1.4041	1.4023	1.4004	1.3985
67	1.3967	1.3949	1.3930	1.3912	1.3894	1.3876	1.3858	1.3840	1.3822	1.3804
68	1.3786	1.3768	1.3750	1.3733	1.3715	1.3697	1.3680	1.3662	1.3645	1.3628
69	1.3610	1.3593	1.3576	1.3559	1.3542	1.3524	1.3507	1.3491	1.3474	1.3457
70	1.3440	1.3423	1.3407	1.3390	1.3373	1.3357	1.3340	1.3324	1.3308	1.3291
71	1.3275	1.3259	1.3243	1.3227	1.3211	1.3195	1.3179	1.3163	1.3147	1.3131
72	1.3115	1.3099	1.3084	1.3068	1.3053	1.3037	1.3022	1.3006	1.2991	1.2975
73	1.2960	1.2945	1.2930	1.2914	1.2899	1.2884	1.2869	1.2854	1.2839	1.2824
74	1.2809	1.2795	1.2780	1.2765	1.2750	1.2736	1.2721	1.2707	1.2692	1.2678
75	1.2663	1.2649	1.2635	1.2620	1.2606	1.2592	1.2578	1.2563	1.2549	1.2535
76	1.2521	1.2507	1.2493	1.2480	1.2466	1.2452	1.2438	1.2424	1.2411	1.2397
77	1.2383	1.2370	1.2356	1.2343	1.2329	1.2316	1.2303	1.2289	1.2276	1.2263
78	1.2250	1.2236	1.2223	1.2210	1.2197	1.2184	1.2171	1.2158	1.2145	1.2132
79	1.2119	1.2107	1.2094	1.2081	1.2068	1.2056	1.2043	1.2030	1.2018	1.2005
80	1.1993	1.1980	1.1968	1.1956	1.1943	1.1931	1.1919	1.1906	1.1894	1.1882
81	1.1870	1.1858	1.1846	1.1834	1.1822	1.1810	1.1798	1.1786	1.1774	1.1762
82	1.1750	1.1739	1.1727	1.1715	1.1703	1.1692	1.1680	1.1669	1.1657	1.1645
83	1.1634	1.1623	1.1611	1.1600	1.1588	1.1577	1.1566	1.1554	1.1543	1.1532
84	1.1521	1.1510	1.1498	1.1487	1.1476	1.1465	1.1454	1.1443	1.1433	1.1421
85	1.1411	1.1400	1.1389	1.1378	1.1367	1.1357	1.1346	1.1335	1.1325	1.1314
86	1.1303	1.1293	1.1282	1.1272	1.1261	1.1251	1.1240	1.1230	1.1220	1.1209
87	1.1199	1.1189	1.1178	1.1168	1.1158	1.1148	1.1138	1.1128	1.1118	1.1107
88	1.1098	1.1088	1.1078	1.1067	1.1057	1.1048	1.1038	1.1028	1.1018	1.1008
89	1.0998	1.0989	1.0979	1.0969	1.0960	1.0950	1.0940	1.0931	1.0921	1.0912
90	1.0902	1.0893	1.0883	1.0874	1.0864	1.0855	1.0845	1.0836	1.0827	1.0817
91	1.0808	1.0799	1.0790	1.0780	1.0771	1.0762	1.0753	1.0744	1.0735	1.0726
92	1.0717	1.0708	1.0699	1.0690	1.0681	1.0672	1.0663	1.0654	1.0645	1.0636
93	1.0627	1.0619	1.0610	1.0601	1.0592	1.0584	1.0575	1.0566	1.0558	1.0549
94	1.0541	1.0532	1.0523	1.0515	1.0506	1.0498	1.0490	1.0481	1.0473	1.0464
95	1.0456	1.0448	1.0439	1.0431	1.0423	1.0414	1.0406	1.0398	1.0390	1.0382
96	1.0373	1.0365	1.0357	1.0349	1.0341	1.0333	1.0325	1.0317	1.0309	1.0301
97	1.0293	1.0285	1.0277	1.0269	1.0261	1.0253	1.0246	1.0238	1.0230	1.0222
98	1.0214	1.0207	1.0199	1.0191	1.0184	1.0176	1.0168	1.0161	1.0153	1.0145
99	1.0138	1.0130	1.0123	1.0115	1.0108	1.0100	1.0093	1.0085	1.0078	1.0071
100	1.0063	1.0056	1.0049	1.0041	1.0034	1.0027	1.0019	1.0012	1.0005	.99977

TABLE 21.1   (continued)

2θ	.0	.1	.2	.3	.4	.5	.6	.7	.8	.9
101	.99905	.99833	.99761	.99690	.99619	.99548	.99477	.99406	.99336	.99265
102	.99195	.99125	.99055	.98985	.98916	.98847	.98778	.98709	.98640	.98571
103	.98503	.98434	.98366	.98298	.98231	.98163	.98095	.98028	.97961	.97894
104	.97827	.97761	.97694	.97628	.97562	.97496	.97430	.97364	.97300	.97234
105	.97169	.97104	.97039	.96974	.96910	.96845	.96781	.96717	.96653	.96589
106	.96526	.96436	.96399	.96336	.96273	.96210	.96148	.96085	.96023	.95961
107	.95899	.95837	.95775	.95714	.95652	.95591	.95530	.95469	.95408	.95348
108	.95287	.95227	.95167	.95107	.95047	.94987	.94927	.94868	.94809	.94750
109	.94690	.94632	.94573	.94514	.94456	.94398	.94339	.94281	.94224	.94166
110	.94108	.94051	.93994	.93936	.93879	.93823	.93766	.93709	.93653	.93596
111	.93540	.93484	.93428	.93373	.93317	.93262	.93206	.93151	.93096	.93041
112	.92986	.92931	.92877	.92823	.92768	.92714	.92660	.92606	.92553	.92499
113	.92446	.92392	.92339	.92286	.92233	.92180	.92127	.92075	.92023	.91970
114	.91918	.91866	.91814	.91762	.91711	.91659	.91608	.91557	.91505	.91454
115	.91404	.91353	.91302	.91252	.91201	.91151	.91101	.91051	.91001	.90951
116	.90902	.90852	.90803	.90754	.90704	.90655	.90606	.90558	.90509	.90461
117	.90412	.90364	.90316	.90268	.90220	.90172	.90124	.90077	.90029	.89982
118	.89935	.89888	.89841	.89794	.89747	.89700	.89654	.89607	.89561	.89515
119	.89469	.89423	.89377	.89331	.89286	.89240	.89195	.89150	.89105	.89060
120	.89015	.88970	.88925	.88881	.88836	.88792	.88748	.88704	.88659	.88616

SOURCE: By permission of B. M. Loeffler, Colorado College, Colorado Springs, Colorado.

not completely monochromatic (meaning of a single wavelength) but shows a spectrum of a few peaks at different wavelengths. For a Cu X-ray tube the range of wavelengths of the K spectrum is as follows:

CuKβ   at   1.39217 Å
CuKα₁   at   1.54051 Å
CuKα₂   at   1.54433 Å

In other words, the CuK spectrum consists of three peaks, two of which (α_1 and α_2) almost overlap in wavelength. Standard filtering methods eliminate almost all the Kβ radiation before the X-ray beam enters the X-ray camera. However, the α_1 and α_2 double peaks remain and may resolve into two arcs or peak reflections from the same internal plane in a structure, but generally only at high angles of diffraction (higher than 90° 2θ, in the region marked "back reflection" in Fig. 21.2a). In the region of lower angles of diffraction, a weighted average of the wavelengths for the α_1 and α_2 spectral lines is used and is referred to as CuKᾱ at 1.5418 Å. Although CuK radiation was used as an example, similar peak distributions (at different λ values) occur for the K spectra of other metals. In solving the Bragg equation for diffraction lines in the drawing of Fig. 21.2b, we use the wavelength of CuKᾱ = 1.54178 Å for all the lines, except for the doublet around the right-hand hole of the film. This doublet represents one and the same lattice plane but has been resolved into two diffraction lines because of the difference in λ for CuKα₁ and CuKα₂ radiation.

Another aspect of the Bragg equation that needs amplification is the angle θ. This is the angle of incidence, as well as of "reflection" (diffraction), analogous to the reflection of light from a mirror (see Fig. 21.3a). In X-ray diffraction there is not just "reflection" from a top surface, as light on a mirror is reflected, but the X-ray beam penetrates inside the structure so that the resulting diffraction effect is the sum of all diffraction from a stack of parallel lattice planes. As noted in Fig. 21.3a, such planes are given Miller index notations, such as *hkl*. Figure 21.3b shows a single lattice plane with respect to the cylindrical film position in a camera. The angle of incidence is θ, the angle of diffraction is θ, and the angle between the diffracted beam and the nondiffracted part of the X-ray beam (exiting through hole 2 in the film) is 2θ. In other words, the distance that is measured on a film (in millimeters) represent 2θ. The formulation of the Bragg equation given earlier refers to θ, which is half the value measured on the film.

With an electronic calculator (with trigonometric functions) it is not a difficult task to solve the Bragg equation for each of the measured lines of the film in Fig. 21.2b. However, a standard table with solutions to the Bragg equation for several wavelengths of X-rays and increments of 0.01° 2θ, published by Fang and Bloss (1966), is available. A sample page of this book of tables is shown in Fig. 21.4. It tabulates many solutions to the Bragg equation as a function of the wavelength used. The film drawn in figure 21.2b was obtained with CuK radiation. All the lines, except for the right-hand doublet, are the result of CuKᾱ radiation (the third column under Cu in Fig. 21.4). The doublet is the result of α_1 and α_2 resolution (columns 1 and 2 under Cu). You must have access to this table (or a good calculator) in order to convert your angular values (2θ values measured in millimeters) to values of *d*

FIGURE 21.3 (a). A parallel X-ray beam irradiating a stack of parallel lattice planes, *hkl.* The angle of incidence (θ) is the same as the angle of diffraction (θ), or "reflection". (b) Schematic cross section of an X-ray powder camera with a single lattice plane in the sample in a diffracting position. The angle that is measured on the film, between the center in hole 2 and the diffraction line is 2θ.

(a)

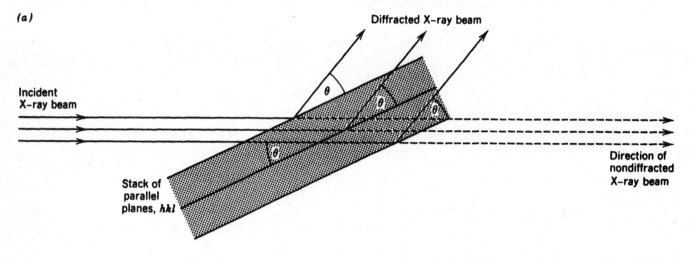

(b)

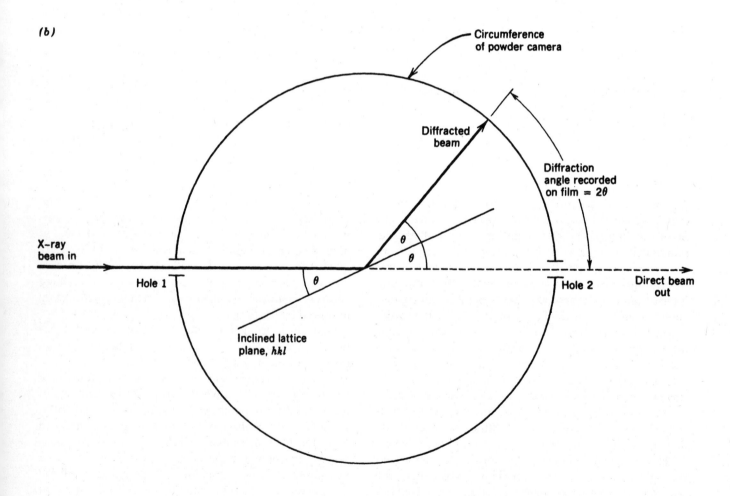

FIGURE 21.4 A sample page from *X-ray Diffraction Tables,* by J. M. Fang and F. D. Bloss. The 2θ column represents the measured angles. Corresponding *d* values are read from the appropriate wavelength columns (e.g., CuKα, CuKα₁ and CuKα₂). (Copyright © 1966, by Southern Illinois University Press, Carbondale, reprinted by permission of the publisher.)

Interplanar spacing (*d*) corresponding to 2θ for various wavelengths

53.01–53.50°

2θ	Sin²θ	Cu Kα₁	Cu Kα₂	Cu Kα	Cu Kβ	Fe Kα₁	Fe Kα₂	Fe Kα	Fe Kβ	Mo Kα₁	Mo Kα₂	Mo Kα	Mo Kβ	Cr Kα₁	Cr Kα₂	Cr Kα	Cr Kβ	W Lα₁
53.01	.19916	1.7260	1.7302	1.7274	1.5598	2.1690	2.1734	2.1705	1.968	.79464	.79944	.79624	.7084	2.5653	2.5696	2.5667	2.336	1.654
53.02	.19923	1.7257	1.7299	1.7271	1.5595	2.1686	2.1731	2.1701	1.968	.79450	.79930	.79611	.7082	2.5648	2.5692	2.5663	2.335	1.654
53.03	.19930	1.7254	1.7296	1.7268	1.5592	2.1683	2.1727	2.1697	1.967	.79436	.79916	.79597	.7081	2.5644	2.5687	2.5658	2.335	1.653
53.04	.19937	1.7251	1.7293	1.7265	1.5589	2.1679	2.1723	2.1694	1.967	.79423	.79902	.79583	.7080	2.5639	2.5683	2.5654	2.334	1.653
53.05	.19944	1.7248	1.7290	1.7262	1.5587	2.1675	2.1719	2.1690	1.967	.79409	.79888	.79569	.7079	2.5635	2.5678	2.5649	2.334	1.653
53.06	.19951	1.7245	1.7287	1.7259	1.5584	2.1671	2.1715	2.1686	1.966	.79395	.79874	.79555	.7077	2.5630	2.5674	2.5645	2.334	1.653
53.07	.19958	1.7242	1.7284	1.7256	1.5581	2.1668	2.1712	2.1682	1.966	.79381	.79860	.79541	.7076	2.5626	2.5669	2.5640	2.333	1.652
53.08	.19965	1.7238	1.7281	1.7253	1.5579	2.1664	2.1708	2.1678	1.966	.79367	.79846	.79527	.7075	2.5621	2.5665	2.5636	2.333	1.652
53.09	.19972	1.7235	1.7278	1.7250	1.5576	2.1660	2.1704	2.1675	1.965	.79353	.79832	.79513	.7074	2.5617	2.5660	2.5631	2.333	1.652
53.10	.19979	1.7232	1.7275	1.7247	1.5573	2.1656	2.1700	2.1671	1.965	.79339	.79818	.79499	.7072	2.5612	2.5656	2.5627	2.332	1.651
53.11	.19986	1.7229	1.7272	1.7244	1.5570	2.1652	2.1696	2.1667	1.965	.79325	.79804	.79485	.7071	2.5608	2.5651	2.5622	2.332	1.651
53.12	.19993	1.7226	1.7269	1.7241	1.5568	2.1649	2.1693	2.1663	1.964	.79312	.79790	.79472	.7070	2.5603	2.5647	2.5618	2.331	1.651
53.13	.20000	1.7223	1.7266	1.7238	1.5565	2.1645	2.1689	2.1659	1.964	.79298	.79776	.79458	.7069	2.5599	2.5642	2.5613	2.331	1.651
53.14	.20007	1.7220	1.7263	1.7235	1.5562	2.1641	2.1685	2.1656	1.964	.79284	.79762	.79444	.7068	2.5594	2.5638	2.5609	2.330	1.650
53.15	.20014	1.7217	1.7260	1.7232	1.5560	2.1637	2.1681	2.1652	1.963	.79270	.79748	.79430	.7066	2.5590	2.5633	2.5604	2.330	1.650
53.16	.20021	1.7214	1.7257	1.7229	1.5557	2.1634	2.1678	2.1648	1.963	.79256	.79735	.79416	.7065	2.5585	2.5629	2.5600	2.330	1.650
53.17	.20028	1.7211	1.7254	1.7226	1.5554	2.1630	2.1674	2.1644	1.962	.79242	.79721	.79402	.7064	2.5581	2.5624	2.5595	2.329	1.649
53.18	.20035	1.7208	1.7251	1.7223	1.5551	2.1626	2.1670	2.1641	1.962	.79229	.79707	.79388	.7063	2.5576	2.5620	2.5591	2.329	1.649
53.19	.20042	1.7205	1.7248	1.7220	1.5549	2.1622	2.1666	2.1637	1.962	.79215	.79693	.79375	.7061	2.5572	2.5615	2.5587	2.328	1.649
53.20	.20049	1.7202	1.7245	1.7217	1.5546	2.1618	2.1662	2.1633	1.961	.79201	.79679	.79361	.7060	2.5568	2.5611	2.5582	2.328	1.649
53.21	.20056	1.7199	1.7242	1.7214	1.5543	2.1615	2.1659	2.1629	1.961	.79187	.79665	.79347	.7059	2.5563	2.5607	2.5578	2.328	1.648
53.22	.20063	1.7196	1.7239	1.7211	1.5541	2.1611	2.1655	2.1626	1.961	.79173	.79651	.79333	.7058	2.5559	2.5602	2.5573	2.327	1.648
53.23	.20070	1.7193	1.7236	1.7208	1.5538	2.1607	2.1651	2.1622	1.961	.79160	.79637	.79319	.7056	2.5555	2.5598	2.5569	2.327	1.648
53.24	.20077	1.7190	1.7233	1.7205	1.5535	2.1603	2.1647	2.1618	1.960	.79146	.79623	.79305	.7055	2.5550	2.5593	2.5564	2.326	1.647
53.25	.20084	1.7187	1.7230	1.7202	1.5532	2.1600	2.1644	2.1614	1.960	.79132	.79610	.79292	.7054	2.5545	2.5589	2.5560	2.326	1.647
53.26	.20091	1.7184	1.7227	1.7199	1.5530	2.1596	2.1640	2.1610	1.959	.79118	.79596	.79278	.7053	2.5541	2.5584	2.5555	2.326	1.647
53.27	.20098	1.7181	1.7224	1.7196	1.5527	2.1592	2.1636	2.1607	1.959	.79105	.79582	.79264	.7052	2.5536	2.5580	2.5551	2.325	1.647
53.28	.20105	1.7178	1.7221	1.7193	1.5524	2.1588	2.1632	2.1603	1.959	.79091	.79568	.79250	.7050	2.5532	2.5575	2.5546	2.325	1.646
53.29	.20112	1.7176	1.7218	1.7190	1.5522	2.1585	2.1629	2.1599	1.958	.79077	.79554	.79236	.7049	2.5528	2.5571	2.5542	2.324	1.646
53.30	.20119	1.7173	1.7215	1.7187	1.5519	2.1581	2.1625	2.1595	1.958	.79063	.79540	.79223	.7048	2.5523	2.5566	2.5538	2.324	1.646
53.31	.20126	1.7170	1.7212	1.7184	1.5516	2.1577	2.1621	2.1592	1.958	.79050	.79527	.79209	.7047	2.5519	2.5562	2.5533	2.324	1.645
53.32	.20133	1.7167	1.7209	1.7181	1.5514	2.1573	2.1617	2.1588	1.957	.79036	.79513	.79195	.7045	2.5514	2.5558	2.5529	2.323	1.645
53.33	.20140	1.7164	1.7206	1.7178	1.5511	2.1570	2.1613	2.1584	1.957	.79022	.79499	.79181	.7044	2.5510	2.5553	2.5524	2.323	1.645
53.34	.20147	1.7161	1.7203	1.7176	1.5508	2.1566	2.1610	2.1580	1.957	.79008	.79485	.79168	.7043	2.5505	2.5549	2.5520	2.322	1.645
53.35	.20154	1.7158	1.7200	1.7172	1.5505	2.1562	2.1606	2.1577	1.956	.78995	.79471	.79154	.7042	2.5501	2.5544	2.5515	2.322	1.644
53.36	.20161	1.7155	1.7197	1.7169	1.5503	2.1558	2.1602	2.1573	1.956	.78981	.79458	.79140	.7041	2.5496	2.5540	2.5511	2.322	1.644
53.37	.20168	1.7152	1.7194	1.7166	1.5500	2.1555	2.1598	2.1569	1.956	.78967	.79444	.79126	.7039	2.5492	2.5535	2.5507	2.321	1.644
53.38	.20175	1.7149	1.7191	1.7163	1.5497	2.1551	2.1595	2.1565	1.955	.78953	.79430	.79113	.7038	2.5488	2.5531	2.5502	2.321	1.643
53.39	.20182	1.7146	1.7188	1.7160	1.5494	2.1547	2.1591	2.1562	1.955	.78940	.79416	.79099	.7037	2.5483	2.5526	2.5498	2.320	1.643
53.40	.20189	1.7143	1.7185	1.7157	1.5492	2.1543	2.1587	2.1558	1.955	.78926	.79402	.79085	.7036	2.5479	2.5522	2.5493	2.320	1.643
53.41	.20196	1.7140	1.7182	1.7154	1.5489	2.1540	2.1583	2.1554	1.954	.78912	.79389	.79071	.7034	2.5474	2.5518	2.5489	2.320	1.643
53.42	.20203	1.7137	1.7179	1.7151	1.5487	2.1536	2.1580	2.1550	1.954	.78899	.79375	.79058	.7033	2.5470	2.5513	2.5484	2.319	1.642
53.43	.20210	1.7134	1.7176	1.7148	1.5484	2.1532	2.1576	2.1547	1.954	.78885	.79361	.79044	.7032	2.5466	2.5509	2.5480	2.319	1.642
53.44	.20217	1.7131	1.7173	1.7145	1.5481	2.1528	2.1572	2.1543	1.953	.78871	.79347	.79030	.7031	2.5461	2.5504	2.5476	2.318	1.642
53.45	.20224	1.7128	1.7170	1.7142	1.5479	2.1525	2.1569	2.1539	1.953	.78858	.79334	.79017	.7030	2.5457	2.5500	2.5471	2.318	1.641
53.46	.20231	1.7125	1.7167	1.7139	1.5476	2.1521	2.1565	2.1536	1.953	.78844	.79320	.79003	.7028	2.5452	2.5496	2.5467	2.318	1.641
53.47	.20238	1.7122	1.7164	1.7136	1.5473	2.1517	2.1561	2.1532	1.952	.78830	.79306	.78989	.7027	2.5448	2.5491	2.5462	2.317	1.641
53.48	.20245	1.7119	1.7161	1.7133	1.5471	2.1514	2.1557	2.1528	1.952	.78817	.79292	.78976	.7026	2.5443	2.5487	2.5458	2.317	1.641
53.49	.20252	1.7116	1.7158	1.7130	1.5468	2.1510	2.1554	2.1524	1.952	.78803	.79279	.78962	.7025	2.5439	2.5482	2.5453	2.316	1.640
53.50	.20259	1.7113	1.7155	1.7127	1.5465	2.1506	2.1550	2.1521	1.951	.78789	.79265	.78948	.7023	2.5435	2.5478	2.5449	2.316	1.640

(interplanar spacings). In the absence of this published table you can use Table 21.1, which lists solutions to the Bragg equation (that is, *d*, the interplanar spacing) as a function of 2θ-value increments of 0.1°. Table 21.1 will give you much lesser resolution than the book of tables by Fang and Bloss (1966), and it also stops at $2\theta = 120.9°$ 2θ. In the assignment you are requested to convert each of the 2θ values to its equivalent value of *d*.

With a listing of *d*s for your unknown, you are close to beginning the process of mineral identification. However, you also need to estimate the intensity of each of your measured diffraction lines. On an actual film, you measure the intensity of a line by the intensity of the blackening of the film. This is done on a relative scale, with the darkest line given the intensity of 100 and the weakest line, whose intensity you can barely estimate, a value of 10 or less. Because Fig. 21.2*b* is a black-line drawing on which relative intensities cannot be measured, their values are given as numbers next to the diffraction lines. With a listing of *d*s and their corresponding intensities as recorded in Table 21.2, part *B*, you are now ready for the various identification steps.

The X-ray diffraction pattern of an inorganic unknown is identified using the *Powder Diffraction File (Inorganic)*, published by the International Center for Diffraction Data. This organization publishes search manuals, which are known as the *Hanawalt Index* of the *Inorganic Phases*, as well as the *Mineral Powder Diffraction File*. In this exercise we will use the *Hanawalt Index*. In this the strongest three lines of all known patterns are arranged in decreasing value of *d*. The *d* value of the *highest-intensity line* ($I = 100$) determines the location of the pattern within one of 45 subgroups in the search manual. Within such a group, data are arranged in the order of decreasing value of the *second* strongest line. The *third* strongest line in the pattern is recorded in the third column. A page of the 1986 *Search Manual for Inorganic Phases* is shown in Fig. 21.5. A page of the Hanawalt (search) section of the *Mineral Powder Diffraction File* (1986) is shown in Fig. 21.6. Both figures show that an additional five of the next strongest lines are entered as well as part of the search manual tabulation. Each pattern is given an identification number (as in the right-hand column of Figs. 21.5 and 21.6). This identification number allows for the location of the complete X-ray diffraction pattern, as well as additional data (e.g., Miller indices for each diffraction line, chemical formula of the substance, its space group, cell dimensions, origin of the material used, and X-ray wavelength employed). Figure 21.7 shows such a compilation of data for the mineral halite. In older files these data were arranged in card form. More recently all such cards are available in microfilm files.

MATERIALS

The drawing of an actual X-ray powder diffraction film (Fig. 21.2*b*) forms the basis for this assignment. The data obtained from the film are to be reported in Table 21.2. Access to a copy of the *X-ray Diffraction Tables* by Fang and Bloss will be very helpful. However, in its absence, the necessary 2θ values can be converted to their appropriate *d*s by using Table 21.1 or by solving the Bragg equation, using an electronic calculator with trigonometric functions. Figures 21.5 and 21.6 suffice for the identification of the unknown. If, however, you wish to see the complete information on the identified unknown, you will need to consult a microfiche or card file of the Powder Diffraction File published by the International Center for Diffraction Data.

ASSIGNMENT

Depending on the wishes of your instructor, you can begin this assignment at step 1 (see below) using the film, or step 2, using a tabulation of published 2θ values (see Table 21.2B on the lower far right.).

1. Using the drawing of the film and metric scale in Fig. 21.2*b*, locate as carefully as possible the right and left centers of the diffraction curves and tabulate them in Table 21.2*A*. Subsequently measure, again carefully, the positions of all the diffraction positions between holes 2 and 1 in Fig. 21.2*b*, and tabulate these 2θ results in Table 21.2*B*. They should compare well with the published data already given in the last two columns of Table 21.2*B*.

2. Using your own tabulation in Table 21.2*B*, or the 2θ values already provided, convert the 2θ values into *d*s; consult the conversion tables of Fang and Bloss. If these tables are not available, use Table 21.1, or use the Bragg equation $n\lambda = 2d\sin\theta$, to solve *d* for a given 2θ measurement. That is, find the sine of half your 2θ measurement to plug into $d = n\lambda/(2\sin\theta)$. Use the CuK$\overline{\alpha}$ wavelength, 1.54178 Å, if there is no resolution of doublets and set *n* as 1. If you wish to obtain the *d* appropriate to a doublet, you must use CuKα_1 and CuKα_2 wavelengths, respectively.

3. Enter the intensity (*I*) for each diffraction line, as given on Fig. 21.2*b*, into Table 21.2*B*.

4. Using the *d* values of the three most intense lines, identify the substance; refer to the two pages of the diffraction search manuals in Figs. 21.5 and 21.6.

5. Now that you have identified the substance, you can consult the complete reference data on it by searching in the Diffraction File for the file card that corresponds to your identification.

FIGURE 21.5 A page from the *Search Manual* (Hanawalt Section) of *the Powder Diffraction File,* for Inorganic Phases, 1986. The *d* values of families of planes that produce the eight strongest diffraction lines from a specific substance are listed. The number of the complete reference file card is given in the right column. (Reproduced by permission of the International Center for Diffraction Data.)

2.01 – 1.86 (± .01)

	d_1	d_2	d_3	d_4	d_5	d_6	d_7	d_8	Substance	File No.	I/I_c
	1.91_3	3.16_6	2.93_6	2.74_x	2.11_4	1.63_4	1.58_4	1.22_2	$ZrBr_4$	15-332	
	1.91_5	3.16_6	2.87_6	1.72_5	1.57_2	1.45_4	2.58_4	2.05_2	Tl_2UO	28-1346	
o	1.90_5	3.16_6	3.57_5	2.49_6	1.61_5	3.04_2	2.82_2	1.78_2	$Co(TcO_4)\cdot4H_2O$	32-316	
	1.87_5	3.16_8	2.27_6	1.24_5	3.26_5	1.65_3	1.45_3	1.34_4	Li_2O_2	19-722	
	1.86_9	3.16_8	2.04_6	3.71_5	2.63_5	1.61_5	2.15_5	1.67_4	$FeIn_2S_4$	35-1065	
	1.85_5	3.16_5	3.31_5	1.96_5	1.63_5	5.85_4	2.68_4	2.93_5	$Na_2Np_2O_7$	18-1231	
•	1.85_5	3.16_5	2.62_5	2.02_5	3.71_5	1.31_5	1.07_5	2.14_5	$ZnSc_2S_4$	29-1392	2.40
	1.96_5	3.15_5	3.00_5	2.66_5	1.88_5	1.71_5	1.62_5	6.30_4	K_3VF_6	28-839	
	1.95_5	3.15_5	2.49_5	1.76_5	1.58_5	1.33_5	3.52_5	1.31_5	K_2PdCl_4	12-412	
	1.94_5	3.15_5	2.73_5	1.65_5	1.26_4	1.23_5	1.58_4	1.37_5	$Dy_4Te_3O_{12}$	22-262	
o	1.94_5	3.15_5	1.65_5	2.97_5	2.22_5	1.87_4	1.62_5	1.57_5	$MnOTe$	22-440	
	1.93_5	3.15_5	2.73_5	1.64_5	1.25_5	1.22_5	1.11_5	0.92_5	GaP	32-397	
•	1.93_3	3.15_5	1.65_5	1.12_5	1.37_5	1.25_5	0.86_5	1.05_5	CaF_2 Fluorite, syn	35-816	
	1.93_5	3.15_5	1.64_5	1.11_5	1.25_5	1.05_5	2.73_5	1.36_5	$(Zn,Hg)S$ Sphalerite, mercurian	22-731	
	1.91_5	3.15_5	2.45_5	2.06_5	1.76_5	1.73_5	3.37_5	1.69_5	ScI_3	18-1156	
i	1.90_5	3.15_x	3.56_5	2.91_5	1.72_5	1.54_5	5.82_5	2.81_5	$ti\text{-}SnWO_4$	27-902	
i	1.89_5	3.15_x	3.00_5	2.63_5	1.83_5	1.62_5	1.58_5	1.56_5	Tl_2TeO_{12}	32-1337	
i	1.88_5	3.15_x	3.30_5	1.49_5	1.10_5	3.58_5	1.94_5	1.65_5	$VOMoO_4$	18-1454	
	2.02_5	3.14_x	1.97_5	3.58_5	1.76_5	1.70_5	3.50_5	1.41_5	BkF_3	23-70	
i	2.01_5	3.14_x	3.70_5	1.85_5	2.61_5	2.13_5	1.59_5	1.36_5	$CdCr_{1.25}In_{0.75}S_4$	32-131	
o	1.99_5	3.14_x	1.95_5	1.17_5	3.19_5	2.10_5	1.46_5	1.27_x	$RbGd(SO_4)_2\cdot4H_2O$	21-1026	
	1.97_5	3.14_x	5.14_5	4.20_5	3.35_5	3.02_5	2.39_5	5.60_5	$CoCO_3TeO_2$/Mroseite	29-309	
	1.97_5	3.14_x	2.02_5	3.58_5	1.76_5	1.70_5	3.50_5	1.41_5	BkF_3	23-70	
o	1.96_5	3.14_x	3.02_5	1.31_5	1.29_5	1.27_x	1.47_5	2.42_5	$PrOC$	20-965	
	1.95_5	3.14_x	3.11_x	1.91_5	2.79_5	2.71_5	2.61_5	1.90_5	Gd_2MoO_6	26-656	
o	1.95_5	3.14_x	2.89_5	2.64_5	1.87_5	1.75_5	1.60_5	1.57_5	$Na_3ThMo_3O_{12}$	30-1248	
i	1.95_5	3.14_x	2.03_5	1.84_5	2.65_5	1.73_4	2.12_4	4.44_5	$K_2Zn_7(BeF_4)_3$	20-861	
o	1.95_5	3.14_x	1.99_x	1.17_5	3.19_5	2.10_5	1.46_5	1.27_x	$RbGd(SO_4)_2\cdot4H_2O$	21-1026	
	1.94_5	3.14_x	2.09_5	1.92_5	1.41_5	2.84_5	2.81_5	2.28_5	$PtSb_{1.3}Sn_{0.7}$	22-784	
i	1.93_5	3.14_x	1.64_5	2.91_5	1.99_5	2.57_5	1.76_5	2.13_5	$(Cu,Hg)_{12}Sb_4Se_{13}$/Hakite	25-297	
	1.93_5	3.14_x	1.64_5	1.25_5	1.11_5	0.92_5	0.86_5	1.05_5	AlP	12-470	
	1.93_5	3.14_x	1.64_5	1.11_5	0.86_5	1.25_5	0.92_5	0.83_5	$\alpha\text{-}NaYF_4$	6-342	
	1.93_5	3.14_x	1.64_5	1.11_5	0.92_5	1.25_5	1.05_5	0.86_5	$NaTlF_4$	18-1252	
	1.92_5	3.14_x	2.73_5	1.94_5	1.64_5	1.37_5	3.19_5	1.66_5	$TbOF$	8-230	
	1.92_5	3.14_x	2.09_5	1.94_5	1.41_5	2.84_5	2.81_5	2.28_5	$PtSb_{1.3}Sn_{0.7}$	22-784	
i	1.92_5	3.14_x	1.64_5	2.72_5	5.50_5	1.36_5	1.25_5	4.24_5	Cu_2SnS_3	27-198	
	1.92_5	3.14_x	1.64_5	2.72_5	1.25_5	1.22_5	1.11_5	1.05_5	$Ce_{0.8}U_{0.2}O_{2.04}$	33-343	3.70
	1.92_5	3.14_x	1.64_5	1.11_5	0.92_5	0.86_5	1.24_5	1.05_5	Ga_2Se_3	5-724	
	1.92_x	3.14_x	1.64_5	0.92_5	1.11_5	1.05_5	0.96_5	1.25_5	$NaCoCdYF_8$	7-261	
o	1.91_5	3.14_x	3.10_x	2.68_5	2.87_5	2.74_5	4.82_x	3.87_x	Pr_3SbO_7	33-1071	
o	1.89_5	3.14_x	2.82_5	1.85_5	1.56_5	2.70_5	1.32_5	1.08_5	Cu_5FeS_4/Idaite, syn	13-161	
	1.88_5	3.14_x	3.43_5	6.51_5	3.24_5	2.97_5	2.53_5	3.47_5	$Rb_{0.125}Ta_{0.375}O$	34-734	
	1.88_5	3.14_x	2.42_5	2.08_5	6.13_5	1.74_5	1.58_5	1.39_5	HfI_3	19-543	
i	1.85_x	3.14_x	3.70_5	2.01_5	2.61_5	2.13_5	1.59_5	1.36_5	$CdCr_{1.25}In_{0.75}S_4$	32-131	
o	1.85_x	3.14_x	2.82_5	1.89_5	1.56_5	2.70_5	1.32_5	1.08_5	Cu_5FeS_4/Idaite, syn	13-161	
i	2.01_x	3.13_x	2.64_5	1.21_5	2.24_5	2.13_5	1.65_5	1.60_5	$CoSn$	2-559	
i	2.00_5	3.13_x	1.84_5	3.67_5	2.60_5	2.12_5	1.59_5	1.58_5	$CdCr_{1.5}In_{0.5}S_4$	32-129	
i	2.00_5	3.13_x	1.83_5	3.66_5	2.99_5	2.38_5	1.64_5	1.58_5	$CsFe_2F_7$	36-868	
i	1.99_5	3.13_x	3.38_5	3.17_5	2.91_5	2.13_5	1.88_5	1.73_5	$La_6Th_2O_{13}$	33-724	
i	1.96_5	3.13_x	2.62_5	2.77_5	1.90_5	1.66_5	1.60_5	1.57_5	Tl_2TeO_3	32-1338	
	1.96_5	3.13_x	2.50_5	3.53_5	1.77_4	1.74_5	1.58_4	1.56_4	Rb_3NbO_8	27-543	
i	1.96_5	3.13_x	2.47_5	1.92_5	1.58_5	3.56_5	3.43_5	1.75_5	Ag_2BaSnS_4	33-1170	
o	1.95_5	3.13_x	1.61_5	1.72_5	2.64_5	1.86_5	2.90_5	3.94_5	$Pm_2Mo_2O_{12}$	33-1088	
i	1.92_5	3.13_x	2.71_5	1.63_5	2.12_5	1.24_5	1.56_5	1.21_5	Cf_2O_3	23-140	
i	1.92_5	3.13_x	2.47_5	1.96_5	1.58_5	3.56_5	3.43_5	1.75_5	Ag_2BaSnS_4	33-1170	
	1.92_5	3.13_x	1.91_5	1.63_5	3.45_5	4.06_5	7.37_5	4.37_5	$Na_7Zr_6F_{31}$	10-177	
	1.92_5	3.13_x	1.89_5	1.61_5	2.41_5	1.66_5	1.33_5	1.24_5	$FeSe$	3-533	
i	1.92_5	3.13_x	1.64_5	2.71_5	1.57_5	1.36_5	0.00_5	0.00_5	Ni_3Se_2	23-1277	
i	1.92_5	3.13_x	1.64_3	1.57_5	1.25_5	2.71_5	4.87_5	2.37_5	Cu_2ZnSnS_4/Kesterite, syn	26-575	
i	1.92_5	3.13_x	1.63_5	2.71_5	1.56_5	2.70_5	1.36_5	1.35_5	$CdAgF_4$	26-273	
	1.91_5	3.13_x	1.92_5	1.63_5	3.45_5	4.06_5	7.37_5	4.37_5	$Na_7Zr_6F_{31}$	10-177	
	1.91_5	3.13_x	1.64_5	2.71_5	6.26_5	4.02_5	3.28_5	2.48_5	$ZnBr_2$	34-621	
o	1.91_x	3.13_x	1.64_5	1.11_5	1.24_5	2.70_5	1.04_5	2.44_5	Cu_3SnS_4/Kuramite	33-501	
	1.90_5	3.13_x	1.64_5	1.93_5	1.62_5	1.34_5	1.24_5	1.10_5	$ZnGeP_2$	33-1471	
	1.89_5	3.13_x	1.92_5	1.61_5	2.41_5	1.66_5	1.33_5	1.24_5	$FeSe$	3-533	
	1.89_5	3.13_x	1.61_5	1.02_5	1.08_5	1.22_5	1.09_5	1.04_5	$Cu_6Zn_3As_4S_{12}$/Nowackiite	25-323	
c	1.88_5	3.13_x	2.57_5	0.81_5	0.80_5	0.78_5	2.26_5	1.95_5	$HfAs$	15-490	
	2.02_5	3.12_x	2.14_5	2.24_5	2.64_5	2.10_5	1.32_5	1.92_5	$HfNi_3$	32-426	6.52
	2.01_5	3.12_x	1.98_5	3.59_5	3.46_5	1.76_5	1.69_5	1.42_5	PmF_3	33-1086	
i	1.99_5	3.12_x	2.99_5	1.83_5	3.66_5	1.75_5	1.58_5	1.56_5	$Tl_{0.50}(W_{0.50}Nb_{0.50})O_3$	22-931	
i	1.99_5	3.12_x	2.98_5	1.83_5	3.65_5	1.56_5	1.75_5	1.58_5	$Tl_{0.50}(W_{0.50}Ta_{0.50})O_3$	22-1484	
	1.99_5	3.12_x	2.87_5	2.14_5	1.85_5	1.75_5	4.24_5	2.59_5	$LaHP_2O_7$	21-442	
i	1.98_5	3.12_x	3.63_5	3.55_5	3.46_5	3.23_5	3.18_5	2.94_5	$Sr_3V_2F_{12}$	28-1265	
	1.98_5	3.12_x	2.76_5	2.48_5	1.71_5	1.40_5	1.63_5	1.52_5	ZnF_2	21-1480	
	1.98_5	3.12_x	2.01_5	3.59_5	3.46_5	1.76_5	1.69_5	1.42_5	PmF_3	33-1086	
	1.98_5	3.12_x	1.94_5	1.16_5	3.18_5	2.33_5	1.46_5	1.27_5	$RbTb(SO_4)_2\cdot4H_2O$	21-1045	
	1.95_5	3.12_x	1.91_5	2.79_5	2.71_5	2.61_5	1.90_5	1.67_5	Gd_2MoO_6	26-656	
	1.94_5	3.12_x	2.65_5	1.34_5	1.32_5	3.49_5	2.47_5	2.21_5	$Rb_2Sr_3(CO_3)_3$	33-1127	
	1.94_5	3.12_x	2.63_5	3.49_5	1.80_5	2.10_5	1.34_5	1.15_5	Cs_2O_3	10-248	
	1.94_5	3.12_x	1.98_5	1.16_5	3.18_5	2.33_5	1.46_5	1.27_5	$RbTb(SO_4)_2\cdot4H_2O$	21-1045	
	1.93_5	3.12_x	1.63_5	1.10_5	1.04_5	2.68_5	1.25_5	1.24_5	UO_2	9-174	
	1.92_5	3.12_x	3.93_5	1.64_5	1.83_5	1.80_5	3.58_5	3.41_5	$NaU_xZr_7F_{17}$	13-45	
	1.92_5	3.12_x	1.64_5	2.71_5	1.25_5	1.22_5	1.58_5	1.11_5	Bi_2WO_6 Russellite	26-1044	
	1.92_5	3.12_x	1.63_5	1.90_5	2.68_5	1.24_5	1.10_5	1.04_5	$DyOF$	19-437	
	1.92_5	3.12_x	1.63_5	1.11_5	2.70_5	1.24_5	1.04_5	2.42_5	Cu_7FeSnS_4	24-366	
i	1.91_5	3.12_x	1.63_5	2.70_5	1.24_5	1.10_5	1.56_5	1.35_5	$Li_4Cl_2N_2$	24-605	
	1.91_5	3.12_x	1.63_5	2.70_5	1.35_5	1.24_5	2.42_5	4.82_5	Cu_2CoSnS_4	26-513	
	1.90_5	3.12_x	1.63_5	1.92_5	2.68_5	1.24_5	1.10_5	1.04_5	$DyOF$	19-437	
	1.89_5	3.12_x	1.74_5	2.05_5	1.71_5	1.15_5	2.76_5	2.48_5	$(NH_4)_2Er_5F_{17}$	29-110	
	1.89_5	3.12_x	1.63_5	3.41_5	1.86_5	1.67_5	2.04_5	1.97_5	$GaGeS_3$	33-554	

FIGURE 21.6 A page from the *Search Manual* (Hanawalt Section) of the *Mineral Powder Diffraction File.* The information given here is similar to that in Fig. 21.5, except that here all data entries are for minerals only. (Reproduced by permission of the International Center for Diffraction Data.)

1.93 – 1.86 (± .01)

											File No.
	1.92_x	11.7_x	10.7_x	1.52_x	5.45_8	3.68_8	3.34_8	2.58_8	Koenenite	$Mg_3Al_4O_{11}\cdot 15H_2O$	11- 492
	1.93_x	8.12_6	2.92_6	2.88_6	2.04_5	1.71_5	4.06_4	3.50_4	Bultfonteinite	$Ca_2SiO_2(OH,F)_4$	8- 223
	1.94_8	7.04_x	2.99_9	3.54_8	2.88_6	3.20_5	2.49_5	5.86_4	Pentahydroborite	$CaB_2O_4\cdot 5H_2O$	14- 339
o	1.92_9	6.20_x	3.25_x	3.38_8	2.95_8	2.19_8	2.06_8	1.29_8	Pseudo-autunite	$(H_3O)_4Ca_2(UO_2)_2(PO_4)_4\cdot 5H_2O$	18-1084
	1.91_8	5.40_x	3.12_5	2.40_4	1.63_4	1.10_4	1.80_3	1.35_3	Sulvanite	Cu_3VS_4	11- 104
i	1.90_8	5.12_x	7.13_9	3.03_8	3.08_7	9.60_6	5.60_6	3.91_5	Coyoteite	$NaFe_3S_5\cdot 2H_2O$	35- 565
o	1.87_x	4.75_x	2.35_x	1.44_7	9.68_6	1.38_6	1.24_6	1.57_5	Lithiophorite	$(Co,Mn)O(OH)$	12- 647
	1.93_8	4.55_x	3.15_8	2.28_8	3.34_7	2.15_6	1.74_6	2.07_4	Gearksutite	$CaAl(F,OH)_5\cdot H_2O$	5- 283
o	1.91_x	4.52_x	2.21_x	1.16_8	0.86_8	0.88_7	1.36_4	1.34_4	Hongshiite	$CuPtAs$	29- 574
•	1.94_x	4.44_x	3.63_5	6.28_5	3.97_5	2.81_4	2.09_4	2.37_3	Metaborite, syn	HBO_2	15- 868
	1.89_x	3.79_x	1.87_8	2.74_7	2.78_6	2.01_6	1.83_6	1.70_6	Osarsite	$(Os,Ru)AsS$	25- 595
	1.89_x	3.79_x	1.87_8	2.74_7	2.78_6	2.01_6	1.83_6	1.70_6	Ruarsite, syn	$RuAsS$	27- 568
	1.87_8	3.79_x	1.89_x	2.74_7	2.78_6	2.01_6	1.83_6	1.70_6	Osarsite	$(Os,Ru)AsS$	25- 595
	1.87_8	3.79_x	1.89_x	2.74_7	2.78_6	2.01_6	1.83_6	1.70_6	Ruarsite, syn	$RuAsS$	27- 568
	1.91_9	3.63_x	3.10_8	3.02_8	2.73_8	2.71_8	2.07_8	2.05_8	Freudenbergite	$(Na,K)_2(Ti,Nb)_6(Fe,Si)_2(O,OH)_{16}$	17- 531
i	1.89_9	3.62_x	2.64_9	3.44_8	6.27_7	1.72_6	1.68_6	7.70_5	Fleischerite, syn	$Pb_3Ge(SO_4)_2(OH)_6\cdot 3H_2O$	29- 771
	1.86_8	3.50_x	2.79_x	2.47_4	1.52_3	1.57_2	1.49_2	1.27_2	Plattnerite	PbO_2	25- 447
	1.87_8	3.49_x	4.88_8	2.74_8	2.45_6	1.97_6	1.54_6	2.40_4	Kennedyite	$Fe_2MgTi_3O_{10}$	13- 353
o	1.90_9	3.47_x	3.09_x	2.05_x	3.13_8	2.97_7	2.43_5	5.55_4	Sulfoborite	$Mg_6B_4O_{10}(SO_4)_2\cdot 9H_2O$	14- 639
i	1.90_8	3.42_x	2.46_7	2.28_7	3.32_6	1.86_6	1.62_6	4.70_5	Brannerite, heated	UTi_2O_6	8- 2
i	1.93_8	3.39_x	2.17_x	2.24_6	2.21_6	2.14_6	1.87_6	1.79_6	Stenonite	$Sr_2AlF_5CO_3$	15- 366
o	1.87_8	3.34_x	5.74_9	3.23_7	2.12_7	3.57_6	1.75_6	7.13_5	Unnamed mineral	$Fe\text{-}TeO$	16- 146
	1.92_8	3.33_x	1.75_x	2.98_8	2.89_8	2.80_6	2.05_6	1.02_5	Unnamed mineral	$(Ni,Fe,Cu)_{0.75}Ir_{0.25}S$	29- 555
	1.92_8	3.32_x	2.17_7	2.57_7	1.66_7	1.48_5	1.44_5	1.39_5	Genthelvite	$Zn_4Be_3(SiO_4)_3S$	29- 224
	1.93_9	3.29_x	3.84_4	2.11_4	5.42_3	3.16_2	1.67_2	1.42_2	Shadlunite	$(Cu,Fe)_8(Pb,Cd)S_8$	25-1426
i	1.93_x	3.28_9	3.15_9	2.09_6	1.11_6	1.66_5	2.44_4	1.82_4	Geffroyite	$(Cu,Fe,Ag)_9(Se,S)_8$	35- 523
	1.93_x	3.27_x	2.98_x	2.90_9	1.97_8	3.58_5	3.41_5	3.40_5	Juanite	$Ca_9(Mg,Fe)_3(Si,Al)_{12}(O,OH)_{36}\cdot 7H_2O$	29- 335
o	1.91_8	3.24_x	3.06_x	1.97_x	1.71_8	1.68_8	1.61_8	3.58_6	Fergusonite-beta-(Ce)	$CeNbO_4$	29- 402
	1.89_9	3.23_x	1.10_4	3.08_3	2.07_3	3.78_2	2.46_2	1.64_2	Manganese-shadlunite	$(Cu,Fe)_8(Mn,Pb)S_8$	25-1425
	1.87_8	3.22_x	1.75_7	1.17_5	1.05_5	3.49_4	3.00_4	2.79_4	Cubanite	$CuFe_2S_3$	9- 324
	1.94_x	3.21_x	2.01_9	3.91_7	2.50_7	1.62_7	1.56_7	1.34_7	Huanghoite	$BaCe(CO_3)_2F$	15- 286
o	1.88_9	3.21_x	3.03_x	2.80_5	2.70_5	2.06_5	1.96_5	1.67_5	Fergusonite-beta-(Nd)	$(Nd,Ce)NbO_4$	35- 703
	1.88_9	3.20_x	2.05_7	1.09_8	1.03_7	3.76_5	1.38_5	1.23_5	Indite	$FeIn_2S_4$	16- 170
c	1.88_x	3.20_x	2.04_8	2.57_7	2.84_6	3.35_5	2.50_4	1.77_4	Palladseite, syn	$Pd_{17}Se_{15}$	29-1437
	1.94_x	3.18_6	2.74_5	1.26_5	1.12_5	3.31_4	2.50_4	1.65_3	Bornite	Cu_5FeS_4	14- 323
c	1.94_x	3.17_4	2.74_4	3.31_3	2.52_2	1.66_1	1.12_1	1.37_1	Bornite	Cu_5FeS_4	34- 135
	1.94_8	3.17_x	1.67_4	1.65_4	1.96_3	1.26_3	1.13_2	5.37_1	Kuramite	$(Cu,Hg)_{5.5}Sn_2S_8$	29- 570
i	1.94_8	3.17_x	1.65_8	2.74_7	4.47_6	2.93_4	2.58_4	1.39_4	Winstanleyite	$TiTe_3O_8$	34-1485
i	1.86_x	3.17_x	2.02_4	1.07_3	6.06_2	5.25_2	3.71_2	3.02_2	Argentopentlandite	$(Fe,Ni)_8Ag_{1.1}S_8$	25- 406
	1.93_9	3.16_x	1.65_8	1.11_6	1.25_5	1.05_5	2.12_3	1.36_3	Gruzdevite	$Cu_6Hg_3Sb_4S_{12}$	35- 659
•	1.93_x	3.15_9	1.65_3	1.12_2	1.37_1	1.25_1	0.86_1	1.05_1	Fluorite, syn	CaF_2	35- 816
	1.93_9	3.15_x	1.64_8	1.11_5	1.25_4	1.05_4	2.73_3	1.36_3	Sphalerite, mercurian	$(Zn,Hg)S$	22- 731
i	1.93_9	3.14_x	1.64_8	2.91_7	1.99_7	2.57_6	1.76_6	2.13_5	Hakite	$(Cu,Hg)_{12}Sb_4Se_{13}$	25- 297
o	1.89_x	3.14_x	2.82_x	1.85_x	1.56_x	2.70_8	1.32_8	1.08_8	Idaite, syn	Cu_5FeS_6	13- 161
o	1.85_x	3.14_x	2.82_x	1.89_x	1.56_x	2.70_8	1.32_8	1.08_8	Idaite, syn	Cu_5FeS_6	13- 161
i	1.92_9	3.13_x	1.64_3	1.57_1	1.25_1	2.71_1	4.87_1	2.37_1	Kesterite, syn	Cu_2ZnSnS_4	26- 575
o	1.91_8	3.13_x	1.64_6	1.11_4	1.24_3	2.70_2	1.04_2	2.44_1	Kuramite	Cu_3SnS_4	33- 501
	1.89_7	3.13_6	1.61_6	1.02_x	1.08_8	1.22_5	1.09_2	1.04_2	Nowackiite	$Cu_6Zn_3As_4S_{12}$	25- 323
	1.92_9	3.12_x	1.64_x	2.71_8	1.25_5	1.22_4	1.58_3	1.11_3	Russellite	Bi_2WO_6	26-1044
	1.94_8	3.11_x	1.60_9	1.25_8	4.77_7	1.69_7	1.56_7	1.09_7	Scheelite	$CaWO_4$	8- 145
o	1.90_x	3.11_8	1.63_4	1.57_4	2.87_3	1.06_3	2.70_2	2.55_2	Chatkalite	$Cu_6FeSn_2S_8$	35- 683
o	1.86_x	3.11_x	3.03_x	2.81_x	3.46_9	1.30_9	1.27_9	1.24_9	Calciocopiapite	$CaFe_4(SO_4)_6(OH)_2\cdot 19H_2O$	27- 77
	1.90_x	3.10_x	1.62_x	2.69_7	1.24_4	1.35_3	4.04_2	1.55_2	Aktashite	$Cu_6Hg_3As_4S_{12}$	25- 298
	1.90_8	3.10_x	1.61_6	1.23_5	1.10_5	1.80_4	1.75_4	1.64_4	Lautite	$CuAsS$	12- 738
	1.89_8	3.08_x	3.16_7	1.61_7	3.63_6	1.22_6	1.09_6	1.02_6	Polhemusite	$(Zn,Hg)S$	31- 870
	1.89_x	3.08_x	1.98_x	2.80_9	3.43_8	3.70_7	2.97_6	2.24_6	Hypercinnabar, syn	HgS	19- 798
	1.89_8	3.07_x	1.60_6	1.08_6	1.22_5	1.87_4	1.32_4	1.02_4	Mooihoekite, syn	$Cu_9Fe_9S_{16}$	25- 286
	1.88_8	3.07_x	1.89_6	1.61_6	1.09_6	0.94_6	1.21_5	2.67_4	Haycockite	$Cu_4(Fe,Ni)_5S_8$	25- 289
	1.88_x	3.06_x	1.87_x	1.60_x	2.66_8	1.59_5	2.63_4	1.53_4	Briartite, syn	Cu_2FeGeS_4	25- 282
	1.87_x	3.06_x	1.88_x	1.60_x	2.66_8	1.59_5	2.63_4	1.53_4	Briartite, syn	Cu_2FeGeS_4	25- 282
i	1.87_8	3.06_x	1.61_6	1.89_5	1.59_3	1.22_3	1.34_2	1.23_2	Gallite, syn	$CuGaS_2$	25- 279
	1.87_8	3.06_x	1.60_6	2.65_3	1.21_3	4.31_2	7.50_1	3.34_1	Renierite	$Cu_3(Fe,Ge)(S,As)_4$	9- 424
	1.87_9	3.06_x	1.60_7	1.08_6	2.64_5	1.21_5	1.02_5	0.94_5	Talnakhite	$Cu_9(Fe,Ni)_8S_{16}$	25- 287
	1.87_8	3.05_x	1.59_8	2.64_6	1.52_4	1.21_4	1.18_4	1.02_4	Stannomicrolite	$Sn_2(Ta,Nb)_2O_7$	23-1441
	1.87_9	3.05_x	1.59_7	1.93_6	2.50_5	2.08_4	1.71_4	2.83_3	Giraudite	$(Cu,Zn,Ag)_{12}(As,Sb)_4(Se,S)_{13}$	35- 525
	1.86_9	3.05_x	1.59_7	1.58_6	1.20_6	1.08_6	1.32_5	1.07_5	Luzonite, syn	Cu_3AsS_4	10- 450
o	1.87_x	3.04_x	1.59_5	1.08_3	2.77_1	1.82_1	1.66_1	1.32_1	Arsenosulvanite	Cu_3AsS_4	25- 265
o	1.86_x	3.04_x	1.59_x	1.01_x	3.17_9	1.37_9	1.02_9	2.02_8	Cesstibtantite	$(Cs,Na)SbTa_4O_{12}$	35- 672
o	1.86_9	3.03_x	1.58_9	2.63_8	6.07_6	3.17_5	1.21_5	1.17_5	Natrobistantite	$(Na,Cs)Bi(Ta,Nb,Sb)_4O_{12}$	35- 706
	1.86_9	3.02_x	1.58_9	1.08_8	1.02_8	1.21_7	2.62_6	1.18_6	Plumbopyrochlore	$(Pb,Ln)_{2-x}(Nb,Ta)_2O_6(OH)$	25- 453
	1.85_8	3.02_x	1.58_7	2.61_4	1.21_3	3.34_2	1.56_2	1.31_2	Sinnerite, syn	$Cu_6As_4S_9$	25- 264
i	1.92_8	3.01_x	1.76_x	5.75_6	2.88_6	1.02_6	2.29_5	1.30_5	Cobalt pentlandite	$(Co,Fe,Ni)_9S_8$	12- 723
	1.85_x	3.01_x	2.83_x	3.45_9	2.14_9	1.31_8	2.56_5	1.16_5	Saryarkite	$(Ca,Y,Th)_2Al_4(SiO_4,PO_4)_4(OH)\cdot 9H_2O$	16- 712
	1.86_8	3.00_x	1.58_6	2.62_4	2.06_4	1.20_4	1.07_4	2.47_3	Tetrahedrite, argentian	$(Cu,Ag,Fe)_{12}Sb_4S_{13}$	11- 101
o	1.90_8	2.99_x	1.77_9	2.04_8	2.77_7	2.27_6	1.44_6	5.88_5	Steigerite, chromian	$(Al,Cr)VO_4\cdot 3H_2O$	29- 20
	1.91_8	2.98_x	5.73_9	3.51_9	2.26_7	1.76_6	1.50_6	1.48_6	Hidalgoite	$PbAl_3AsO_4SO_4(OH)_6$	6- 380
	1.90_9	2.96_x	3.51_9	2.20_8	1.76_8	5.71_7	1.29_6	2.75_5	Kemmlitzite-(Ln)	$(Sr,Ln)Al_3(OH)_6(AsO_4)SO_4$	22-1248
	1.85_8	2.95_x	1.81_7	2.24_5	1.01_5	2.04_4	1.56_4	1.48_4	Calcite, manganoan	$(Ca,Mn)CO_3$	2- 714
	1.89_9	2.94_x	2.18_9	1.74_8	1.43_6	3.49_4	1.47_4	1.49_3	Woodhouseite	$CaAl_3(PO_4)(SO_4)(OH)_6$	4- 670
	1.89_8	2.93_x	5.63_9	2.16_8	3.48_7	2.21_7	1.74_7	1.43_6	Florencite-(Ce)	$CeAl_3(PO_4)_2(OH)_6$	8- 143
o	1.85_9	2.93_x	2.86_x	2.14_9	2.06_8	2.80_7	3.23_6	1.99_6	Tyretskite-1Tc	$Ca_2B_5O_9(OH)_3$	26- 2
	1.92_9	2.92_x	3.59_9	2.09_9	2.05_8	1.70_7	1.32_7	1.46_5	Hydroxylbastnaesite-(Ce)	$(Ce,La)CO_3(OH)$	17- 503
	1.87_8	2.91_x	2.84_9	2.82_9	3.99_7	2.67_7	3.43_6	3.21_6	Svabite	$Ca_5(AsO_4)_3(OH,Cl,F)$	19- 215
i	1.86_x	2.91_x	3.42_9	4.44_9	2.97_8	2.13_8	2.09_6	2.22_5	Stillwellite	$CeBSiO_5$	26- 349
	1.89_8	2.89_x	1.73_9	1.59_9	1.20_8	2.10_7	1.28_6	2.69_4	Kettnerite	$CaBi(CO_3)OF$	25- 126
	1.86_8	2.84_x	3.48_8	2.81_8	4.12_6	3.93_6	3.21_6	3.11_6	Britholite-(Ce), heated	$(Ca,Ce,Th)_{4.87}(P,Si,Al)_{2.92}O_{11.90}(OH,F)_1$	17- 724
	1.94_9	2.80_x	3.48_8	4.70_8	1.66_8	1.77_7	1.71_7	1.84_6	Cappelenite	$Ba(Y,Ln)_6(Si_3B_6O_{24})$	27- 42
	1.92_9	2.78_x	2.67_8	1.23_8	2.26_7	2.01_7	1.63_6	1.98_5	Kurchatovite	$Ca(Mg,Mn)B_2O_5$	19- 648
•	1.86_x	2.78_x	2.51_7	4.81_6	2.23_6	1.82_5	2.95_4	1.74_4	Millerite	NiS	12- 41
	1.85_8	2.76_x	2.38_8	2.28_8	4.33_7	1.79_7	3.21_6	3.58_5	Tucekite	$Ni_9Sb_2S_8$	29- 927

FIGURE 21.7 The JCPDS card number 5-628 for synthetic halite. (Reproduced by permission of the International Center for Diffraction Data.)

5-628

	d Å	Int	hkl	d Å	Int	hkl
NaCl	3.26	13	111			
	2.821	100	200			
Sodium Chloride Halite, syn	1.994	55	220			
	1.701	2	311			
	1.628	15	222			
Rad. CuKα₁ **λ** 1.5405 **Filter** Ni **d-sp**	1.410	6	400			
Cut off **Int.** Diffractometer **I/I**cor. 4.40	1.294	1	331			
Ref. Swanson, Fuyat, *Natl. Bur. Stand. (U.S.), Circ. 539*, **2** 41 (1953)	1.261	11	420			
	1.1515	7	422			
Sys. Cubic **S.G.** Fm3m (225)	1.0855	1	511			
a 5.6402 **b** **c** **A** **C**	0.9969	2	440			
α **β** **γ** **Z** 4	0.9533	1	531			
Ref. Ibid.	0.9401	3	600			
	0.8917	4	620			
Dx 2.16 **D**m 2.17 **mp** 804°	0.8601	1	533			
εα **nωβ** 1.542 **εγ** **Sign** **2V**	0.8503	3	622			
Ref. *Dana's System of Mineralogy, 7th Ed.*, **2** 4	0.8141	2	444			
Color Colorless						

Color Colorless

X-ray pattern at 26 C. An ACS reagent grade sample recrystallized twice from hydrochloric acid. Merck Index, 8th Ed., p. 956. Halite group. halite subgroup. $F_{17} = 92.7(.0108,17)$. PSC: cF8.

Student Name

TABLE 21.2 Record of Data Obtained from the X-Ray Powder Diffraction Film Pictured in Fig. 21.2b. Refer to Fig. 21.2b

A. Location of Left and Right Centers

Line Number	Measurements		Total $x'_1 + x_1$	Center $\dfrac{x'_1 + x'_1}{2}$	Difference Between Left and Right Center (in millimeters)	Mineral Identification:
	Left (x')	Right (x)				JCPD file no.:
1						
2						
3						
13 (α_1)						
13 (α_2)						

B. Measurement of Diffraction Lines Going from Left Center to Right Center

Line Number	Measurement of Right Arc (x)	Value of Left Center	Subtract Left Center from x = 2θ	Conversion to d	Intensity (Copy from Fig. 21.2b)	Published Data 2θ	
						(degrees)	d
1						28.28	3.155
2						47.04	1.9316
3						55.81	1.6471
4						68.74	1.3656
5						75.92	1.2533
6						87.46	1.1152
7						94.31	1.0514
8						105.91	0.9658
9						113.20	0.9234
10						126.36	0.8638
11						135.44	0.8331
12						138.78	0.8236
13 (α_1)						—[a]	—
13 (α_2)						—	—

[a]Not available

Identification of an Unknown by X-Ray Powder Diffractometer Tracing

PURPOSE OF EXERCISE

To understand the steps involved in the identification of an unknown mineral using the diffractometer tracing of its powder diffraction effects. As in exercise 21, you will need to refer to the Powder Diffraction File of the International Center for Diffraction Data.

FURTHER READING

See listing in exercise 21, p. 295.

Background Information: If you have completed exercise 21 before beginning this assignment, you will find that identification of unknown crystalline materials is a great deal faster by diffractometer technique than it is by film. This is because a diffractometer tracing provides you with a graphical display of each peak position, relative to a direct reading 2θ scale, as well as a reasonably quantitative, direct reading, relative intensity scale.

A powder X-ray diffractometer is a great deal more complex and expensive than a powder camera mounted on an X-ray generator. A diffractometer, in conjunction with an X-ray generator, consists of a goniometer (a device that measures the angular location in terms of 2θ for a diffraction peak), an X-ray counting device (such as a Geiger, a scintillation, or a proportional counter for measurement of peak intensity), and an electronic readout system (see Fig. 22.1). On nonautomated powder diffractometers the graphical result is a diffractometer chart obtained over a 2θ region of about 6° to 80°, during a time period of about 45 minutes. On an automated diffractometer the same results are printed out on an $8\frac{1}{2} \times 11$-inch page in about 15 minutes. In either case, the final diagram shows peak locations with respect to a horizontal 2θ scale, as well as relative intensities of the peaks in terms of a vertical scale. To search the diffraction files for identification of an unknown, an investigator needs at the minimum to convert the 2θ values of the three most intense peaks on the graph to d values [using the *X-ray Diffraction Tables* of Fang and Bloss (1966), or, if these are not available, by using Table 21.1, or by solving the Bragg equation as outlined in section 2 of assignment 21]. It is strongly suggested, however, that to be certain the identification is completely unambiguous, the investigator should compare the ds of another ten or so peaks in the pattern with the published pattern on which the identification is based. Even though the identification of an unknown is based on matching of the three most intense X-ray diffraction lines, all other lines in the unknown pattern and those of the selected matching reference pattern should show good agreement in d values and intensity.

MATERIALS

The diffractometer tracing in Fig. 22.2 and Table 22.1 for data tabulation. A 90° triangle is useful for locating peak positions accurately on the diffractometer tracing with respect to the horizontal 2θ scale. You need access to the 2θ-to-d conversion tables of Fang and Bloss (1966), *X-ray Diffraction Tables*. If these are not available, the 2θ values can be converted to their appropriate ds by solving the Bragg equation with the help of an electronic calculator. Figures 22.3 and 22.4 should allow for unambiguous identification of the unknown. To check the ds and the intensities of all the diffraction lines on the pattern, you will need access to a microfiche or card edition of the file of the International Center for Diffraction Data.

ASSIGNMENT

1. Using the diffractometer tracing in Fig. 22.2, assign 2θ values to each of the peaks. Carefully locate each peak position with reference to the horizontal 2θ scale. Write the 2θ appropriate to the peak next to it on the figure. Number the peaks from left to right. Enter the peak numbers and 2θ values into Table 22.1.

2. Read the relative intensities of all the peaks, by measuring the height of the peak on the vertical scale and subtracting the background value in the area of the peak. Assign the value of 100 to the most intense peak. If the height of the tallest peak is y divisions (where y is some number less that 100), multiply all the other peaks by the ratio of 100/y to obtain their values relative to 100. Enter these relative peak heights into Table 22.1.

3. Convert the 2θ angles to d values using the *X-ray Diffraction Tables* of Fang and Bloss, or Table 21.1, or calculate each d, using the Bragg equation as outlined in section 2 of the assignment in exercise 21. Because the pattern shows no $\alpha_1 - \alpha_2$ doublets, use the CuKα column in Fang and Bloss.

4. Using the d values of the three most intense peaks, identify the substance with Figs. 22.3 or 22.4.

FIGURE 22.1 Schematic diagram of some of the experimental configuration of an X-ray powder diffractometer.

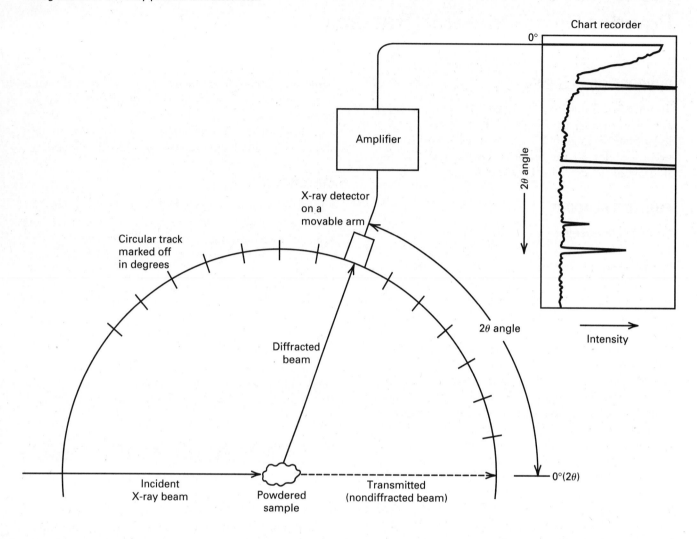

5. After identification, make sure the additional five *d* values listed in Figs. 22.3 or 22.4 show good agreement with the data from your pattern.

6. If search files from the International Center for Diffraction Data are available, locate the complete reference card on the substance and make sure *all* peaks and their intensities of both the "unknown" and the substance you identified it to be show good correspondence.

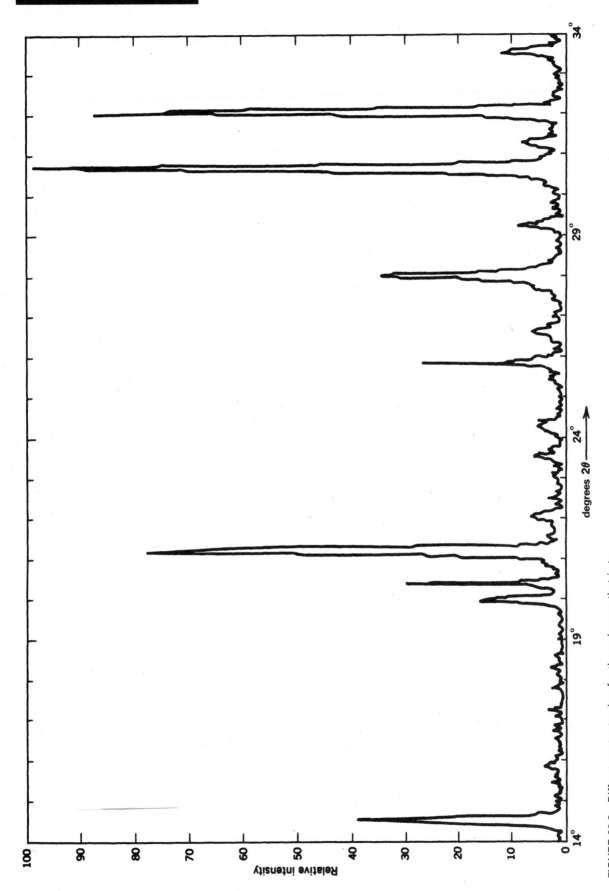

FIGURE 22.2 Diffractometer tracing for the unknown that is to be identified in this exercise. The X-ray radiation was produced by a Cu target X-ray tube. The 2θ-angle scale is horizontal. The relative intensity scale is vertical.

TABLE 22.1 Record of Measurements Obtained from the Diffractogram in Fig. 22.2. Refer to the diffractogram in Fig. 22.2.

Line Number[a]	2θ angle	d	I

Published 2θ values for the three most intense lines in the pattern of Fig. 22.2.

2θ	I
30.62	100
31.97	90
21.06	75

[a]Line numbers go consecutively from left to right.

Mineral identified as: _____

Diffraction file no.: _____

FIGURE 22.3 A page from the *Search Manual* (Hanawalt Section) of the *Powder Diffraction File,* for Inorganic Phases, 1986. The *d* values for families of planes that produce the eight strongest diffraction lines from a specific substance are listed. The number of the complete reference file card is given in the right column. (Reproduced by permission of the International Center for Diffraction Data.)

2.94 – 2.90 (± .01)

	d	d	d	d	d	d	d	d	Substance	File No.	I/Ic
	2.90	2.82	2.80	3.98	3.47	2.68	1.88	2.03	Ca₃(AsO₄)₂(OH) Johnbaumite	33–265	
i	2.90	2.82	2.58	2.23	1.69	4.47	3.81	3.16	Tl₂Te₃	24–1310	
	2.90	2.82	2.33	1.21	1.49	1.34	1.32	2.78	GdSm	20–436	
	2.90	2.82	2.31	2.48	3.11	3.00	2.61	2.36	K₄H₂(CO₃)₃·1.5H₂O	20–886	
	2.89	2.82	3.06	2.86	2.29	2.98	2.55	2.71	Ca₃Si₃O₁₂·H₂O	14–35	
	2.89	2.82	2.84	2.01	1.66	1.72	2.47	1.78	CaY₂O₄	19–265	
	2.89	2.82	2.78	2.14	2.05	1.99	1.84	1.83	(Ca,Sr)₂B₂O₈(OH)₂Cl/Hilgardite-1Tc, strontian	11–405	
	2.89	2.82	2.75	2.80	4.69	4.53	2.07	1.96	La₂CuIrO₆	18–673	
	2.89	2.82	2.02	1.66	2.55	5.16	5.02	1.47	K₃As	4–676	
•	2.95	2.81	4.23	2.69	8.47	2.56	1.71	1.93	LuFeMgO₄	36–965	
	2.95	2.81	3.39	2.90	2.17	1.99	1.62	3.06	K₄CaSi₃O₉	27–1346	
c	2.95	2.81	2.79	1.92	3.06	2.18	2.16	3.17	Bi₂Rh	28–172	
•	2.95	2.81	2.14	1.99	2.19	2.59	1.47	6.57	K₂MgF₄	23–469	
o	2.94	2.81	2.55	2.75	2.45	2.34	2.26	1.96	Ba₂MgCl₆	15–292	
c	2.94	2.81	1.88	5.63	1.72	1.47	3.45	1.27	Bi₂Cs	27–323	
	2.93	2.81	8.43	2.23	6.18	4.22	2.10	3.32	K₄Fe(CN)₆·3H₂O/Kafehydrocyanite, syn	14–695	1.80
•	2.93	2.81	4.05	2.03	2.31	1.64	1.84	2.43	Ba₂MgSi₂O₈	10–74	
	2.93	2.81	3.05	1.72	3.11	2.85	1.87	2.50	Pb₂Sr₂SiO₇	30–728	
•	2.92	2.81	4.16	2.43	2.63	3.03	3.80	2.49	Li₂CO₃	22–1141	0.90
•	2.91	2.81	4.05	5.30	2.44	4.48	3.39	2.57	Na₂CO₃·7H₂O	25–816	
•	2.91	2.81	4.04	2.02	2.31	1.83	1.64	2.42	NaBaPO₄	33–1210	
i	2.91	2.81	3.33	3.18	4.21	1.93	1.82	2.02	Eu₂(PO₄)₂F	26–627	
i	2.91	2.81	3.03	1.80	2.63	2.55	1.77	1.75	Ge₃P₂O₃	26–689	
	2.91	2.81	2.96	2.25	3.97	3.48	2.56	2.46	KBF(OH)₃	30–918	
	2.91	2.81	2.62	2.43	1.70	1.63	1.31	1.30	AuPb₃	26–711	
	2.90	2.81	3.39	2.95	2.17	1.99	1.62	3.06	K₄CaSi₃O₉	27–1346	
	2.90	2.81	3.18	3.31	2.16	4.20	2.01	4.04	Eu₂(PO₄)₂OH	22–613	
	2.90	2.81	2.65	1.59	1.47	1.09	0.83	1.73	Am	15–657	
i	2.90	2.81	2.02	3.45	1.66	2.45	2.20	1.64	Ba₂Sc₂O₅	31–161	
	2.89	2.81	4.46	3.36	2.09	1.92	4.21	3.14	CsAl(SeO₄)₂·12H₂O	28–278	
	2.89	2.81	3.22	6.79	3.99	2.41	2.01	1.79	(Na,Ca,Mn)₃Zr(SiO₄)₂F/Lavenite	14–586	
c	2.94	2.80	2.51	2.20	2.65	1.47	2.72	2.48	Ba₃Zn₃Fe₁₂O₂₂	25–1193	
	2.93	2.80	2.64	3.91	2.95	5.88	2.86	2.43	PbFe₈O₁₀	33–757	
i	2.91	2.80	1.98	1.63	3.78	1.70	2.14	1.45	K₂NiO₂	27–423	
	2.90	2.80	4.23	2.82	5.60	2.28	2.02	1.79	Al₄Ba	10–250	
	2.90	2.80	2.95	2.03	1.67	1.66	2.01	1.70	SrYb₂O₄	20–1222	
o	2.90	2.80	1.90	3.61	3.19	2.03	1.81	1.85	Ca₂Nd₂Sr(PO₄)₄(SiO₄)F₂	30–268	
	2.89	2.80	3.21	2.60	4.41	3.58	3.51	3.26	K₂S₄	30–992	
i	2.95	2.79	2.66	4.12	2.69	3.47	3.39	3.11	Ba₁.₅Sb₁₀O₁₀.₄	31–119	
o	2.95	2.79	2.03	1.74	1.64	4.37	3.73	3.12	BaLa₂Ti₂O₈	35–164	
c	2.94	2.79	2.96	3.16	2.16	1.53	1.91	1.61	CaD₂	31–266	3.08
c	2.93	2.79	4.60	5.05	2.91	3.23	3.97	3.78	Ba₃P₁₄	33–172	1.88
•	2.92	2.79	4.21	6.12	4.36	3.44	3.19	2.45	α-LiAlSi₂O₆/Spodumene	33–786	
	2.92	2.79	2.57	2.30	3.60	3.18	2.83	2.62	Ca₂SiO₄·CaCl₂	36–558	
•	2.92	2.79	2.02	2.06	1.48	1.44	1.21	1.66	Sn/Tin, syn	4–673	
	2.91	2.79	2.94	2.81	3.16	2.58	0.00	0.00	Sr₄Ga₂O₇	21–1181	
	2.91	2.79	2.74	3.74	5.96	3.93	2.09	2.01	Pb₃O₂Cl₂	6–475	
	2.91	2.79	1.96	1.57	1.48	1.27	2.58	1.65	Sr₄N₆	17–797	
i	2.90	2.79	2.78	1.45	2.94	2.02	2.00	1.66	SrLu₂O₄	32–1242	1.90
i	2.90	2.79	1.91	2.01	3.17	3.30	3.60	1.95	Ca₂Sr(PO₄)₆(OH)₂	34–477	
i	2.89	2.79	10.8	4.10	2.67	2.48	2.36	1.67	α₃-Ca₃Al₂O₇·19H₂O	14–628	
	2.89	2.79	2.59	2.68	2.67	2.40	2.29	2.10	Ca₂(Al,Fe)Al₂(SiO₄)₃OH/Clinozoisite	21–128	
	2.89	2.79	2.15	2.75	1.63	1.61	1.70	2.84	Sr₃Al₂O₆	10–65	
i	2.89	2.79	1.98	1.63	6.34	1.68	2.11	1.44	Sr₂Fe₀.₅Ga₀.₅O₄	32–1238	
i	2.89	2.79	1.91	3.58	3.16	2.00	3.29	1.95	Ca₂Sr(PO₄)₆(OH)	34–478	
	2.95	2.78	2.89	3.11	4.44	3.30	2.42	5.57	Ba₂NaCe₂Fe(Ti,Nb)₂Si₈O₂₆(OH,F)·H₂O/Joaquinit	36–385	
i	2.93	2.78	3.84	2.81	5.00	4.79	2.80	8.51	Na₂SiO₃·9H₂O	19–1239	
i	2.93	2.78	2.92	2.89	3.33	3.76	2.15	1.91	Pb₂Mn₂Ge₂O₇	34–1362	
	2.92	2.78	2.57	2.13	2.00	5.71	4.21	2.27	PtSn₄	4–744	
c	2.92	2.78	2.41	2.55	3.56	3.20	3.17	2.15	CaCu	25–1138	
•	2.92	2.78	2.02	2.94	3.26	2.12	4.81	2.37	KSrPO₄	33–1045	
c	2.89	2.78	2.99	2.44	2.14	4.78	3.67	3.10	Sb₂Sr	30–98	6.03
	2.89	2.78	2.95	3.11	4.44	3.30	2.42	5.57	Ba₂NaCe₂Fe(Ti,Nb)₂Si₈O₂₆(OH,F)·H₂O/Joaquinit	36–385	
i	2.89	2.78	2.66	2.17	2.32	1.87	2.54	1.67	Na₂AlH(CO₃)₂F₄/Barentsite	35–693	
	2.89	2.78	2.19	3.10	2.68	2.62	3.46	3.26	KCa₂FSiO₄	27–1343	
	2.95	2.77	4.72	2.53	2.46	2.36	2.27	2.24	K₂MnO₄	24–875	
	2.95	2.77	3.51	2.48	1.20	1.17	1.76	1.72	La₃Si₂	19–660	
	2.94	2.77	4.01	2.17	1.63	3.44	2.61	2.40	Ba(BrO₃)₂	20–120	
	2.93	2.77	3.69	2.69	4.68	2.55	2.52	3.51	H₃Ag₂NO₃P	16–511	
	2.92	2.77	3.75	3.43	3.08	2.33	1.78	1.54	Ag₂V₄O₁₁	20–1385	
•	2.92	2.77	2.01	1.62	2.20	1.60	1.38	1.29	CaLa₄Ti₄O₁₃	36–1278	
	2.92	2.77	2.01	1.46	1.66	1.43	2.06	0.00	InSb	19–579	
c	2.90	2.77	5.55	1.85	1.70	2.40	1.45	1.25	Bi₂Rb	28–175	
i	2.90	2.77	3.22	4.33	5.87	3.00	2.67	2.51	K₂Ba(Ge₄O₉)₂	31–975	
	2.94	2.76	4.26	2.09	1.82	3.38	1.29	5.80	DyPO₄·1.5H₂O	21–316	
o	2.94	2.76	2.62	2.33	3.23	3.18	3.03	2.90	Rb₃BiBr₆	20–986	
	2.93	2.76	3.05	2.40	2.30	2.16	2.85	2.58	Ca₂Yb₂O₅	19–262	
	2.91	2.76	3.06	2.94	2.56	3.12	4.06	3.68	Cu₃(PO₄)₂	21–298	
o	2.91	2.76	2.00	3.23	3.11	2.26	2.22	1.96	SrLa₂Ti₂O₈	36–211	
	2.91	2.76	1.52	3.13	2.14	1.90	1.69	1.66	YbD₂	9–256	
	2.90	2.76	2.01	3.93	3.62	2.18	1.64	1.61	NaCaVO₄	14–54	
i	2.90	2.76	2.00	2.27	3.65	1.62	1.83	1.60	Ca₂SiO₄	23–1045	
	2.90	2.76	1.62	2.00	1.70	1.60	1.45	1.28	β-TlFeO₂	22–490	
i	2.95	2.75	3.52	3.46	2.33	3.86	2.63	4.13	PbAs₂S₄/Sartorite	11–76	
	2.95	2.75	2.01	2.27	1.63	3.75	1.59	1.55	KCaPO₄	33–1002	
	2.95	2.75	2.00	1.47	1.29	1.60	2.06	1.38	InSb	22–67	
	2.94	2.75	3.31	1.95	1.82	2.00	3.64	3.07	Cd₂Cl(VO₄)₃	19–185	
i	2.94	2.75	2.67	3.87	3.14	3.20	1.84	4.55	Pb₃Mg(Si₂O₇)₃	32–525	1.50
	2.94	2.75	2.67	3.14	3.20	4.54	1.84	2.81	Pb₃Ni(Si₂O₇)₃	32–527	1.60
	2.93	2.75	3.64	2.57	3.40	2.88	1.83	1.74	Al₂Th₃	7–308	

FIGURE 22.4 A page from the *Search Manual* (Hanawalt Section) of the *Mineral Powder Diffraction File.* The information here is similar to that in Fig. 22.3, except that here all data entries are for minerals only. (Reproduced by permission of the International Center for Diffraction Data.)

2.94 – 2.90 (± .01)

File No.

								Mineral	Formula	File No.
•	2.93₈	3.32₆	3.60₆	3.04₇	3.56₄	2.35₄	3.75₇	4.69₇	Carnallite	KMgCl₃·6H₂O	24– 869
i	2.93₇	3.31₉	3.13₆	1.91₅	1.76₄	1.63₄	2.27₇	1.30₆	Wurtzite-2H, syn	α-ZnS	5– 492
	2.93₇	3.31₉	1.91₆	1.76₄	3.13₄	1.63₄	2.27₇	1.60₇	Wurtzite-2H, syn	ZnS	10– 434
i	2.91₈	3.30₆	3.44₆	2.75₄	3.76₄	3.02₄	2.28₄	2.07₅	Anderite	AgPbSb₃S₆	35– 596
i	2.91₇	3.30₆	2.76₆	3.02₇	1.77₅	2.02₅	1.73₄	4.26₅	Paulmooreite	Pb₂As₂O₅	33– 726
i	2.89₇	3.30₅	3.42₅	2.74₄	3.72₄	3.00₄	2.86₄	1.88₅	Anderite, cuprian	Ag₃CuPb₄Sb₁₂S₂₄	13– 462
	2.90₈	3.28₆	4.15₆	3.93₅	3.82₅	1.94₅	7.80₇	5.84₆	Chiavennite	CaMnBe₂Si₅O₁₃(OH)₂·2H₂O	35– 602
o	2.89₇	3.28₅	2.79₅	2.59₅	1.91₅	1.53₅	4.46₅	3.57₆	Unnamed mineral	Ba·Ti·Si·O	17– 504
i	2.92₈	3.27₅	3.74₅	3.49₄	4.13₄	2.74₄	3.08₅	2.77₅	Menoghinite	CuPb₁₃Sb₇S₂₄	29– 559
•	2.92₈	3.27₅	3.03₅	3.96₄	2.07₄	2.03₄	2.62₄	1.94₅	Fabianite	CaB₃O₅(OH)	15– 631
	2.90₈	3.27₅	2.98₅	1.93₄	1.97₅	3.58₅	3.41₅	3.40₅	Juanite	Ca₄(Mg,Fe)₇(Si,Al)₇(O,OH)₃₀·7H₂O	29– 335
i	2.95₈	3.26₅	3.81₅	2.86₄	3.35₄	3.88₄	2.15₄	4.22₅	Semseyite	Pb₉Sb₈S₂₁	22–1130
o	2.94₈	3.26₅	1.70₅	2.50₄	3.63₄	1.63₄	1.36₄	2.19₅	Ilmenorutile	Fe₄(Nb,Ta)₂Ti₉₋ₓO₃	31– 646
i	2.91₈	3.25₅	2.19₅	2.97₄	4.93₄	2.61₄	2.98₄	1.93₅	Viitaniemiite	NaCaAl(PO₄)ₓF(OH)ₓ	35– 598
c	2.90₈	3.24₅	3.02₅	3.44₄	2.25₄	3.73₄	2.14₄	3.62₅	Bustamite, ferroan	(Ca,Mn)₃Si₃O₉	33– 292
o	2.92₈	3.23₅	4.38₅	3.73₄	2.76₄	5.06₄	4.13₄	3.57₅	Roeddorite	(Na,K)₂(Mg,Fe)₅(Al,Si)₁₂O₃₀	23– 76
i	2.92₈	3.23₅	2.77₅	4.13₅	3.75₄	7.13₄	5.09₄	5.53₅	Osumilite	(K,Na,Ca)(Mg,Fe)₂(Al,Fe)₃(Si,Al)₁₂O₃₀·H₂O	25– 658
	2.90₈	3.23₅	5.12₅	2.69₄	2.62₄	1.65₄	4.25₄	2.30₅	Descleizite	(Zn,Cu)PbVO₄(OH)	12– 537
	2.95₈	3.21₅	2.44₅	2.14₄	1.57₄	4.20₄	5.17₄	1.37₅	Rusakovite	(Fe,Al)₅(VO₄,PO₄)₂(OH)₉·3H₂O	14– 60
i	2.92₈	3.21₅	3.02₅	2.91₄	2.49₄	1.63₄	2.57₄	2.14₄	Kanoite	(Mn,Mg)₂(Si₂O₆)	29– 865
i	2.91₈	3.21₅	3.26₅	3.87₄	3.77₄	2.62₄	3.61₄	5.87₅	Plagionite	Pb₅Sb₈S₁₇	22–1129
i	2.91₈	3.21₅	3.02₅	2.92₄	2.49₄	1.63₄	2.57₄	2.14₄	Kanoite	(Mn,Mg)₂(Si₂O₆)	29– 865
•	2.94₈	3.20₅	3.69₅	2.36₄	2.11₄	3.04₄	1.67₄	2.38₅	Topaz	Al₂SiO₄(F,OH)₂	12– 765
	2.93₈	3.20₅	2.58₅	1.80₄	3.50₄	2.71₄	2.15₄	4.54₅	Carminite	PbFe₂(AsO₄)₂(OH)₂	12– 278
	2.91₈	3.20₅	3.40₅	2.63₄	2.19₄	2.10₄	1.68₄	1.57₅	Shcherbakovite	NaK(Ba,K)Ti₂(Si₂O₇)₂	31–1324
o	2.90₈	3.20₅	6.11₅	2.37₄	2.31₄	1.92₄	2.49₄	3.86₅	Unnamed mineral	K–Ca–CO₃	25– 627
	2.90₈	3.20₅	4.09₅	1.84₄	1.33₄	1.52₄	4.51₄	3.52₅	Sogdianite	(K,Na)₂Li₃(Li,Fe,Al)₃ZrSi₁₂O₃₀	21– 501
c	2.95₈	3.19₅	3.22₅	2.88₄	2.93₄	3.70₄	3.05₄	3.03₅	Baratovite	Li₃KCa₇(Ti,Zr)₂(Si₆O₁₈)₂F₂	33– 811
	2.90₈	3.18₅	1.78₅	2.22₄	2.51₄	2.40₄	1.46₄	1.05₅	Leightonite	K₂Ca₂Cu(SO₄)₄·2H₂O	15– 128
	2.90₈	3.17₅	2.79₅	2.01₄	1.91₄	1.47₄	2.32₄	2.15₅	Belovite	Sr₃(Ce,Na,Ca)₂(PO₄)₃OH	31–1350
i	2.93₈	3.15₅	3.23₅	4.52₄	2.72₄	2.66₄	2.26₄	2.70₄	Bayldonite	Cu₃Pb(AsO₄)₂(OH)₂	26–1410
	2.93₈	3.13₅	1.90₅	1.64₄	1.57₄	2.72₄	1.22₄	1.19₄	Formanite, heated	LnTaO₄	26–1478
	2.92₈	3.13₅	2.69₅	4.06₄	1.78₄	5.29₄	2.49₄	4.32₅	Kermesite	Sb₂OS₂	11– 91
i	2.95₈	3.11₅	2.93₅	1.78₄	4.35₄	1.98₄	2.68₄	2.14₅	Okanoganite	(Na,Ca)₃Ln₁₇Si₆B₃O₂₇F₁₄	35– 483
c	2.94₈	3.11₅	3.22₅	2.84₄	3.67₄	3.41₄	2.58₄	2.89₅	Tripleidite	Mn₁.₅Fe₀.₅PO₄(OH)	26–1239
i	2.93₈	3.11₅	3.04₅	2.66₄	8.84₄	2.69₄	4.51₄	3.58₅	Kulanite	Ba(Fe,Mn,Mg)₂Al₂(PO₄)₃(OH)₃	29– 170
	2.94₈	3.10₅	3.19₅	1.80₄	3.41₄	2.58₄	2.31₄	2.15₅	Tripleidite	(Mn,Fe)₂PO₄(OH)	26–1240
	2.94₈	3.10₅	3.00₅	2.16₄	3.44₄	1.65₄	1.43₄	1.07₅	Manganbabingtonite	Ca₂(Mn,Fe)FeSi₅O₁₄OH	26– 313
	2.93₈	3.09₅	3.18₅	1.79₄	3.37₄	2.57₄	2.29₄	2.14₅	Wolfeite	(Fe,Mn)₂(PO₄)(OH)	5– 612
i	2.92₈	3.09₅	2.65₅	8.81₄	3.03₄	2.68₄	4.49₄	2.69₅	Penikisite	Ba(Mg,Fe)₂Al₂(PO₄)₃(OH)₃	29– 169
	2.94₈	3.08₅	2.79₅	3.44₄	3.26₅	2.24₅	2.08₄	1.88₄	Kalborsite	K₆BAl₄Si₆O₂₀(OH)₄Cl	33– 999
	2.91₈	3.08₅	1.64₅	4.69₄	4.81₅	2.36₄	4.20₅	4.27₅	Canasite	(Na,K)₆Ca₅Si₁₂O₃₀(OH,F)₄	13– 553
	2.90₈	3.08₅	1.76₅	3.06₄	3.31₅	3.27₄	2.16₄	2.73₅	Pectolite	NaCa₂HSi₃O₉	33–1223
o	2.92₈	3.07₅	2.59₅	1.84₄	1.50₄	3.68₄	1.90₄	11.0₅	Samarskite, heated	(Y,U,Fe)(Nb,Ta,Ti)₂O₆	4– 617
i	2.94₈	3.06₅	1.89₅	3.96₄	2.63₄	2.48₄	1.82₄	2.20₅	Rosenbuschite	(Na,Ca)₃(Fe,Ti,Zr)(SiO₄)₂F	14– 447
	2.92₈	3.06₅	4.21₅	3.52₄	3.24₄	6.15₄	6.51₄	2.18₅	Kasolite	Pb(UO₂)SiO₄·H₂O	29– 788
o	2.93₈	3.04₅	2.99₅	3.32₄	2.08₄	1.56₄	1.52₄	1.97₄	Mimetite, phosphatian	Pb₅Cl(As,P)O₄)₃	13– 124
i	2.92₈	3.04₅	3.35₅	2.78₄	2.91₄	3.83₄	1.91₄	2.56₅	Reinhardbraunsite	Ca₅(SiO₄)₂(OH)₂	29– 380
i	2.89₈	3.03₅	3.19₅	2.52₄	1.91₄	1.75₄	2.75₄	2.63₅	Rustumite	Ca₁₀(Si₂O₇)₂SiO₄Cl₂(OH)₂	18– 305
o	2.95₈	3.02₅	3.09₅	5.50₄	2.97₄	2.66₄	4.43₄	2.80₅	Yttropyrochlore	(Y,Ce,Nd,Th)(Nb,Ti,Ta)₂O₆	18– 765
i	2.91₈	3.02₅	2.90₅	3.21₄	2.58₄	1.63₄	1.49₄	1.39₄	Pigeonite	(Fe,Mg,Ca)SiO₃	13– 421
i	2.90₈	3.02₅	3.21₅	2.91₄	2.58₄	1.63₄	1.49₄	1.39₄	Pigeonite	(Fe,Mg,Ca)SiO₃	13– 421
	2.94₈	3.01₅	2.81₅	1.75₄	2.25₄	1.98₄	1.66₄	1.63₅	Sahlinite	Pb₁₄(AsO₄)₂O₉Cl₄	22– 664
	2.89₈	3.01₅	5.75₅	3.20₄	3.93₄	3.08₄	2.82₄	5.01₅	Realgar, high, syn	β-AsS	25– 57
i	2.95₈	3.00₅	4.84₅	3.78₄	3.70₄	2.50₄	2.88₄	1.73₅	Huebnerite, syn	MnWO₄	13– 434
	2.94₈	3.00₅	3.06₅	3.20₄	2.88₄	2.53₄	2.20₄	2.11₄	Roeblingite	Ca₂Pb₂(SO₄)₂Si₆O₁₄(OH)₁₀	18– 292
	2.94₈	3.00₅	1.58₅	2.65₄	1.70₄	1.86₄	3.71₄	2.21₅	Tantalaeschynite-(Y), heated	(Y,Ce,Ca)(Ta,Ti,Nb)₂O₆	26– 1
o	2.90₈	3.00₅	2.89₅	4.18₄	2.42₄	2.09₄	2.08₄	4.16₅	Arcanite, syn	K₂SO₄	5– 613
i	2.91₈	2.99₅	3.07₅	1.57₄	1.51₄	2.59₄	1.85₄	2.77₅	Aeschynite-(Y), syn	YTiNbO₆	20–1401
i	2.92₈	2.99₅	2.29₅	1.93₄	4.96₄	1.50₄	5.77₄	1.90₅	Alunite	(K,Na)Al₃(SO₄)₂(OH)₆	14– 136
o	2.91₈	2.98₅	2.52₅	2.49₄	2.20₄	2.02₄	1.41₄	2.54₅	Diopside, manganoan	(Ca,Mn)(Mg,Fe,Mn)Si₂O₆	22– 534
i	2.95₈	2.97₅	4.78₅	3.76₄	3.67₄	2.49₄	2.86₄	2.39₅	Wolframite, syn	FeMn(WO₄)₂	12– 727
i	2.94₈	2.97₅	3.11₅	2.93₄	1.78₄	1.98₄	2.68₄	2.63₅	Okanoganite	(Na,Ca)₃Ln₁₇Si₆B₃O₂₇F₁₄	35– 483
	2.93₈	2.97₅	1.83₅	11.7₄	6.00₄	3.13₄	4.21₄	4.00₅	Federite	(K,Na)₂.₅(Ca,Na)₂Si₁₄O₃₆(OH,F)₂·H₂O	19– 466
	2.89₈	2.94₅	4.43₅	3.29₄	2.61₄	1.39₄	1.87₄	3.05₅	Joaquinite	Ba₂NaCe₂FeTi₂Si₈O₂₆(OH)	26–1034
i	2.91₈	2.93₅	3.73₅	4.69₄	3.62₄	2.47₄	2.46₄	2.86₅	Sanmartinite, syn	ZnWO₄	15– 774
i	2.90₈	2.92₅	4.37₅	3.79₄	2.73₄	2.51₄	4.66₄	2.45₅	Pumpellyite	Ca₂MgAl₂(Si₂O₇)₂(OH)₂·H₂O	25– 156
•	2.93₈	2.91₅	3.73₅	4.69₄	3.62₄	2.47₄	2.46₄	2.86₅	Sanmartinite, syn	ZnWO₄	15– 774
c	2.89₈	2.91₅	3.24₅	3.02₄	3.44₄	2.25₄	3.73₄	2.14₄	Bustamite, ferroan	(Ca,Mn)₃Si₃O₉	33– 292
c	2.91₈	2.89₅	3.24₅	3.02₄	3.44₄	2.25₄	3.73₄	2.14₄	Bustamite, ferroan	(Ca,Mn)₃Si₃O₉	33– 292
i	2.95₈	2.88₅	2.78₅	3.02₄	2.84₄	2.71₄	4.54₄	3.30₅	Iimoriite	Y₂(SiO₄)(CO₃)	35– 640
i	2.89₈	2.87₅	5.90₅	6.60₄	4.41₄	4.72₄	4.37₄	2.94₅	Mesolite	Na₂Ca₂Al₆Si₉O₃₀·9H₂O	24–1064
	2.95₈	2.86₅	1.76₅	1.14₄	3.31₄	2.77₄	2.42₄	1.55₅	Nordite-(La)	Na₃LnSrMnSi₆O₁₇	27– 672
o	2.93₈	2.86₅	2.14₅	1.85₄	2.06₄	2.80₄	3.23₄	1.99₅	Tyretskite-1Tc	Ca₂B₅O₉(OH)₅	2– 2
	2.91₈	2.86₅	2.61₅	1.72₄	1.19₄	2.64₄	1.86₄	1.14₅	Vysotskite	((Pd,Ni)S)	15– 151
c	2.90₈	2.86₅	2.74₅	3.71₄	3.51₄	3.24₄	2.84₄	5.56₅	Kentrolite, ferroan	Pb₂(Mn,Fe)₂Si₂O₉	20– 586
c	2.95₈	2.84₅	2.86₅	3.08₄	2.54₄	2.90₄	3.82₄	3.29₅	Latiumite	(Ca,K)₄(Si,Al)₅O₁₁(SO₄,CO₃)	25–1202
	2.91₈	2.84₅	2.82₅	1.87₄	3.99₄	2.67₄	3.43₄	3.21₅	Svabite	Ca₅(AsO₄)₃(OH,Cl,F)	19– 215
	2.94₈	2.82₅	2.10₅	4.25₄	4.05₄	3.35₄	3.19₄	2.58₅	Eudialyte, yttrian	Na₄(Ca,Ln)₂ZrSi₆O₁₇(OH)₂	25– 814
c	2.93₈	2.82₅	2.51₅	1.80₄	1.75₄	1.81₄	1.52₄	1.54₅	Zirkelite	CaZrTi₂O₇	34– 167
•	2.92₈	2.82₅	2.03₅	1.93₄	1.47₇	3.20₄	1.83₄	1.97₅	Strontium-apatite, syn	Sr₅(PO₄)₃OH	33–1348
	2.90₈	2.82₅	2.80₅	3.98₄	3.47₄	2.68₄	1.88₄	2.03₅	Johnbaumite	Ca₅(AsO₄)₃(OH)	33– 265
	2.89₈	2.82₅	2.78₅	2.14₄	2.05₄	1.99₄	1.84₄	1.83₅	Hilgardite-1Tc, strontian	(Ca,Sr)₂B₅O₈(OH)₂Cl	14– 695
	2.93₈	2.81₅	8.43₅	2.23₄	6.18₄	4.22₄	2.10₄	3.32₅	Kafehydrocyanite, syn	K₄Fe(CN)₆·3H₂O	14– 695
	2.89₈	2.81₅	3.22₅	6.79₄	3.99₄	2.41₄	2.01₄	1.79₅	Lovenite	(Na,Ca,Mn)₃Zr(SiO₄)₂F	14– 586
•	2.92₈	2.79₅	4.21₅	6.12₄	4.36₄	3.44₄	3.19₄	2.45₅	Spodumene	α-LiAlSi₂O₆	33– 786
i	2.92₈	2.79₅	2.02₅	2.06₄	1.48₄	1.44₄	1.21₄	1.66₅	Tin, syn	Sn	4– 673
i	2.89₈	2.79₅	2.59₅	2.68₄	2.67₄	2.40₄	2.29₄	2.10₅	Clinozoisite	Ca₂(Al,Fe)Al₂(SiO₄)₃OH	21– 128
	2.89₈	2.78₅	2.66₅	2.17₄	2.32₄	1.87₄	2.54₄	1.67₅	Barentsite	Na₇AlH₂(CO₃)₄F₂	35– 693
i	2.95₈	2.75₅	3.52₅	3.46₄	2.33₄	3.86₄	2.63₄	4.13₅	Sartorite	PbAs₂S₄	11– 76

Determination of the Unit Cell Size of an Isometric Mineral from Its X-ray Powder Diffraction Pattern

PURPOSE OF EXERCISE

To understand the use of X-ray powder diffraction data for the determination of the unit cell size (in angstrom or nanometer units) of an isometric crystalline substance. This approach is restricted to isometric substances because only in materials of the highest symmetry can powder diffraction data be used for the unambiguous evaluation of unit cell size.

FURTHER READING

See listing in exercise 21, p. 295.

Background Information: In exercise 21 the 2θ angles and corresponding d values were obtained for an isometric mineral. The question now is "Can such d values be related to the Miller indices of the parallel stacks of planes responsible for the diffraction?" Furthermore, if we know the Miller indices, "Can d and hkl be related to unit cell size?" The answer to both these questions is yes, because there are equations that relate d, hkl, and unit cell size. For orthorhombic symmetry this equation is as follows,

$$\frac{1}{d_{hkl}^2} = \frac{h^2}{a^2} + \frac{k^2}{b^2} + \frac{l^2}{c^2}$$

where d is a specific interplanar spacing, a, b, and c are the unit cell edges, and hkl is the Miller index of a specific stack of parallel planes responsible for the diffraction measured at the corresponding d. This relationship becomes much more complicated in nonorthogonal systems (such as monoclinic, and triclinic) but reduces to a very simple expression for the isometric system. Because there is only one cell edge, a, in isometric crystals, the expression becomes

$$\frac{1}{d_{hkl}^2} = \frac{1}{a^2}(h^2 + k^2 + l^2)$$

The foregoing equation relates the measured value of d to a specific unit cell edge and Miller index. Another way of saying this is that spacings in the isometric system are proportional to $(h^2 + k^2 + l^2)^{-1/2}$ and that they differ from one isometric substance to the next only in scale, which is a function of the value of a. This means that the spacing of a specific plane is a linear function of the cell edge, and this relationship can be expressed by a graph as in Fig. 23.1.

As will be outlined in the assignment, the d values (for the substance measured in exercise 21 and transferred to Table 23.2) must be plotted on a strip of paper, according to the horizontal scale of Fig. 23.2. This strip is moved vertically on the graph (keeping it parallel to the horizontal axis) until all marks (representing the various observed d values) coincide simultaneously with a number of the inclined lines. The lines on the graph give the appropriate Miller index* (or indices) of the diffraction planes, and the horizontal intersection on the vertical axis (= cell edge) gives an approximate unit cell size.

The best fit for the d values of gold in Fig. 23.1 shows that, by no means, are all possible hkls (as plotted on the graph) present in the gold pattern. The absence of quite a number of possible hkls is due to (1) random absences, caused by such low X-ray diffraction intensity on the film that a specific line may be missed, and (2) systematic absences, which are the result of space group elements (these are known as "space group extinctions"). X-ray "reflections" are systematically absent because of lattice centering, as occurs in F, I, R, and C (or B, or A) lattices; and because lattice nodes are interleaved owing to glide planes and screw axes. The space group of gold is $Fm3m$, which is equivalent to $F4/m\overline{3}2/m$. Because the lattice type is all-face-centered (F) instead of primitive (P), specific absences ("extinctions") occur in the listing of possible hkls.

Specifically for an F cell, hkl is present only when $h + k = 2n$, $k + l = 2n$, and $h + l = 2n$, where n is an integer including zero. These conditions are equivalent to requiring that h, k, and l be all odd or all even. If you inspect the hkl listing in Fig. 23.1 for gold on the strip, you will note that this is indeed the case.

This very short introduction to some aspects of space group extinctions should serve as a warning that you must not expect all lines (and Miller indices) that are part of the graphical construction in Figs. 23.1 and 23.2 to be present in your own pattern. Indeed, the mineral used in exercise 21 (for which the data are now recorded in Table 23.2) has the same space group as gold, $F4/m\overline{3}2/m$.

As noted earlier, once you have determined a good line fit on the graph, you can read an approximate unit cell edge on the left vertical scale. Subsequently, using the relationship

$$d_{hkl} = \frac{a}{\sqrt{h^2 + k^2 + l^2}}$$

*By now, you may have noticed that the Miller index notation for X-ray diffraction lines consists of hkl values without enclosing parentheses, or brackets. This allows for recognition of X-ray related Miller indices, as distinct from Miller indices for morphological forms.

FIGURE 23.1 An example of the graphical solution of *hkl*s and the unit cell edge for an isometric substance. The strip that is overlaid on the graph shows the measured *d* values for gold, as well as their Miller indices at an *a* value of 4.07 Å. (From E. W. Nuffield, 1966, *X-ray Diffraction Methods,* Wiley, Hoboken, New Jersey, p. 120.)

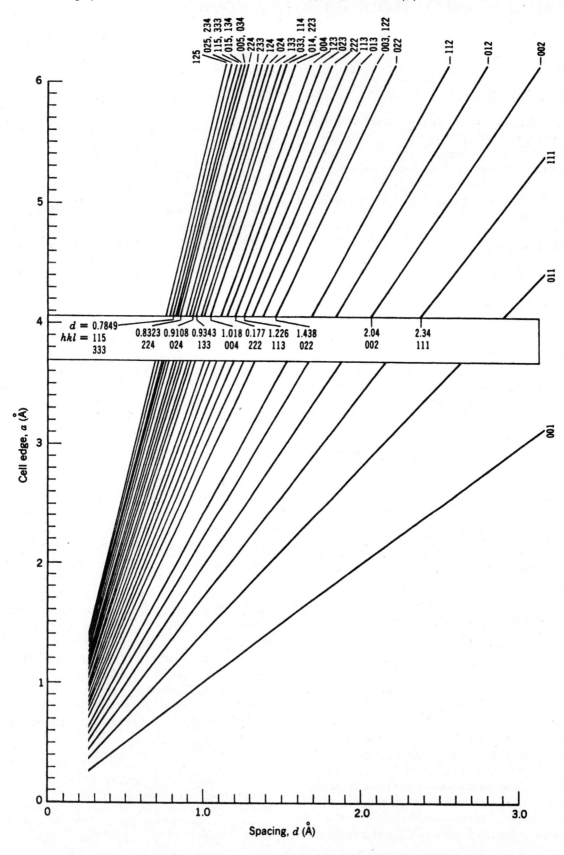

FIGURE 23.2 The enlarged upper portion of the graph in Fig. 23.1 for use in this exercise.

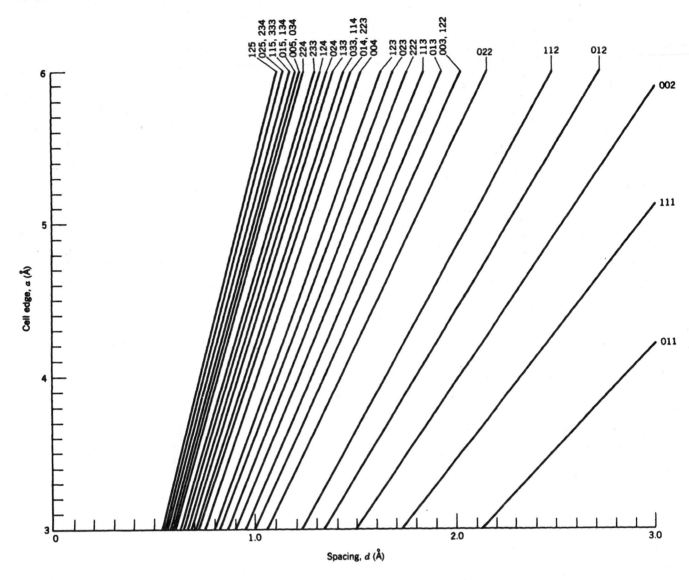

you can calculate an *a* value for each of the *d*s for which you have (with the graph) assigned a corresponding *hkl*. This procedure, instead of giving you one graphically determined *a*, gives you as many calculated *a* values as you have *d* values. Some of these calculated *a* values will be more precise than others. Those determined on the basis of lines at high 2θ values will be more precise than those calculated for low 2θ "reflections". In Table 23.2 you are asked to average the *a*s calculated for the three highest 2θ lines (smallest *d* values) for which you have a Miller index. This allows for a comparison of your graphically determined *a* and calculated *a*. Table 23.1 gives a partial listing of solutions to the quadratic form $h^2 + k^2 + l^2$, which will help you to solve the equation. An exhaustive list of these solutions is given in Azaroff and Buerger (1958), *The Powder Method*, McGraw Hill Book Co., New York, Appendix 1, pp. 269–273.

MATERIALS

If you have measured the film in Fig. 21.2*b* (exercise 21) you should transfer the *d* and intensity data from Table 21.2, part *B* to the third column in Table 23.2. If you have not made your own measurements, you can use the published *d* values as listed in Table 23.2 (second column). You will need a strip of paper on which to plot these *d*s, on a scale identical to the horizontal scale of Fig. 23.2. A pocket calculator with square root function will be helpful.

ASSIGNMENT

This assignment is based on the X-ray diffraction data (*d* values and their intensities) that were obtained for the isometric mineral in exercise 21.

1. Transfer the ds you obtained as a result of your measurement of the film in Fig. 21.2b (as listed in Table 21.2) to the third column in Table 23.2 or use the published d values as listed in the second column of Table 23.2.

2. Use either set of d values and scale the ds on a horizontal paper strip to exactly the same scale as that given in Fig. 23.2 for the horizontal axis of the graph. Notice that the d value of your first line (line 1) will lie to the far right and that decreasing values of d will plot to the left of this first position.

3. Move the carefully marked strip vertically on the graph in Fig. 23.2 (keeping it parallel to the horizontal axis) until many of your marks coincide with many of the inclined lines. When you have such coincidences, you can read off the Miller index that corresponds to each of the d values. The horizontal intersection on the vertical axis will give you a reasonably good value for the unit cell size. Keep in mind that because the substance for which you are resolving the Miller indices (from its interplanar spacings) has a space group with lattice centering, not all possible hkl values will occur in the pattern. The space group of the substance is $F4/m\bar{3}2/m$, which results in all hkls that are not all odd or all even being absent.

4. In Table 23.2 list the appropriate hkl (or hkls if two or three coincide) for each d.

5. In Table 23.2 record the graphically determined unit cell edge.

6. Using the relationship

$$d_{hkl} = \frac{a}{\sqrt{h^2 + k^2 + l^2}}$$

determine an a value for each of the ds for which you now have a Miller index. List these a values in Table 23.2.

7. Average the a values for three lines with the highest 2θ values (that is, the lowest ds) in the pattern. This averaged value will be different from the value you determined graphically. Unit cell edges calculated from high-angle lines in a diffraction pattern tend to be more accurate than those determined from other (lower-angle) lines.

TABLE 23.1 Some Solutions to the Quadratic Form $N = h^2 + k^2 + l^2$ for Isometric Minerals

N	hkl	N	hkl
1	1, 0, 0	26	5, 1, 0; 4, 3, 1
2	1, 1, 0	27	5, 1, 1; 3, 3, 3
3	1, 1, 1	28	
4	2, 0, 0	29	5, 2, 0; 4, 3, 2
5	2, 1, 0	30	5, 2, 1
6	2, 1, 1	31	
7		32	4, 4, 0
8	2, 2, 0	33	5, 2, 2; 4, 4, 1
9	3, 0, 0; 2, 2, 1	34	5, 3, 0; 4, 3, 3
10	3, 1, 0	35	5, 3, 1
11	3, 1, 1	36	6, 0, 0; 4, 4, 2
12	2, 2, 2	37	6, 1, 0
13	3, 2, 0	38	6, 1, 1; 5, 3, 2
14	3, 2, 1	39	
15		40	6, 2, 0
16	4, 0, 0	41	6, 2, 1; 5, 4, 0; 4, 4, 3
17	4, 1, 0; 3, 2, 2	42	5, 4, 1
18	4, 1, 1; 3, 3, 0	43	5, 3, 3
19	3, 3, 1	44	6, 2, 2
20	4, 2, 0	45	6, 3, 0; 5, 4, 2
21	4, 2, 1	46	6, 3, 1
22	3, 3, 2	47	
23		48	4, 4, 4
24	4, 2, 2	49	7, 0, 0; 6, 3, 2
25	5, 0, 0; 4, 3, 0	50	7, 1, 0; 5, 5, 0; 5, 4, 3

TABLE 23.2 Tabulation of Data, *hkl* Assignments, and Unit Cell Size Determinations. Refer to Table 21.2

Line Number	Published[a] *d*	Your Own Measurements of *d*	I[b]	hkl[c]	a[d]
1	3.155		90		
2	1.9316		100		
3	1.6471		40		
4	1.3656		10		
5	1.2533		10		
6	1.1152		20		
7	1.0514		10		
8	0.9658		<10		
9	0.9234		<10		
10	0.8638		15		
11	0.8331		<10		
12	0.8236		<10		
13	n.a.				

[a]From last column in Table 21.2.
[b]From values marked on Fig. 21.2*b*.
[c]As obtained from best fit with the graph in Fig. 23.2.
[d]A value for *a* for each of the *d*s using the quadratic equation and Table 23.1.
n.a. = not available.

Graphically determined *a*: _____

Average of three highest 2θ values; *a* = _____

A. Introduction to the Study of Minerals in Hand Specimens

B. Rock-Forming Silicates: Common Framework Silicates and their Physical Properties

PURPOSE OF EXERCISE

To gain (1) an understanding of how best to study hand specimens of minerals, and (2) an overview of the most diagnostic physical properties and common occurrences of framework silicates.

FURTHER READING AND CD-ROM INSTRUCTIONS

Klein, C. and Dutrow, B., 2008, *Manual of Mineral Science*, 23rd ed., Wiley, Hoboken, New Jersey, pp. 19–36.

CD-ROM, entitled *Mineralogy Tutorials*, version 3.0, that accompanies *Manual of Mineral Science*, 23rd ed., click on "Module IV", and subsequently the button "Silicates". Then click "Tectosilicates" on the next screen.

Nesse, W. D., 2000, *Introduction to Mineralogy*, Oxford University Press, New York, pp. 201–234.

Perkins, D., 2002, *Mineralogy*, 2nd ed., Prentice Hall, Upper Saddle River, New Jersey, pp. 298–310.

Wenk, H. R. and Bulakh, A., 2004, *Minerals: Their Constitution and Origin*, Cambridge University Press, New York, New York, pp. 313–331; 496–508.

Deer, W. A., Howie, R. A. and Zussman, J., 1992, *An Introduction to The Rock-Forming Minerals*, 2nd ed., John Wiley and Sons, Hoboken, New Jersey, pp. 391–520.

Background Information

a. *The study of minerals in hand specimen*. In exercises 1 and 2 you were introduced to various diagnostic physical tests and a chemical test that, when employed together, should lead to a relatively clear-cut identification of an unknown mineral in hand specimen. It is now appropriate to give you some helpful hints on how best to approach hand specimen mineral identification.

Time and effort must be spent in training the eye and other senses about the various properties of minerals. This means that a large variety of minerals must be looked at and handled. Through careful inspection as well as handling of specimens, you will develop the ability to assess the overall geometrical shape of a crystal or mineral specimen. You can then draw specific conclusions about crystal form, cleavage, and the type of aggregation of intergrown mineral grains (details of this are given in exercise 1): At the same time you can assess whether the specimen is metallic or nonmetallic in luster, and what its color is.

Some soft minerals may have a "greasy" feel, and "hefting" (lifting) of a specimen will provide a relative estimate of its specific gravity. Some of the most common minerals such as quartz, feldspar, and calcite have nearly the same specific gravity. Relative to them, gypsum may appear light and barite rather heavy. For a few minerals taste is diagnostic, and occasionally odor is a useful property.

All these assessments can be made without specific tests such as those that evaluate hardness and streak. Indeed, you, the student of mineralogy, must make every effort to stimulate and cultivate your powers of observation. This will take time and commitment; you should observe, study, and handle as many minerals as possible in the laboratory, or elsewhere in your institution. In a mineralogy or hand specimen petrology class or in a sequence of such courses, the laboratory is commonly arranged so that the student can study materials in hand specimens in two different types of collections. One collection might be a *reference* or *display collection* with identification numbers and labels. This collection is generally made of reasonably good- to high-quality specimens and is meant for visual inspection only, not for hardness, streak, and other somewhat destructive testing. The reference collection should be used to study better-quality examples and a larger variety of the same minerals that were tested in the collection of unknowns. Time spent on the reference collection will give you a broader understanding and appreciation for the variation within a specific mineral or mineral group, but you will also learn to recognize the importance of mineral associations. In well-labeled collections the minerals that are associated with the hand specimen on display are generally identified as well. Observations of such associations are commonly helpful in mineral identification. In your independent study of minerals and associations, expand the scope of your observations to any materials on display outside the laboratory. Geology departments commonly have hallway displays of minerals, and sometimes a small mineral museum as well. Additional mineral exhibits may be found in local museums dealing with natural history.

The other collection is commonly a *collection of unknowns*, without identification labels, on which the student can exercise his or her powers of identification. This second collection can be used for specific gravity, cleavage, hardness, and streak tests, all of which lead to some degradation of the original specimen. If both types of collections are available, you should develop your specific testing and mineral recognition skills on the unknowns, with the aid of determinative tables in textbooks.

In this and the subsequent six exercises (exercises 25 through 30), properties of many of the most common rock-forming and most important ore-forming minerals will be highlighted, and only properties of these minerals that are the most diagnostic will be listed. For example, if cleavage is not mentioned, either it is not observed or it is too imperfect to be important. Because almost all minerals are brittle, this property will not be given; however, if the mineral is not brittle but sectile or malleable, these properties will be listed and should therefore be carefully noted. If the streak is not given, it is understood to be white or nearly so, as is the case for most nonmetallic minerals. All metallic minerals are opaque.

The goal of this and the subsequent six exercises is to help you improve your ability in mineral identification. In order to gain a broad base and versatility in mineral recognition, you must never rely on only the properties listed in this exercise manual. While studying a specific mineral, or various members of a mineral group, you must also consult the complete description of that mineral (or mineral group) in the text assigned for the course. Reading the full mineral description in the textbook while at the same time handling one or several specimens of a specific mineral in a collection is the best way to familiarize yourself with a broad range of properties. If some (or all) specimens in your laboratory assignments are given as "unknowns," you will find it most expeditious to use the mineral identification tables provided in several textbooks (e.g., Chapter 22 in *Manual of Mineral Science*, 23rd ed.; Appendix B in *Mineralogy* by Nesse; Appendix B in *Mineralogy* by Perkins; and Appendix 1 in *Minerals, Their Constitution and Origin* by Wenk and Bulakh). Before using such tables, familiarize yourself with the introductory statement to the tables, outlining the manner in which the tables are set up.

You will find that after making various diagnostic tests (as outlined in the first two exercises) and after consulting descriptions in the textbook and comparing these with the specimens to which you have access, it is helpful to make a listing of various aspects of each of the minerals. Such a listing for each mineral can be made on 3 × 5 inch index cards. If you have been provided with a collection of unknown specimens that you are to "curate" for the duration of the course, you can file such cards with the appropriate specimens. Figure 24.1 gives suggestions for the types of entries that you might make. Or you may wish to use a notebook

FIGURE 24.1 Example of the type of listing of important physical properties for the mineral quartz. The most diagnostic properties have been circled.

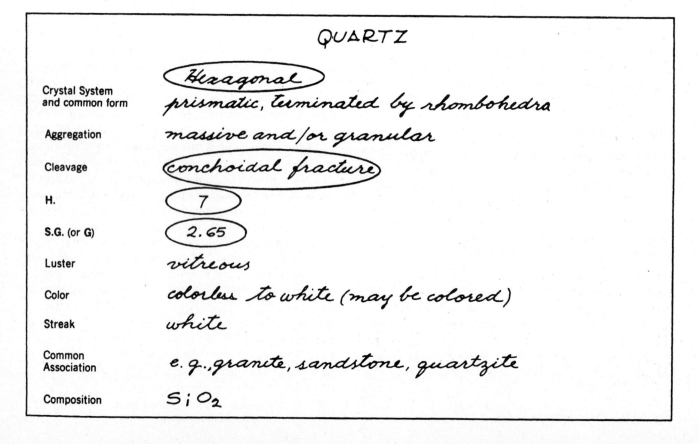

for such listings. In such a notebook you can rule a series of parallel vertical columns, and, to avoid writing the list of properties on each page, they may be written on the edge of the first left-hand page. This vertical strip can be photocopied and the resulting strips can be glued on the various pages on which you plan to record your observations of physical properties. When a property is particularly diagnostic, it can be circled or underscored. It is not worthwhile to repeat in tabular form the complete description given in a textbook. With some experience you will soon learn what should or should not be included in your tabulations. Although you cannot determine the chemical composition of a mineral, it is very important that you become familiar with chemical formulas and ranges of compositions of solid-solution series in minerals. The chemical composition of the minerals you may be studying in the laboratory is not included in the present exercise, nor in the following six exercises. Nor is there any reference to the amount of substitution of elements that may occur in the minerals that you study in the laboratory. For all such information you must consult the text that your instructor assigned for the course. It is therefore a good learning experience to enter a statement about composition on your index card or in your tabulation.

It is not necessary to try to commit to memory all the diagnostic properties; this is not only impossible but tiresome as well. You will acquire a sense of the important physical properties by repeated observation and handling of specimens. But you will need to memorize such things as chemical formulas and compositional ranges as well as common mineral associations.

b. *Diagnostic properties of the framework silicates.* The study of hand specimens of minerals begins with framework silicates because they represent several of the most common rock-forming silicates; they include *quartz*, the *feldspars*, *feldspathoids*, and *zeolites*. These framework silicates are also known as tectosilicates, from the Greek root *tekton*, meaning builder. It is instructive to supplement any laboratory session dealing with specific minerals and mineral groups with representative models of the appropriate crystal structures (if available). Descriptions and illustrations of crystal structures are not repeated here but are available in the various textbooks referenced above.

MATERIALS

Representative collections of the framework silicates, which include quartz (and its microcrystalline and granular varieties such as chalcedony, chert, and so on), feldspars (microcline, orthoclase, sanidine, and members of the plagioclase series), felspathoids (such as leucite, nepheline, and sodalite), scapolite, analcime, and zeolites (such as natrolite, chabazite, and stilbite). Basic testing tools for hardness, specific gravity, and streak as well as a hand lens, a binocular microscope or both are needed. Index cards or a notebook must be available for abstracting the most important data for each mineral (see Fig. 24.1).

ASSIGNMENT

Hand specimen study of the various specimens available of the framework silicates. Consult the listing of highly diagnostic physical properties that follows, as well as determinative tables and mineral descriptions in the textbook. The last two pages in this exercise provide two different types of tables (Tables 24.1 and 24.2) in which you can enter your observations on physical properties, as well a some other data. Your instructor may also use the last table for testing your observational skills of unknown samples in the laboratory.

DIAGNOSTIC ASPECTS OF MINERALS BELONGING TO THE FRAMEWORK SILICATE GROUP

QUARTZ

Crystal form:	Hexagonal, prismatic with horizontal striations on prism faces; terminations are usually a combination of a positive and a negative rhombohedron.
H:	7
G:	2.65
Fracture:	Conchoidal
Luster:	Vitreous
Color:	Usually colorless or white (milky). Colored varieties are *amethyst*—purple; *rose quartz*—light pink; *smoky quartz*—smoky brown to black; *citrine*—light yellow.
Occurrence:	A common rock-forming mineral. Examples of rock types in which it is abundant: granite, granodiorite, pegmatites, quartz–mica schists, quartzites, and sandstones. Quartz

is absent from SiO_2-poor igneous rocks (subsiliceous) in which feldspathoids (e.g., nepheline or leucite) occur; in other words, if nepheline or leucite is recognized in the rock, quartz cannot be present.

Microcrystalline varieties (fibrous): *Chalcedony* is the general term for fibrous, very fine-grained quartz

carnelian–a red-brown chalcedony

chrysoprase–an apple-green chalcedony

onyx–a layered chalcedony with parallel layers

agate–various layers of chalcedony with different colors

silicified wood–wood that has been replaced by clouded agate

Microcrystalline varieties (granular): *chert*–light-colored, in bedded deposits of very fine grained quartz

flint–dark-colored siliceous nodules are common in chalk beds; there is no sharp distinction between *chert* and *flint*

jasper–very fine-grained quartz colored red by traces of hematite

CRISTOBALITE

Crystal form:	Small octahedrons in small cavities in lavas.
H:	$6\frac{1}{2}$
G:	2.32
Color:	Colorless
Occurrence:	Occurs as linings of spherical aggregates in small cavities of siliceous lavas; present also in the matrix of such rocks, but can be identified with certainty only in a thin section under a microscope.

OPAL

Crystal form:	Often botryoidal and stalactitic, massive.
H:	5–6 (less than quartz)
Luster:	Vitreous, somewhat resinous
Color:	Colorless, white, pale shades of yellow, green, gray, red, and blue. Brilliant *play of color* in precious opal.
Occurrence:	Deposited from meteoric waters or by hot springs and as such may be found as linings in cavities.
Varieties:	*common opal*–milky white and other colors without internal reflections
	precious opal–brilliant internal play of colors
	hyalite–clear and colorless opal with a globular or botryoidal surface
	geyserite or *siliceous sinter*–deposited from hot springs, as in Yellowstone National Park
	wood opal–silicified fossil wood with opal as the petrifying material
	diatomite–very fine-grained, resembling chalk in appearance. The result of accumulation on the seafloor of siliceous tests of diatoms.

Feldspars: Microcline, Orthoclase, and Sanidine

The three polymorphs of $KAlSi_3O_8$, microcline, orthoclase, and sanidine, all have essentially the same composition, but they have subtly different atomic structures as a function of their temperature of crystallization. Microcline is the lowest-temperature form with triclinic symmetry; orthoclase is a medium-temperature form, with monoclinic symmetry; and sanidine is the high-temperature form, also with monoclinic symmetry. Microcline and orthoclase are common in plutonic igneous rocks and in metamorphic assemblages, as well as in sedimentary rocks. Sanidine is restricted in occurrence to high-temperature extrusive rocks. All three polymorphs have similar hardness (**H** = 6), perfect {001} and good {010} cleavage (at approximately right angles to each other), and **G** ranging from 2.54 to 2.62.

MICROCLINE

Crystal form:	Triclinic, usually short prismatic, frequently twinned according to Carlsbad law. Common in coarse, cleavable masses. Crystals and cleavage identical to those of orthoclase.
Cleavage:	{001} perfect, {010} good at 89°30' to each other
Color:	White to pale yellow. Green variety known as *amazonstone.*
Occurrence:	Common constituent of granites and syenites that have cooled slowly; in sedimentary sandstone, arkose, and conglomerate and in metamorphic gneiss. Large crystals and cleavage masses in pegmatites.
Remarks:	Cannot be distinguished from orthoclase in hand specimen. However, if the feldspar is deep green or contains abundant perthite lamallae, it is likely to be microcline.

ORTHOCLASE

Color:	Colorless, white, gray and flesh red (owing to traces of hematite).
Occurrence:	Major constituent of granites, granodiorites, and syenites that cooled at reasonably fast rates.
Remarks:	Lack of striations (representing albite twin lamallae) on the cleavage faces distinguishes it from members of the plagioclase feldspar series.

SANIDINE

Remarks:	Sanidine can be positively identified only by optical and X-ray techniques. However, colorless to white monoclinic phenocrysts in extrusive igneous rocks such as rhyolites are characteristically sanidine. Sanidine, being the high-temperature polymorph of $KAlSi_3O_8$, is characteristic of extrusive igneous rocks that cooled quickly from an initially high eruption temperature.

THE PLAGIOCLASE FELDSPAR SERIES: ALBITE TO ANORTHITE

Crystal form:	Distinct crystals are rare. Usually in twinned, cleavable masses and as irregular grains in igneous rocks.
Twinning:	Polysynthetic *albite twinning* is the diagnostic property in this series. The {001} crystal face or cleavage direction will show finely spaced parallel groovings or striations. The striations are commonly very finely spaced, so that a hand lens or binocular microscope should be used to identify them. Albite twinning occurs in all members of the plagioclase series.
Cleavage:	{001} perfect and {010} good, at about 86° to each other
H:	6
Color:	From colorless to white to light and dark gray. A general darkening occurs in specimens that become more anorthite-rich; although albite is commonly white, more Anrich members are generally various shades of gray. Good *play of colors* is diagnostic of labradorite and andesine.
Occurrence:	As a major rock-forming mineral group, plagioclase feldspars are even more common than potash feldspar. They occur in igneous, metamorphic, and more rarely sedimentary rocks.

The classification of igneous rocks is based largely on the kind and amount of feldspar present. As a result, the greater the percentage of SiO_2 in a rock, the fewer the dark minerals, the greater the amount of potash feldspar, and the more sodic the plagioclase; and conversely, the lower the percentage of SiO_2, the greater the percentage of dark minerals and the more calcic the plagioclase.

Varieties: Members of the series are defined in terms of their anorthite (An) content: *albite* (An_{0-10}), *oligoclase* (An_{10-30}), *andesine* (An_{30-50}), *labradorite* (An_{50-70}), *bytownite* (An_{70-90}), and *anorthite* (An_{90-100}).

Moonstone is a variety of albite and oligoclase that shows an opalescent play of colors

Feldspathoids

As the name implies, feldspathoids are chemically and structurally closely related to the feldspars. Feldspathoids tend to form from igneous melts that are rich in alkalis (Na and K) and poor in SiO_2. Feldspathoids contain only about two-thirds as much silica as alkali feldspar. They generally show much less well-developed cleavage than the feldspars do. They have a hardness that ranges from $5\frac{1}{2}$ to 6, and leucite and nepheline have light colors.

LEUCITE

Crystal form:	Usually in trapezohedral crystals embedded in a fine-grained matrix of silica-poor lavas.
H:	$5\frac{1}{2}-6$
G:	2.47 (considerably lower than that of quartz or feldspar)
Luster:	Vitreous to dull; translucent
Color:	White to gray
Occurrence:	Abundant as phenocrysts in the fine-grained matrix of some recent silica-deficient lava flows.
Remarks:	If quartz has been identified in the rock, there cannot be leucite, and vice versa. Feldspathoids are found only in SiO_2-poor rocks, that is, without quartz.

NEPHELINE

Crystal form:	Rarely in crystals; generally massive, compact, or as embedded grains.
Cleavage:	$\{10\bar{1}0\}$ distinct
H:	$5\frac{1}{2}-6$
G:	2.60–2.65
Luster:	Vitreous in clear material; greasy in the massive variety.
Occurrence:	A rock-forming mineral in SiO_2-deficient intrusive and extrusive igneous rocks.
Remarks:	If you have positively identified quartz as part of the assemblage, there cannot be any nepheline, and vice versa. It is distinguished from quartz by the lower hardness and from plagioclase by the lack of albite twins.

SODALITE

Crystal form:	Most commonly massive; crystals rare.
Cleavage:	{011} poor
H:	$5\frac{1}{2}-6$
G:	2.15–2.3
Luster:	Vitreous
Color:	Usually blue, but also white and gray, or green. Translucent.
Occurrence:	Associated with nepheline and other feldspathoids in nepheline syenite, trachyte, and phonolite.
Remarks:	Usually recognized by its blue color. Because of the common association of nepheline and sodalite, as soon as you see this distinctly blue mineral, look for nepheline and realize that you are dealing with an SiO_2-deficient assemblage. Therefore quartz cannot be present.

LAZURITE

Crystal form:	Usually massive or in foliated cleavable masses.
Cleavage:	{011} imperfect
H:	$5-5\frac{1}{2}$
G:	2.4–2.45
Luster:	Vitreous, translucent
Color:	Deep azure blue, greenish blue
Occurrence:	A rare mineral occurring usually in crystalline limestone as a product of contact metamorphism. *Lapis lazuli* is a mixture of lazurite, with small amounts of calcite, pyroxene, and other silicates, and commonly some pyrite.
Remarks:	The blue color and association with pyrite is characteristic.

SCAPOLITE

Crystal form:	Commonly as coarse prismatic crystals with a faint fibrous appearance.
Cleavage:	{100} and {110} imperfect but distinct
H:	5–6
G:	2.55–2.74
Luster:	Vitreous when fresh
Color:	White, gray, pale green; transparent to translucent.
Occurrence:	Common in crystalline schists, gneisses, amphibolites, and granulite facies metamorphic rocks. Also in crystalline limestones as a product of contact metamorphism. Common associations: diopside, garnet, apatite, and amphibole.
Remarks:	Characterized by crystals with a square cross section and four cleavage directions at 45°. When massive, resembles feldspar but has a characteristic fibrous appearance on the cleavage surfaces.

ANALCIME

Crystal form:	Usually in trapezohedral crystals or in combinations of cubes and trapezohedral truncations. Also massive.
H:	$5-5\frac{1}{2}$
G:	2.27
Color:	Colorless to white; transparent to translucent.
Occurrence:	A primary mineral in some igneous rocks and also the product of hydrothermal action in the filling of cavities in basaltic flows. Examples are analcime basalts and as a lining of vesicles in association with calcite and zeolites.
Remarks:	Usually recognized by free-growing crystals and vitreous luster. Crystals may resemble leucite, but remember that leucite crystals are always embedded in the rock matrix.

Zeolites

The zeolites constitute a large group of hydrous framework silicates in which the tetrahedral (SiO_4 and AlO_4) linkages are much more open than those of feldspars and feldspathoids. These openings, commonly forming interconnecting spaces and channels, house variable amounts of H_2O. Because of these structural features, all zeolites have a hardness that is considerably less than that of quartz and feldspar (common **H** ranges from $3\frac{1}{2}$ to 5) and S.G. values that are relatively low as well (**G** ranging from 2.05 to 2.25).

NATROLITE

Crystal form:	Usually in radiating groups of acicular, vertically striated crystals. Also fibrous, massive, granular, and compact.
Cleavage:	{110} perfect
H:	$5-5\frac{1}{2}$
G:	2.25 (low compared with quartz and feldspar)

Color:	Colorless to white; transparent to translucent.
Occurrence:	Characteristically found lining cavities in basaltic rocks associated with other zeolites and calcite.

CHABAZITE

Crystal form:	Usually in rhombohedral crystals.
Cleavage:	Rhombohedral $\{10\bar{1}1\}$ poor
H:	4–5
G:	2.05–2.15 (low as compared with quartz and feldspar)
Luster:	Vitreous
Color:	White, pink, red; transparent to translucent.
Occurrence:	Usually in association with other zeolites, lining cavities in basalts.
Remarks:	Generally recognized by rhombohedral crystals, and distinguished from calcite by its poorer cleavage and lack of effervescence in HCl.

STILBITE

Crystal form:	Usually in crystals that are tabular parallel to {010} or in sheaflike aggregates.
Cleavage:	{010} perfect
H:	$3\frac{1}{2}-4$
G:	2.1–2.2 (low as compared with quartz and feldspar)
Luster:	Vitreous; pearly on {010}.
Color:	White, rarely yellow, brown, or red; translucent.
Occurrence:	Associated with other zeolites in cavities in basaltic rocks.
Remarks:	Characterized chiefly by its cleavage, pearly luster on cleavage faces, and common sheaflike groups of crystals.

TABLE 24.1   Tabulation of physical properties and subsequent mineral identification for three unknowns.

Specimen number: . . . .

Crystal Morphology/Symmetry/Twinning/Striations:

Cleavage/Parting/Fracture:

Hardness:

Luster:

Streak:

Color(s)

Tenacity:

Specific gravity (heft):

Other distinguishing features (effervescence in HCl, magnetism, smell, exsolution lamellae, etc.):

Name:

Mineral classification; Formula:

Association:

Specimen number: . . . .

Crystal Morphology/Symmetry/Twinning/Striations:

Cleavage/Parting/Fracture:

Hardness:

Luster:

Streak:

Color(s)

Tenacity:

Specific gravity (heft):

Other distinguishing features (effervescence in HCl, magnetism, smell, exsolution lamellae, etc.):

Name:

Mineral classification; Formula:

Association:

Specimen number: . . . .

Crystal Morphology/Symmetry/Twinning/Striations:

Cleavage/Parting/Fracture:

Hardness:

Luster:

Streak:

Color(s)

Tenacity:

Specific gravity (heft):

Other distinguishing features (effervescence in HCl, magnetism, smell, exsolution lamellae, etc.):

Name:

Mineral classification; Formula:

Association:

Please make photocopies of this form if more copies are needed.

Copyright John Wiley & Sons, Inc.

333

TABLE 24.2 Tabulation of physical properties and subsequent mineral identification for ten unknowns.

Specimen number	Luster (metallic versus non-metallic)	Hardness	Cleavage or fracture	Crystal form	Color	Other (streak, texture, striations)	Mineral Name

Please make photocopies of this form if more copies are needed.

Rock-Forming Silicates: Common Chain Silicates and their Physical Properties

PURPOSE OF EXERCISE

An overview of some of the most diagnostic physical properties and common occurrences of chain silicates.

FURTHER READING AND CD-ROM INSTRUCTIONS

Klein, C. and Dutrow, B., 2008, *Manual of Mineral Science,* **23rd ed., Wiley, Hoboken, New Jersey, pp. 505–519.**

CD-ROM, entitled *Mineralogy Tutorials,* **version 3.0, that accompanies** *Manual of Mineral Science,* **23rd ed., click on "Module IV", and subsequently click the button "Silicates". On the next screen click "Inosilicates".**

Nesse, W. D., 2000, *Introduction to Mineralogy,* Oxford University Press, New York, pp. 261–331.

Perkins, D., 2002, *Mineralogy,* 2nd ed., Prentice Hall, Upper Saddle River, New Jersey, pp. 321–331.

Wenk, H. R. and Bulakh, A., 2004, *Minerals: Their Constitution and Origin,* Cambridge University Press, New York, New York, pp. 470–483.

Deer, W. A., Howie, R. A. and Zussman, J., 1992, *An Introduction to The Rock-Forming Minerals,* 2nd ed., John Wiley and Sons, Hoboken, New Jersey, pp. 143–232.

Background Information: The study of hand specimens of minerals continues with the chain silicate group because this group includes the very common mineral groups known as *pyroxenes* and *amphiboles,* as well as *pyroxenoids.*

Associations of quartz and feldspar and pyroxenes and amphiboles are very common in many rock types. The chain silicates are also known as the inosilicates, from the Greek root *inos,* meaning thread or fiber.

MATERIALS

Representative collections of members of the chain silicates, which include, among the pyroxenes, enstatite, ferrosilite, pigeonite, diopside, hedenbergite, augite, jadeite, aegirine, and spodumene; among the pyroxenoids, wollastonite, rhodonite, and pectolite; and among the amphiboles, anthophyllite, cummingtonite and grunerite, tremolite, actinolite, hornblende, glaucophane, and riebeckite. Basic testing tools for hardness, specific gravity, and streak as well as a hand lens, a binocular microscope, or both are needed. Index cards or a notebook must be available for abstracting the most important data for each mineral (see Fig. 24.1). Tables 25.1 and 25.2 can also be used.

ASSIGNMENT

Hand specimen study of the various specimens available of the chain silicates. Consult the listing of highly diagnostic physical properties that follows, as well as determinative tables and mineral descriptions in the various textbooks listed above.

DIAGNOSTIC ASPECTS OF MINERALS BELONGING TO THE CHAIN SILICATE GROUP

Pyroxenes

Because the internal structures of members of the pyroxene group, of the pyroxenoid group, and of the amphibole group are based on the presence of infinitely extending chains, all three major groups share some common characteristics, such as prismatic cleavage, hardness, and luster. For pyroxenes, if they do not occur in euhedral crystals, the prismatic cleavage angles of about 92° to 93° and 87° to 88° are the most diagnostic property. In other words, these prismatic cleavages (such as {210} and {110}) show approximately rectangular intersections. (It is commonly necessary to use a hand lens or a binocular microscope to observe the cleavage directions and to estimate the angles between them.) Although there is generally some elongation along the *c* axis in pyroxene crystals and cleavage fragments, the elongation is not highly pronounced; indeed, crystals and cleavage fragments tend to be stubby (instead of elongate or even fibrous as in the amphiboles). The average hardness of pyroxenes ranges from about 5 to 6. Specific gravity shows a considerable range, because of the extensive solid solution, from about 3.2 to 3.6.

ENSTATITE–FERROSILITE (ORTHOPYROXENE SERIES)

Crystal form and cleavage:	Most commonly massive, showing typical pyroxene (about 90°) cleavages identified as {210}.
Color:	Light beige in enstatite; somewhat darker brown with a submetallic, bronzelike luster in hypersthene. The common brownish and olive-green colors intensify with increasing Fe content of members of this series.

Occurrence: Orthopyroxenes are common constituents of mafic igneous types such as gabbros, peridotites; norites, and basalts. They are also present in pyroxenites and in metamorphic rocks that have originated at high temperature and pressure, such as granulite facies rocks. Commonly found in coexistence with clinopyroxenes in all the foregoing rock types. Also present in iron and stony meterorites.

Remarks: In hand specimen it is generally impossible to distinguish the various species within the orthopyroxene series. Darker color commonly indicates higher Fe content. If a greenish pyroxene coexists with an orthopyroxene, it is most likely diopside or augite.

PIGEONITE

Remarks: A monoclinic pyroxene that is close to the compositional range of the orthopyroxenes, but somewhat more calcic in composition. Only careful optical or X-ray techniques will allow its identification. It is common in high-temperature, rapidly cooled lavas.

DIOPSIDE–HEDENBERGITE AND AUGITE (CLINOPYROXENE SERIES)

Crystal form: In prismatic, monoclinic crystals with square or eight-sided cross sections.

Cleavage: Typical pyroxene cleavage of about 90° for {110}.

Color: White to light green for essentially pure end-member diopside ($CaMgSi_2O_6$); darker green varieties reflect increasing Fe content. Augite is commonly very dark green to black.

Occurrence: Diopside and hedenbergite are common in metamorphosed Ca-rich rocks, such as metamorphosed siliceous limestones and dolomites. Both can also be part of igneous assemblages. Augite is the most common pyroxene and may be a major constituent of dark-colored igneous rocks such as basaltic lavas and plutonic gabbros.

Remarks: If the pyroxene is white or light green and is part of a calcite-rich metamorphic assemblage, it is probably close to diopside in composition. A green pyroxene (commonly in coexistence with an orthopyroxene) in igneous rocks tends toward hedenbergite or augite. Very dark to black pyroxenes are augite.

JADEITE

Remarks: Rarely in crystals. Usually in granular, compact, or massive apple-green to emerald-green aggregates. May also be white with irregular spots of green. Some aggregates show fibrous texture. Jadeite is one of the main constituents of the ornamental stone known as *jade*.

AEGIRINE

Crystal form: Slender prismatic; often in fibrous aggregates.

Color: Brown to green

Occurrence: Most commonly found in rocks rich in Na and poor in SiO_2 such as nepheline syenite.

Remarks: Slender prismatic crystal habit, brown to green color, and mineral associations in SiO_2-poor rocks are diagnostic.

SPODUMENE

Crystal form: Prismatic crystals commonly deeply striated vertically. Often in coarse crystals and cleavable masses.

Cleavage: Perfect {110} cleavage at about 90° and well-developed parting on {100}.

Color: White, gray, pink, yellow, and green.

Occurrence: A major constituent of Li-rich pegmatites; therefore commonly in coarse cleavable masses.

Remarks: The commonly coarse size of its grains, light color (white through pink and gray), and good cleavage (as well as parting) are characteristic. Association with other Li-rich minerals such as lepidolite (pink Li-mica) is highly diagnostic.

Pyroxenoids

The pyroxenoid minerals, like those of the pyroxene group, are based on the architecture of infinitely extending SiO_3 chains. However, the chains in pyroxenoids have a much lower symmetry than those of pyroxenes. As a result, they are triclinic, instead of monoclinic or orthorhombic as are the pyroxenes.

WOLLASTONITE

Crystal form: Commonly massive, cleavable to fibrous. Rarely in tabular, triclinic crystals with either {001} or {100} prominent.

Cleavage: {100} and {001} perfect, at 84° angles; {$\overline{1}01$} good.

H: $5–5\frac{1}{2}$

G: 2.8–2.9

Color: Colorless, white, or gray. Luster vitreous, pearly on cleavage surfaces.

Occurrence: Chiefly a contact-metamorphosed mineral in crystalline limestones and marbles. Therefore, in association with calcite, tremolite, diopside, and so on.

Remarks: Resembles tremolite but distinguished from it by the 84° cleavage angle. Decomposed by HCl without effervescence.

RHODONITE

Crystal form:	Triclinic crystals commonly tabular or parallel to {001}. Generally massive, cleavable to compact.
Cleavage:	{110} and {1$\bar{1}$0} perfect at 90° to each other.
H:	5$\frac{1}{2}$–6
G:	3.4–3.7
Color:	Rose red, pink to brown; frequently with a black exterior owing to manganese oxide.
Occurrence:	Common in metamorphosed Mn-rich deposits and assemblages, resulting in assemblages with rhodochrosite, spessartine garnet, and black manganese oxides.
Remarks:	Characterized by its pink color and near 90° cleavages. Distinguished from rhodochrosite by greater hardness and insolubility in HCl.

PECTOLITE

Crystal form:	Usually in aggregates of acicular crystals; radiating with a fibrous appearance. Also in compact masses.
Cleavage:	{001} and {100} perfect
H:	5
G:	About 2.8
Color:	Colorless, white to gray. Luster vitreous to silky.
Occurrence:	Most commonly as linings in basalt cavities, in association with various zeolites.
Remarks:	Characterized by two directions of perfect cleavage, yielding sharp, acicular fragments, which will puncture the skin if not handled carefully. May resemble wollastonite.

Amphiboles

The atomic structures of the amphibole group minerals are based on the packing of infinitely extending chains (as in pyroxenes) that are twice the width of those in pyroxene structures. The amphibole chains are, therefore, known as "double chains," and because of their doubled width (over the chains in pyroxenes), the excellent cleavage around the chains shows distinctly different angles from those of the pyroxenes. The excellent prismatic cleavage (in orthoamphiboles described as {210} and in clinoamphiboles described as {110}) results in angles of about 56° and 124° (as compared with approximately right-angle cleavage in pyroxenes). It is commonly necessary to use a hand lens or a binocular microscope to observe the cleavage directions and to estimate the angles between them. In addition to these highly diagnostic cleavage angles, you will find that the general habit of amphiboles is more elongate (at times, bladed or even fibrous) than that of the more stubby pyroxenes. All amphiboles have a hardness somewhere between **H** = 5 and 6, thus overlapping with the hardness of pyroxenes. The specific gravity of the amphibole group ranges from 2.85 to 3.6 as a result of the extensive chemical substitutions in the group.

ANTHOPHYLLITE

Crystal form:	Rarely in distinct crystals. Commonly lamellar or fibrous.
Color:	Gray, beige, and clove brown; also some shades of greenish brown.
Occurrence:	A metamorphic mineral in Mg-rich rocks such as ultrabasic igneous rocks and impure dolomitic shales. May be associated with cordierite in gneisses and schists.
Remarks:	Clove-brown color and fibrous to lamellar habit are the most diagnostic properties. Cannot be distinguished with certainty from cummingtonite or grunerite without optical or X-ray tests.

CUMMINGTONITE – GRUNERITE

Crystal form:	Rarely in distinct crystals. Commonly fibrous or lamellar, often radiated. The rare abestiform variety is known as *amosite*.
Color:	Various shades of light to dark brown.
Occurrence:	Cummingtonite (the more magnesian member of this clinoamphibole series) is common in regionally metamorphosed rocks such as amphibolites. Commonly coexists with a greenish amphibole which may be actinolite or hornblende. Grunerite (the more Fe-rich member of this series) is the most common in metamorphosed iron-formation assemblages, in association with, for example, magnetite.
Remarks:	Light brown color, fibrous and radiating habits are most diagnostic. Both cummingtonite and grunerite may be difficult to distinguish from anthophyllite without optical or X-ray tests. The coexistence with magnetite in metamorphosed iron-formations is highly diagnostic of grunerite.

TREMOLITE – ACTINOLITE – FERROACTINOLITE AND HORNBLENDE

The Ca-rich clinoamphiboles tremolite, actinolite, and ferroactinolite are all part of the Ca–Mg–Fe series. Tremolite is the essentially Fe-free end-member, actinolite generally contains several weight percent FeO, and ferroactinolite is the Ca–Fe end-member. Increasing coloration in this series (from white in tremolite, through light green in actinolite, to dark green in ferroactinolite) is a function of increasing Fe content. Hornblende is the most common Ca-rich clinoamphibole, with additional Al as well as alkali atoms in its structure. Hornblende is commonly very dark green to black.

TREMOLITE

Crystal form:	Prismatic crystals uncommon. Commonly bladed and in radiating columnar aggregates.
Color:	White to greenish white

Occurrence: Common in metamorphosed dolomitic limestones. May be associated with diopside. May be asbestiform.

ACTINOLITE

Crystal form: Commonly in columnar aggregates

Color: Various shades of light to dark green

Occurrence: A common mineral in greenschist facies metamorphic rocks.

HORNBLENDE

Crystal form: Prismatic, monoclinic crystals; also in massive coarse- to fine-grained aggregates.

Color: Various shades of dark green to black

Occurrence: Hornblende is the most common species of amphibole, occurring as a major constituent in both igneous and metamorphic rocks. It is particularly characteristic of medium-grained metamorphic rocks known as amphibolites. May occur as a medium-temperature reaction rim around higher temperature augite, both in igneous and metamorphic assemblages.

Remarks: Any of the medium to dark green to black amphiboles are easily distinguished from similar pyroxenes by the typical amphibole cleavage angles.

GLAUCOPHANE

Crystal form: May be in slender acicular crystals; most commonly aggregated.

Color: Blue to lavender blue

Occurrence: In metamorphic rocks such as schists, eclogite, and marble. May be associated with jadeite. A major constituent of glaucophane schists in the Franciscan Formation of California.

RIEBECKITE

Crystal form: Slender acicular crystals and aggregated. The asbestiform variety is known as *crocidolite*.

Color: Blue to dark blue, and blue-black

Occurrence: The nonasbestiform variety of riebeckite is a common accessory mineral in igneous rocks, such as granites, syenites, nepheline syenites, and pegmatites. Crocidolite, the asbestiform variety, is mainly restricted to sedimentary iron-formations in South Africa and Western Australia; here in association with chert, magnetite, and hematite.

TABLE 25.1. Tabulation of physical properties and subsequent mineral identification for three unknowns.

Specimen number: . . . .

Crystal Morphology/Symmetry/Twinning/Striations:

Cleavage/Parting/Fracture:	Color(s)
Hardness:	Tenacity:
Luster:	Specific gravity (heft):
Streak:	Other distinguishing features (effervescence in HCl, magnetism, smell, exsolution lamellae, etc.):

Name:

Mineral classification; Formula:

Association:

Specimen number: . . . .

Crystal Morphology/Symmetry/Twinning/Striations:

Cleavage/Parting/Fracture:	Color(s)
Hardness:	Tenacity:
Luster:	Specific gravity (heft):
Streak:	Other distinguishing features (effervescence in HCl, magnetism, smell, exsolution lamellae, etc.):

Name:

Mineral classification; Formula:

Association:

Specimen number: . . . .

Crystal Morphology/Symmetry/Twinning/Striations:

Cleavage/Parting/Fracture:	Color(s)
Hardness:	Tenacity:
Luster:	Specific gravity (heft):
Streak:	Other distinguishing features (effervescence in HCl, magnetism, smell, exsolution lamellae, etc.):

Name:

Mineral classification; Formula:

Association:

TABLE 25.2. Tabulation of physical properties and subsequent mineral identification for ten unknowns.

Specimen number	Luster (metallic versus non-metallic)	Hardness	Cleavage or fracture	Crystal form	Color	Other (streak, texture, striations)	Mineral Name

Please make photocopies of this form if more copies are needed.

Rock-Forming Silicates: Common Layer Silicates and their Physical Properties

PURPOSE OF EXERCISE

An overview of some of the most diagnostic physical properties and common occurrences of layer silicates.

FURTHER READING AND CD-ROM INSTRUCTIONS

Klein, C. and Dutrow, B., 2008, *Manual of Mineral Science*, 23rd ed., Wiley, Hoboken, New Jersey, pp. 519–534.

CD-ROM, entitled *Mineralogy Tutorials*, version 3.0, that accompanies *Manual of Mineral Science*, 23rd ed., click on "Module IV", and subsequently click on the button "Silicates". On the next screen click on "Phyllosilicates". You may also want to view, in Module I, "Architecture of Layer Silicates".

Nesse, W. D., 2000, *Introduction to Mineralogy*, Oxford University Press, New York, pp. 235–260.

Perkins, D., 2002, *Mineralogy*, 2nd ed., Prentice Hall, Upper Saddle River, New Jersey, pp. 312–321.

Wenk, H. R. and Bulakh, A., 2004, *Minerals: Their Constitution and Origin*, Cambridge University Press, New York, New York, pp. 448–469.

Deer, W. A., Howie, R. A. and Zussman, J., 1992, *An Introduction to The Rock-Forming Minerals*, 2nd ed., John Wiley and Sons, Hoboken, New Jersey, pp. 279–384.

Background Information: The study of hand specimens of minerals continues with the layer silicate group, which includes the very common mineral groups known as the *serpentines*, the *clays*, the *micas*, and the *chlorites*. Any of these layer silicates can be major constituents of various rock types or accessory minerals in a wide range of associations. The layer silicates are also known as the phyllosilicates, from the Greek word *phyllon*, meaning leaf.

MATERIALS

Representative collections of members of the layer silicates, which include, among the serpentines, antigorite and chrysotile; among the clay mineral group, kaolinite; talc; among the micas, muscovite, phlogopite, biotite, and lepidolite; chlorite, apophyllite, prehnite, and chrysocolla. Basic tools for testing hardness, specific gravity, and streak as well as a hand lens, a binocular microscope, or both are needed. Index cards or a notebook must be available for abstracting the most important data for each mineral (see Fig. 24.1). Tables 26.1 and 26.2 can also be used.

ASSIGNMENT

Hand specimen study of the available specimens of various layer silicates. Consult the listing of highly diagnostic physical properties that follows, as well as the determinative tables and mineral descriptions in one or several textbooks listed above.

DIAGNOSTIC ASPECTS OF MINERALS BELONGING TO THE LAYER SILICATE GROUP

The atomic structure of the layer silicates is based on infinitely extending sheets of SiO_4 and AlO_4 tetrahedra. The flat nature of these layers is responsible for the most diagnostic properties of this mineral group, namely their flaky or platy habit and pronounced planar cleavage. They are generally soft, are of relatively low specific gravity, and may show flexibility or even elasticity of the cleavage lamellae.

Serpentine Minerals: Antigorite and Chrysotile

Both antigorite and chrysotile, polymorphs of $Mg_3Si_2O_5(OH)_4$, have a range of hardness, **H** = 3 to 5, but usually 4, and they are commonly intergrown with each other in hand specimens. *Antigorite* is commonly massive and fine-grained, whereas *chrysotile* is fibrous. Both range in color from light to darker green.

ANTIGORITE

Crystal form:	Massive, fine-grained
Luster:	Greasy, waxlike in massive varieties
Color:	Often variegated, showing mottling of lighter and darker shades of green. Translucent.
Occurrence:	Most common as an alteration product of magnesium-rich silicates, such as olivine, pyroxene, and amphibole. The rock known as *serpentine* commonly has antigorite as its major constituent.

CHRYSOTILE

Crystal form:	Fibrous variety of antigorite
Luster:	Silky owing to its fibrous habit
Color:	Light green to almost white
Occurrence:	Most commonly intergrown with antigorite

Remarks: This fibrous variety is the most common asbestos mineral. The characteristic morphology of all asbestos minerals, in their natural form, is a parallel-sided fiber with a length-to-diameter ratio of three or greater.

Clay Minerals

Clay is a rock term, and like most rocks clay is made up of a number of different minerals in varying proportions; these are clay minerals as well as additional, very fine-grained feldspar, quartz, mica, and so on. The term *clay* implies a very fine grain size and, indeed, all clay minerals are so fine-grained that it is impossible to see grain sizes with the naked eye, or even with a hand lens or binocular microscope. The hardness and specific gravity of clay minerals are low.

KAOLINITE

Crystal form: Usually in compact or friable masses of very fine grain size.

Cleavage: Basal {001} perfect

H: 2

G: 2.6

Luster: Usually dull and earthy; may have a soapy feel.

Color: White, but often variously colored by impurities such as iron, giving it a brownish or reddish color.

Occurrence: Kaolinite is commonly the main constituent of kaolin or clay. A major component of soils. When pure and present in mineable quantities, it is referred to as *china clay* or *kaolin*.

Remarks: Usually recognized by its very fine grain size, low degree of hardness, and claylike character. Only X-ray tests will distinguish it from other clay minerals such as *montmorillonite* or *illite*.

TALC

Crystal form: Usually foliated and in radiating foliated groups. When compact and massive, known as *soapstone*.

Cleavage: Basal {001} perfect. Thin folia somewhat flexible but not elastic.

H: 1; sectile. Will leave a mark on cloth. Greasy feel.

Luster: Pearly to greasy

Color: Apple green, gray, white, or silver white. In soapstone often dark gray or green. Translucent.

Occurrence: A secondary mineral resulting from the alteration of Mg-rich silicates such as olivine, pyroxenes, and amphiboles. Characteristically part of low-grade metamorphic rock types, such as talc schists.

Remarks: Characterized by its foliated habit, cleavage, softness, and greasy feel. Cannot be distinguished in hand specimen from *pyrophyllite*.

Micas

Crystals of the mica group minerals are usually tabular with prominent basal planes. Crystals as well as granular intergrowths are characterized by perfect {001}, basal cleavage. The three most common varieties of mica, *muscovite, phlogopite,* and *biotite,* are distinguished from one another on the basis of their color. Muscovite is generally colorless; phlogopite has a bronzelike, brownish color; biotite is very dark green to black.

MUSCOVITE

Crystal form: Distinct crystals rare; usually with prominent {001}. Generally foliated in large to small sheets. May be in scales that are aggregated into plumose or globular forms. Also very fine-grained and massive.

Cleavage: Basal {001} perfect, allowing the mineral to split into very thin sheets. Folia flexible and elastic.

H: $2-2\frac{1}{2}$

Color: Colorless and transparent in thin sheets. In thicker blocks translucent, with light shades of yellow, brown, green, red. Luster vitreous to silky or pearly.

Occurrence: Common in granites and granite pegmatites. Also a major constituent of metamorphic rocks such as mica schists, and common in chlorite–muscovite–feldspar schists. When present as fibrous aggregates of minute scales with a silky luster, known as *sericite*.

Remarks: Characterized by its highly perfect cleavage and light color.

PHLOGOPITE

All properties, except for color, like those of muscovite.

Color: Yellowish to bronze brown, green, white, often with copperlike reflections from cleavage surfaces. Transparent in thin sheets. Luster vitreous to pearly.

Occurrence: A common constituent of metamorphosed Mg-rich limestones, dolomites, and ultramafic rocks. Also found in kimberlites.

Remarks: The yellowish-brown color distinguishes it from muscovite and biotite.

BIOTITE

All properties, except for color, are like those of muscovite and phlogopite.

Color: Usually dark green, brown to black. Thin sheets usually have a smoky color, whereas those of muscovite are almost colorless. Splendent luster.

Occurrence: A common constituent of many rock types. In igneous rocks such as granite, diorite, gabbro, peridotite, and granite pegmatite; also in lavas. In metamorphic rocks such as biotite schists and feldspar–garnet–biotite gneisses. As an accessory mineral in detrital sedimentary rocks such as arenites and greywackes.

Remarks: Distinguished from muscovite and phlogopite by its dark, often black, color

LEPIDOLITE

Crystal form: Crystals may be small plates or prisms with hexagonal outline. Commonly in coarse- to fine-grained scaly aggregates.

Cleavage: Characteristic, perfect {001} mica cleavage

Color: Pink and lilac to grayish white. Luster pearly.

Occurrence: Lepidolite is a comparatively rare mineral, found in pegmatites, usually associated with other Li-bearing minerals such as pink and green tourmaline, amblygonite, and spodumene.

Remarks: Its most obvious characteristics are the typical micaceous cleavage and lilac to pink color.

CHLORITE

A number of species belong to the chlorite group, all of which have similar chemical, crystallographic, and physical properties. Without quantitative chemical analysis or careful study of the optical and X-ray properties, it is impossible to distinguish between these species. The following description is for the most common type of chlorite.

Crystal form: Crystals not common; when they occur they are similar in habit to mica crystals. Usually foliated, massive, or in aggregates of minute scales. Also in finely dissiminated particles.

Cleavage: {001} perfect. Folia flexible but not elastic.

H: $2-2\frac{1}{2}$

G: 2.6–3.3

Color: Green of various shades is most common. Rarely yellow, white, or rose red. Transparent to translucent.

Occurrence: Chlorite, a common mineral in metamorphic rocks, is the diagnostic mineral of the greenschist facies. It is a common constituent of pelitic (aluminous) schists and occurs in association with actinolite and epidote. It is also a common constituent of igneous rocks, where it has formed as an alteration product (often as rims or "coronas") about Mg–Fe silicates such as pyroxenes, amphiboles, biotite, and garnet.

Remarks: Characterized by its green color, by its micaceous habit and cleavage, and by the fact that the folia are not elastic.

APOPHYLLITE

Crystal form: In tetragonal crystals with prisms, pyramids, and pinacoid. Sometimes resembling isometric crystals but differences in the lusters of the prism faces and of the basal pinacoid show the crystals to be tetragonal.

Cleavage: {001} perfect

H: $4\frac{1}{2}-5$

G: 2.3–2.4

Color: Colorless, white, or grayish; may show pale shades of green, yellow, rose. Luster of base pearly, other faces vitreous. Translucent to transparent.

Occurrence: Occurs as a secondary mineral lining cavities in basalt and related rocks, associated with zeolites, calcite, and pectolite.

Remarks: Usually recognized by its crystals, color, luster, and basal cleavage.

PREHNITE

Crystal form: Distinct crystals rare. Usually reniform, or stalactitic, and in rounded groups of tabular crystals.

H: $6-6\frac{1}{2}$

G: 2.8–2.95

Color: Usually light green, passing into white. Translucent.

Occurrence: Occurs as a secondary mineral lining cavities in basalt and related rocks. Associated with zeolites, pectolite, and calcite.

Remarks: Characterized by its green color and crystalline aggregates forming reniform surfaces. Resembles hemimorphite but is of lower specific gravity.

CHRYSOCOLLA

Crystal form: Massive, compact, and in some cases earthy. Individual specimens inhomogeneous.

H: Highly variable between 2 and 4

G: 2–2.4

Cleavage: None, conchoidal fracture

Color: Green to greenish blue; brown to black when impure. Luster vitreous to earthy.

Occurrence: In oxidized zones of copper deposits and may be associated with malachite, azurite, cuprite, or native copper.

Remarks: Characterized by its green to blue color and conchoidal fracture. Distinguished from turquoise by inferior hardness.

TABLE 26.1. Tabulation of physical properties and subsequent mineral identification for three unknowns.

Specimen number: . . . .

Crystal Morphology/Symmetry/Twinning/Striations:

Cleavage/Parting/Fracture:

Hardness:

Luster:

Streak:

Color(s)

Tenacity:

Specific gravity (heft):

Other distinguishing features (effervescence in HCl, magnetism, smell, exsolution lamellae, etc.):

Name:

Mineral classification; Formula:

Association:

Specimen number: . . . .

Crystal Morphology/Symmetry/Twinning/Striations:

Cleavage/Parting/Fracture:

Hardness:

Luster:

Streak:

Color(s)

Tenacity:

Specific gravity (heft):

Other distinguishing features (effervescence in HCl, magnetism, smell, exsolution lamellae, etc.):

Name:

Mineral classification; Formula:

Association:

Specimen number: . . . .

Crystal Morphology/Symmetry/Twinning/Striations:

Cleavage/Parting/Fracture:

Hardness:

Luster:

Streak:

Color(s)

Tenacity:

Specific gravity (heft):

Other distinguishing features (effervescence in HCl, magnetism, smell, exsolution lamellae, etc.):

Name:

Mineral classification; Formula:

Association:

Student Name

TABLE 26.2. Tabulation of physical properties and subsequent mineral identification for ten unknowns.

Specimen number	Luster (metallic versus non-metallic)	Hardness	Cleavage or fracture	Crystal form	Color	Other (streak, texture, striations)	Mineral Name

Please make photocopies of this form if more copies are needed.

Rock-Forming Silicates: Common Silicates whose Structures contain Independent Tetrahedra, Double Tetrahedra, or Tetrahedral Rings, and their Physical Properties

PURPOSE OF EXERCISE

An overview of the most diagnostic physical properties and common occurrence of several rock-forming silicate groups: (1) those with independent tetrahedra in their atomic structures, (2) those with double tetrahedra, and (3) those with ring structures.

FURTHER READING AND CD-ROM INSTRUCTIONS

Klein, C. and Dutrow, B., 2008, *Manual of Mineral Science*, 23rd ed., Wiley, Hoboken, New Jersey, pp. 483–505.

CD-ROM, entitled *Mineralogy Tutorials*, version 3.0, that accompanies *Manual of Mineral Science*, 23rd ed., click on "Module IV", and subsequently the button "Silicates". On the next screen click "Nesosilicates".

Nesse, W. D., 2000, *Introduction to Mineralogy*, Oxford University Press, New York, pp. 306–325.

Perkins, D., 2002, *Mineralogy*, 2nd ed., Prentice Hall, Upper Saddle River, New Jersey, pp. 331–346.

Wenk, H. R. and Bulakh, A., 2004, *Minerals: Their Constitution and Origin*, Cambridge University Press, New York, New York, pp. 425–447.

Deer, W. A., Howie, R. A. and Zussman, J., 1992, *An Introduction to The Rock-Forming Minerals*, 2nd ed., John Wiley and Sons, Hoboken, New Jersey, pp. 3–138.

Background Information: The study of hand specimens of minerals continues with silicates having independent tetrahedra, known as nesosilicates, from the Greek word

nesos, meaning island; with those with double tetrahedra, known as sorosilicates, from the Greek word *soros*, meaning group; and those with rings of tetrahedra, known as cyclosilicates, from the Greek word *kyklos*, meaning circle. These structural groups include several major mineral groups, among them the *olivines*, *garnets*, *epidote*, and the aluminosilicates *sillimanite*, *kyanite*, and *andalusite*.

MATERIALS

Representative collections of the nesosilicates, which include olivine, garnet, zircon, andalusite, sillimanite, kyanite, staurolite, topaz, datolite, titanite, and chloritoid; of the sorosilicates, which include hemimorphite, epidote, and vesuvianite; and of the cyclosilicates, which include beryl, cordierite, and tourmaline. Basic testing tools for hardness, specific gravity, and streak as well as a hand lens, a binocular microscope, or both are needed. Index cards or a notebook must be available for abstracting the most important data for each mineral (see Fig. 24.1 and also Tables 27.1 and 27.2).

ASSIGNMENT

Hand specimen study of the various specimens available of the nesosilicates, sorosilicates, and cyclosilicates. Consult the listing of highly diagnostic physical properties that follows, as well as determinative tables and mineral descriptions in the above listed textbooks.

OLIVINE

Crystal form:	Usually occurs as embedded grains or granular masses.
Cleavage:	None, conchoidal fracture
H:	$6\frac{1}{2}$–7
G:	3.27–4.37, increasing as a function of Fe content.
Color:	Pale yellow-green to green in Mg-rich varieties such as *forsterite;* becomes darker brownish green with increasing Fe content (toward

fayalite). Luster vitreous. Transparent to translucent. Gem quality material known as *peridot*.

Occurrence:	A common mineral that may be the chief constituent of a rock, as in *dunite*. Occurs in dark-colored igneous rocks such as gabbro, peridotite, and basalt.
Remarks:	Distinguished by its glassy luster, conchoidal fracture, green color, and granular nature. Olivine is readily altered to serpentine minerals such as antigorite. Olivines in

metamorphosed igneous rocks may show *coronas,* which are concentric alteration rims of pyroxene and amphibole.

Garnets

Garnets, members of the nesosilicate group, are very common rock-forming minerals, especially in metamorphic rocks. Their general formula can be written as $A_3B_2(SiO_4)_3$, in which the A site houses Mg, Fe, Mn, or Ca, and in which the B site is the location for Al, Fe, and Cr. These elemental substitutions result in six end-members, which are, with partial chemical formulas indicating the A and B site occupancies,

Pyrope – Mg_3Al_2
Almandine – Fe_3Al_2
Spessartine – Mn_3Al_2
Grossular – Ca_3Al_2
Andradite – Ca_3Fe_2
Uvarovite – Ca_3Cr_2

Of these, the most common metamorphic garnets are almandine, pyrope, grossular, and andradite.

All garnets have very similar isometric crystal forms (see below), similar hardness (ranging from $H = 6\frac{1}{2}$ to $7\frac{1}{2}$), and a broad range of specific gravity ($G = 3.5$ to 4.3) as a result of their extensive compositional substitutions. Here follow the most diagnostic properties for hand specimen identification.

Crystal form: Isometric crystals, commonly showing dodecahedrons or trapezohedrons, or in combination. Hexoctahedrons are much rarer. Usually distinctly crystallized. May also occur in rounded grains, massive or granular.

Luster: Vitreous to resinous. Transparent to translucent.

Color: *Pyrope*–deep red to nearly black

Almandine–fine deep red to brownish red

Spessartine–brownish to red

Grossular–white, green, yellow, cinnamon brown, pale red

Andradite–various shades of yellow, green brown, to black

Uvarovite–emerald green

Occurrence: Garnet is a common and widely distributed mineral, occuring abundantly in some metamorphic rocks and as an accessory constituent in some igneous rocks. Because the colors of several garnet end-members overlap, color is generally not a truly diagnostic property. Instead a careful study of the assemblage in which the garnet occurs is commonly very helpful in arriving at a correct evaluation of garnet type.

Pyrope occurs in ultrabasic rocks such as peridotites and kimberlites.

Almandine is the common garnet in metamorphic rocks as in, for example, garnet–biotite schists. It is also a widespread detrital mineral in sedimentary rocks.

Spessartine occurs in Mn-containing skarn deposits; commonly in association with Mn-oxides, rhodonite, rhodochrosite, and so on.

Grossular is a product of contact or regional metamorphism of impure limestones.

Andradite occurs in environments that are more iron-rich than grossular but is also the result of metamorphic reactions in impure (Fe-containing) limestones.

Uvarovite is very rare and occurs in association with serpentine and chromite.

Remarks: Garnets are usually recognized by their characteristic isometric crystals, their hardness, their color, and their mineral association.

ZIRCON

Crystal form: Usually in crystals, commonly medium- to fine-grained. Tetragonal, consisting of prisms, pyramids, and dipyramids. Commonly doubly terminated. May also be in irregular grains.

H: $7\frac{1}{2}$

G: 4.68

Color: Commonly some shade of brown; also colorless, green, or red. Usually translucent. Luster adamantine.

Occurrence: An accessory mineral in all types of igneous rocks, especially common in the more silicic types such as granite, granodiorite, and syenite. Also found in crystalline limestone, gneisses, and schists. It is a common accessory mineral in many sediments and in the heavy mineral fraction of sandstones. Occurs as rounded grains in stream and beach sands.

Remarks: Usually recognized by its characteristic crystals, color, luster, hardness, and high specific gravity.

Andalusite, Sillimanite, and Kyanite

Andalusite, sillimanite, and kyanite are polymorphs of Al_2SiO_5. All three result from metamorphic reactions in argillaceous (= pelitic, meaning aluminum-rich) rocks, and with the aid of a pressure–temperature phase diagram for Al_2SiO_5 one can interpret the different temperature and pressure ranges for the formation of the three polymorphs.

ANDALUSITE

Crystal form: Orthorhombic, usually in coarse, nearly square prisms.

Cleavage: Prismatic {110} distinct

H: $7\frac{1}{2}$. Fine-grained pseudomorphs of mica after andalusite are common, having a hardness of $2\frac{1}{2}$.

G: 3.16–3.20

Color: Flesh red, reddish brown, olive green. Transparent to translucent. The variety *chiastolite* has dark-colored carbonaceous inclusions arranged in a regular manner that makes a cruciform design.

Occurrence: Commonly formed in the contact-metamorphosed aureoles of igneous rocks within argillaceous rocks. May coexist with cordierite. It is also found in association with kyanite or sillimanite in regionally metamorphosed assemblages.

Remarks: Characterized by the nearly square prisms and hardness. Chiastolite is easily recognized by the symmetrically arranged inclusions.

SILLIMANITE

Crystal form: Occurs in long, slender, orthorhombic crystals without distinct terminations. Commonly in parallel groups; also fibrous, called *fibrolite*.

Cleavage: {010} perfect

H: 6–7

G: 3.23

Color: Brown, pale green, white. Transparent to translucent.

Occurrence: In Al-rich rocks metamorphosed at high temperature, such as cordierite–sillimanite gneiss, or quartz–almandine–biotite–sillimanite schist. In silica-poor rocks may be associated with corundum.

Remarks: Characterized by slender crystals with one direction of cleavage. May resemble other fibrous silicates such as tremolite or wollastonite.

KYANITE

Crystal form: Triclinic. Usually in long, tabular crystals, rarely terminated. Also bladed aggregates.

Cleavage: {100} perfect

H: 5 parallel to the length of crystals, 7 at right angles to this direction.

G: 3.55–3.66

Color: Usually blue, often of a darker shade toward the center of the crystal. Also, in some instances, white, gray, or green. Color may be irregular in streaks and patches. Luster vitreous to pearly.

Occurrence: Common in regionally metamorphosed Al-rich rocks in association with, for example, garnet, staurolite, and corundum.

Remarks: Characterized by its bladed crystals, good cleavage, blue color, and different hardnesses in different directions.

STAUROLITE

Crystal form: Monoclinic prismatic crystals commonly in cruciform twins. Usually in crystals or twinned crystals; rarely massive.

H: $7–7\frac{1}{2}$

G: 3.65–3.75

Color: Red-brown to brownish black. Translucent. Luster vitreous to resinous when fresh, dull to earthy when altered or impure.

Occurrence: Occurs in regionally metamorphosed Al-rich schists and gneisses. In high-grade metamorphic rocks may be associated with kyanite and garnet; in lower-grade metamorphic rocks occurs with chloritoid.

Remarks: Recognized by its characteristic crystals and cruciform twins and by its generally brownish color.

TOPAZ

Crystal form: Common in orthorhombic, prismatic crystals terminated by dipyramids, prisms, and basal pinacoid. Vertical faces frequently striated. Usually in crystals but also in crystalline masses; granular, coarse, or fine.

Cleavage: Perfect basal {001}

H: 8

G: 3.4–3.6

Color: Colorless, yellow, pink, wine yellow, bluish, greenish. Transparent to translucent. Vitreous luster.

Occurrence: In cavities in rhyolitic lavas and granite. Also in pegmatites. Associated with tourmaline, cassiterite, apatite, and fluorite.

Remarks: Recognized chiefly by its crystals, basal cleavage, high degree of hardness, and high specific gravity.

DATOLITE

Crystal form: Usually in crystals. Monoclinic with nearly equidimensional habit and often complex in development. Also coarse- to fine-granular.

H: $5–5\frac{1}{2}$

G: 2.8–3.0

Color: White, often with a faint greenish tinge. Transparent to translucent. Vitreous luster.

Occurrence: It is a secondary mineral found usually in cavities in basaltic lavas and similar rocks. Associated with zeolites, prehnite, apophyllite, and calcite.

Remarks: Characterized by its glassy luster, pale green color, and equidimensional crystals with many faces.

TITANITE

Crystal form: Common as wedge-shaped, monoclinic crystals. May be lamellar or massive.

Cleavage: {110} distinct. Parting on {110} may be present.

H: $5–5\frac{1}{2}$

G: 3.4–3.55

Color: Gray, brown, green, yellow, black. Transparent to translucent. Luster resinous to adamantine.

Occurrence: It is a common accessory, in small crystals, in granites, granodiorites, diorites, syenites, and nepheline syenites.

Remarks: Characterized by its wedge-shaped crystals, high luster, and typically brownish color.

CHLORITOID

Crystal form:	Seldom in distinct tabular crystals. Usually coarsely foliated, massive. Also in thin scales or plates.
Cleavage:	Good basal {001} cleavage producing brittle flakes. This cleavage is not as perfect as in the micas or in chlorite.
H:	$6\frac{1}{2}$ (much harder than chlorite)
G:	3.5–3.8
Color:	Dark green, greenish gray, often grass green in very thin plates. Luster pearly.
Occurrence:	A relatively common constituent of low- to medium-grade regionally metamorphosed iron-rich argillaceous rocks. Commonly as porphyroblasts in association with muscovite, chlorite, staurolite, garnet, and kyanite.
Remarks:	Chloritoid is very difficult to distinguish from chlorite, with which it is commonly associated. Optical or X-ray study or both are necessary for unambiguous identification.

HEMIMORPHITE

Crystal form:	Orthorhombic crystals usually tabular parallel to {010}; prism faces terminated above by domes and a pedion, below by a pyramid, forming polar crystals. Often in divergent crystal groups, forming rounded or coxcomb masses. Also mammillary, stalactitic, massive, and granular.
Cleavage:	Prismatic {110} perfect
H:	$4\frac{1}{2}$–5
G:	3.4–3.5 (relatively heavy for a nonmetallic, light-colored mineral)
Color:	White, in some cases with a faint bluish or greenish tinge; also yellow to brown.
Occurrence:	A secondary mineral found in the oxidized portions of zinc deposits, associated with smithsonite, sphalerite, cerussite, anglesite, and galena.
Remarks:	Characterized by the grouping of crystals. Resembles prehnite but has a higher specific gravity.

Epidote

The epidote group of sorosilicates includes several species, the most common of which belong to the complete solid/solution series between monoclinic *clinozoisite* and *epidote,* and a considerably rarer species, *allanite*. In hand specimens it is nearly impossible to distinguish between the various members of the clinozoisite–epidote series and, therefore, all members of this series are here referred to as epidote.

EPIDOTE

Crystal form:	Usually coarse- to fine-granular; also fibrous. Monoclinic crystals are usually elongated parallel to the *b* axis with a prominent development of faces of the [010] zone,

that is, the zone parallel to *b*. Also striated parallel to *b*.

Cleavage:	Basal {001} perfect
H:	6–7
G:	3.25–3.45
Color:	Pistachio green to yellowish green to black; clinozoisite is paler green to gray. Luster vitreous. Transparent to translucent.
Occurrence:	A common mineral in regionally metamorphosed rocks formed as part of epidote–amphibolite facies metamorphism. May also form as a retrograde mineral as a reaction product of plagioclase, pyroxene, and amphibole. Common in association with actinolite, chlorite, diopside, grossular and andradite garnets, vasuvianite, and calcite.
Remarks:	Characterized by its unique pistachio-green color, one perfect cleavage, and relatively high hardness.

ALLANITE

Crystal form:	Commonly massive in embedded grains
H:	$5\frac{1}{2}$–6
G:	3.5–4.2
Color:	Brown to pitch black. Often coated with a yellow-brown alteration product. Submetallic to pitchy and resinous luster. Subtranslucent, will transmit light only on thin edges.
Occurrence:	A minor constituent of many igneous rocks, such as granite, syenite, diorite, and pegmatites. Frequently associated with epidote.
Remarks:	The hardness and specific gravity of allanite are variable because it occurs in various stages of metamictization, owing to self-irradiation by radioactive elements in its atomic structure. Allanite is characterized by its black color, pitchy luster, slight radioactivity, and occurrence in granitic rocks.

VESUVIANITE

Crystal form:	Tetragonal, prismatic crystals that are commonly vertically striated. Also in striated columnar aggregates, granular, and massive.
Cleavage:	Prismatic {010} poor
H:	$6\frac{1}{2}$
G:	3.35–3.45
Color:	Usually green or brown; also yellow, blue, red. Subtranslucent to translucent. Vitreous to resinous luster.
Occurrence:	Common in contact metamorphosed assemblages resulting from reactions in and recrystallization of impure limestones and dolomites. Associated with grossular and andradite garnet, wollastonite, and diopside.
Remarks:	Brown tetragonal prisms and striated columnar masses are highly characteristic. When in massive form vesuvianite may resemble some garnets, but it has a lower specific gravity.

BERYL

Crystal form:	Hexagonal with a well-developed prismatic habit. Commonly vertically striated and grooved. Pyramidal forms are rare. Crystals may be large in size, with somewhat rough faces.
Cleavage:	Basal {0001} imperfect
H:	$7\frac{1}{2}$–8
G:	2.65–2.8
Color:	Commonly bluish green or light yellow, may be deep emerald green, gold yellow, pink, white, or colorless. Transparent to translucent. Vitreous luster. *Aquamarine* is the pale greenish-blue transparent gem variety. *Morganite,* or *rose beryl,* is pale pink to deep rose. *Emerald* is a deep green transparent gem beryl.
Occurrence:	Common in granitic rocks and pegmatites. Also in mica schists and associated with tin ores.
Remarks:	Usually recognized by its hexagonal crystal form and color. Distinguished from apatite by much greater hardness and from quartz by higher specific gravity.

CORDIERITE

Crystal form:	Short prismatic, orthorhombic crystals are uncommon. Most commonly as embedded granular grains and massive.
Cleavage:	Pinacoidal {010} poor
H:	7–$7\frac{1}{2}$
G:	2.60–2.66
Color:	Various shades of blue to bluish gray. Also colorless, white, gray, light yellow to brown. Transparent to translucent. Vitreous luster.
Occurrence:	Common in regionally and contact metamorphosed argillaceous rocks. Associated with sillimanite, spinel, anthophyllite, and garnet in schists and gneisses. Also present in granites and pegmatites.
Remarks:	Easily recognized when it exhibits the typical blue to bluish-gray color. When it occurs as colorless or gray grains in a rock, it closely resembles quartz, from which it can be distinguished only by optical tests. Distinguished from corundum by lower hardness.

TOURMALINE

Crystal form:	Commonly in hexagonal prismatic crystals with a prominent trigonal prism and subordinate hexagonal prism. Prism faces vertically striated. The prism faces may round into each other, giving the crystals a cross section like a spherical triangle. When doubly terminated, crystals usually show different forms at the opposite ends of the crystal. Also massive compact; coarse- to fine-granular, either radiating or parallel.
Cleavage:	None; conchoidal fracture.
H:	7–$7\frac{1}{2}$
G:	3.0–3.25
Color:	Many colors as a result of chemical differences. Most common is *schorl,* an Fe-rich species, which is black. *Dravite,* a brown variety, is Mg-rich. Rarer gem varieties are pink, green, or blue. Rarely white or colorless. A single crystal may show several different colors arranged in concentric zones about the *c* axis. Crystals may also have layers of different colors transverse to the length. When the outside zone is green and the inner zone pink, the variety is known as *watermelon tourmaline.* Luster vitreous to resinous.
Occurrence:	Most common in granite pegmatites or the rocks immediately surrounding them. Also an accessory mineral in igneous and metamorphic rocks. In most pegmatites the common variety is *schorl,* which is black. Lithium-bearing pegmatites, where in association with lepidolite, beryl, and apatite, are the host for light-colored tourmalines.
Remarks:	Usually recognized by the rounded triangular cross sections of crystals, striations on prism faces, and conchoidal fracture. Distinguished from hornblende by lack of cleavage.

TABLE 27.1. Tabulation of physical properties and subsequent mineral identification for three unknowns.

Specimen number: . . . .

Crystal Morphology/Symmetry/Twinning/Striations:

Cleavage/Parting/Fracture:

Hardness:

Luster:

Streak:

Color(s)

Tenacity:

Specific gravity (heft):

Other distinguishing features
(effervescence in HCl,
magnetism, smell, exsolution
lamellae, etc.):

Name:

Mineral classification; Formula:

Association:

Specimen number: . . . .

Crystal Morphology/Symmetry/Twinning/Striations:

Cleavage/Parting/Fracture:

Hardness:

Luster:

Streak:

Color(s)

Tenacity:

Specific gravity (heft):

Other distinguishing features
(effervescence in HCl,
magnetism, smell, exsolution
lamellae, etc.):

Name:

Mineral classification; Formula:

Association:

Specimen number: . . . .

Crystal Morphology/Symmetry/Twinning/Striations:

Cleavage/Parting/Fracture:

Hardness:

Luster:

Streak:

Color(s)

Tenacity:

Specific gravity (heft):

Other distinguishing features
(effervescence in HCl,
magnetism, smell, exsolution
lamellae, etc.):

Name:

Mineral classification; Formula:

Association:

Please make photocopies of this form if more copies are needed.

Student Name

TABLE 27.2. Tabulation of physical properties and subsequent mineral identification for ten unknowns.

Specimen number	Luster (metallic versus non-metallic)	Hardness	Cleavage or fracture	Crystal form	Color	Other (streak, texture, striations)	Mineral Name

Please make photocopies of this form if more copies are needed.

Common Oxides, Hydroxides and Halides, and their Physical Properties

PURPOSE OF EXERCISE

An overview of the most diagnostic physical properties and common occurrences of the rock-forming oxides, hydroxides, and halides.

FURTHER READING AND CD-ROM INSTRUCTIONS

Klein, C. and Dutrow, B., 2008, *Manual of Mineral Science*, 23rd ed., Wiley, Hoboken, New Jersey, pp. 368–398.

CD-ROM, entitled *Mineralogy Tutorials*, version 3.0, that accompanies *Manual of Mineral Science*, 23rd ed., click on "Module IV", and subsequently click on the button "Oxides and Hydroxides". Find halides by clicking on the button "Sulfides, Halides" on that same screen.

Nesse, W. D., 2000, *Introduction to Mineralogy*, Oxford University Press, New York, pp. 356–377.

Perkins, D., 2002, *Mineralogy*, 2nd ed., Prentice Hall, Upper Saddle River, New Jersey, pp. 364–376; pp. 361–364.

Wenk, H. R. and Bulakh, A., 2004, *Minerals: Their Constitution and Origin*, Cambridge University Press, New York, New York, pp. 406–424; pp. 347–358.

Deer, W. A., Howie, R. A. and Zussman, J., 1992, *An Introduction to The Rock-Forming Minerals*, 2nd ed., John Wiley and Sons, Hoboken, New Jersey, pp. 532–578; pp. 672–677.

Background Information: The study of hand specimens continues with (1) the *oxides*, which occur generally as accessory minerals in igneous and metamorphic rocks and as resistant detrital grains in sediments, (2) the *hydroxides*, which occur mainly as secondary alteration and weathering products, and (3) the *halides*, which represent diverse geologic origins.

MATERIALS

Representative collections of the oxides, which include cuprite, corundum, hematite, ilmenite, rutile, pyrolusite, cassiterite, uraninite, spinel, magnetite, and chromite; of the hydroxides, which include brucite, manganite, romanechite, goethite, limonite, and bauxite; and of the halides, including halite, sylvite, and fluorite. Basic testing tools for hardness, specific gravity, and streak as well as a hand lens, a binocular microscope, or both are needed. Index cards or a notebook should be available for abstracting the most important data for each mineral (see Fig. 24.1 and also Tables 28.1 and 28.2).

ASSIGNMENT

Hand specimen study of various specimens of the oxides, hydroxides, and halides. Consult the listing of highly diagnostic physical properties that follows, as well as the determinative tables and mineral descriptions in the above listed textbooks.

Oxides

The oxides are a group of minerals that are relatively hard, dense, and refractory and generally occur as accessory minerals in igneous and metamorphic rocks and as resistant detrital grains in sediments. Several oxide minerals are major ore minerals; examples are hematite and magnetite (for iron ore), chromite (chromium ore), romanechite and manganite (manganese ore), cassiterite (tin ore), and uraninite (uranium ore). Several of the oxides have submetallic to metallic luster and are opaque; however, not all oxides have these properties.

CUPRITE

Crystal form:	Commonly in isometric crystals that show cube, octahedron, and dodecahedron and combinations thereof.
H:	$3\frac{1}{2}$–4 (relatively soft)
G:	6.1

Color:	Various shades of red; ruby red in transparent crystals, called "ruby copper." Luster metallic–adamantine in clear crystallized varieties. Streak brownish red.
Occurrence:	Found, as a supergene mineral, in the oxidized portions of copper deposits, associated with limonite and secondary copper minerals such as native copper, malachite, azurite, and chrysocolla.
Remarks:	Usually distinguished by its red color, isometric crystal form, high luster, brown streak, and association with limonite.

CORUNDUM

Crystal form:	Prismatic or tabular hexagonal crystals are common. May display hexagonal dipyramids that are rounded into barrel shapes with deep horizontal striations. Usually rudely crystallized or massive; coarse- or fine-granular.

Cleavage:	Only parting on {0001} and {10$\bar{1}$1}.
H:	9 (Because corundum may alter to mica make sure a fresh surface is used in hardness testing.)
G:	4.02
Color:	Usually some shade of brown, pink, or blue; but may be colorless or any color. *Ruby* is red gem corundum; *sapphire* is gem corundum of any other color. *Emery* is a black granular corundum intimately mixed with magnetite and hematite. Luster adamantine to vitreous. Transparent to translucent.
Occurrence:	A common accessory mineral in some metamorphic rocks, such as crystalline limestone, mica schist, or gneiss. Also present in SiO_2–poor igneous rocks such as syenites and nepheline syenites. Occurs also as rolled pebbles in detrital soil and stream sands.
Remarks:	Characterized chiefly by its great hardness, high luster, specific gravity, and parting.

HEMATITE

Crystal form:	If in crystals, they are usually thick- to thin-tabular on {0001} with hexagonal outline. Thin plates may be grouped in rosettes. Also in botryoidal to reniform shapes with radiating structures, *kidney ore.* May also be micaceous and foliated and referred to as *specular* hematite. Also earthy.
Cleavage:	Parting on {10$\bar{1}$1} and {0001}
H:	5$\frac{1}{2}$–6$\frac{1}{2}$ (for crystals)
G:	5.26 (for crystals)
Color:	Reddish brown to black. Red earthy variety known as *red ocher,* and platy and metallic variety known as *specularite.* Luster metallic in crystals and dull in earthy varieties. Streak light to dark red. Translucent.
Occurrence:	An accessory constituent of many rock types and the most abundant iron ore mineral in Precambrian banded iron-formations. It may occur as a sublimation product in volcanic rocks and as an accessory constituent in granites. Also present in metamorphic assemblages, and a minor constituent of red sandstones wherein, as a cementing material, it binds quartz grains together. Also in oolitic sedimentary iron deposits.
Remarks:	Distinguished mainly by its characteristic red streak, reddish-brown to black color, and earthy to metallic luster.

ILMENITE

Crystal form:	Most commonly massive and compact. Also in grains or as sand.
H:	5$\frac{1}{2}$–6
G:	4.7
Color:	Iron black with metallic to submetallic luster. Streak black to brownish red. Opaque.
Occurrence:	A common accessory in igneous rocks. May occur as large masses in gabbros, diorites,

and anorthosites, commonly associated with magnetite. As a constituent of black sand, it is associated with magnetite, rutile, and zircon.

Remarks:	Distinguished from hematite by its streak and from magnetite by its lack of strong magnetism.

RUTILE

Crystal form:	Commonly in tetragonal, prismatic crystals with dipyramid terminations and vertically striated prism faces. Frequently in elbow twins. Habit of crystals may be slender acicular. Also compact, massive.
Cleavage:	Prismatic {110} distinct
H:	6–6$\frac{1}{2}$
G:	4.18–4.25
Color:	Red, reddish brown to black. Streak pale brown. Usually subtranslucent, may be transparent. Luster adamantine to submetallic.
Occurrence:	Found in granite, granite pegmatite, mica schist, gneiss, metamorphic limestone, and dolomite. May be included in quartz and micas as fine-grained slender crystals.
Remarks:	Characterized by its adamantine luster and red color. It has a lower specific gravity than cassiterite.

PYROLUSITE

Crystal form:	Usually in radiating fibers or columns. Also granular massive; often in reniform coats and dendritic shapes finely overgrown with other Mn-oxides and hydroxides. Frequently pseudomorphous after manganite.
Cleavage:	Prismatic {110} perfect; fracture splintery.
H:	1–2 (often soils the fingers); for coarsely crystalline *polianite* the hardness is 6–6$\frac{1}{2}$.
G:	4.75
Color:	Iron black. Streak also iron black. Opaque. Luster metallic.
Occurrence:	It is the most common manganese ore mineral and is widespread in its occurrence. It is found as nodular deposits in bogs, on lake bottoms, and on the floors of oceans. Also occurs in veins associated with quartz and various metallic minerals.
Remarks:	Characterized by its black color and black streak and low degree of hardness.

CASSITERITE

Crystal form:	When in crystals may show two tetragonal prisms and dipyramids. Often in elbow-shaped twins. Usually massive, granular; also in reniform shapes with radiating fibrous appearance, known as *wood tin.*
Cleavage:	Prismatic {010} imperfect
H:	6–7
G:	6.8–7.1 (unusually high for a nonmetallic mineral)

Color: Usually brown or black; rarely yellow or white. Streak white. Translucent, rarely transparent. Luster adamantine to submetallic or dull.

Occurrence: Present as an accessory mineral of igneous rocks and pegmatites, but most commonly found in high-temperature hydrothermal veins in or near granitic rocks. Vein associations may include tourmaline, topaz, fluorite, apatite, molybdenite, and arsenopyrite. May also occur as rolled pebbles in placer deposits, know as *stream tin.*

Remarks: Recognized by its high specific gravity, adamantine luster, and light streak.

URANINITE

Crystal form: Most commonly as massive or botryoidal forms with a banded structure, known as *pitchblende.*

H: $5\frac{1}{2}$

G: 6.5–9, for pitchblende. Crystals show **G** 7.5–9.7.

Color: Black. Streak brownish black. Luster submetallic to pitchlike, dull.

Occurrence: A primary constituent of granitic rocks and pegmatites, and in high-temperature hydrothermal veins associated with cassiterite, chalcopyrite, and arsenopyrite. Also in association, at low temperatures, with secondary uranium minerals.

Remarks: Characterized chiefly by its pitchy luster, high specific gravity, color, streak, and strong radioactivity (as detected by a Geiger counter or scintillation counter).

Spinels

The minerals of the spinel group show extensive solid solution between various end-member compositions. The most common members of this group are *spinel,* $MgAl_2O_4$, *magnetite,* $FeFe_2O_4$ (or Fe_3O_4), and *chromite,* $FeCr_2O_4$.

SPINEL

Crystal form: Isometric. Commonly in octahedral crystals or in twinned octahedrons (spinel twins). Dodecahedron may be present as small truncations but other forms are rare. Also massive and in irregular grains.

H: 8

G: 3.5–4.1

Color: Various shades of white, red, lavender, blue, green, brown, and black. Nonmetallic. Luster vitreous. Streak white. Usually translucent, but may be clear and transparent.

Occurrence: A common high-temperature mineral occurring in contact-metamorphosed limestones and metamorphic rocks poor in SiO_2. Metamorphic assemblages may contain phlogopite, pyrrhotite, and graphite.

Also an accessory mineral in many dark igneous rocks. Also found as rolled pebbles in stream sands.

Remarks: Recognized by its hardness, octahedral crystal form, and vitreous luster. Iron-rich spinel is distinguished from magnetite by being nonmagnetic and having a white streak.

MAGNETITE

Crystal form: Isometric. Frequently in octahedral crystals, more rarely in dodecahedrons. Other forms rare. Usually granular massive, coarse- or fine-grained.

Cleavage: None. Octahedral parting in some specimens.

H: 6

G: 5.18

Color: Iron black. Luster metallic. Opaque. Streak black.

Magnetism: Strong. May act as a natural magnet, known as *lodestone.*

Occurrence: Magnetite is a common mineral found disseminated as an accessory in most igneous rocks. In some igneous rock types magnetite may be the chief constituent of the rock and may form large ore zones; these are commonly titaniferous. Magnetite is also found in metamorphic assemblages. It is a major constituent of Precambrian banded iron-formations, in association with chert and hematite. Also found in black sands along seashores. Often closely intergrown with corundum, forming *emery.*

Remarks: Characterized by its strong magnetism, black color, and hardness of 6.

CHROMITE

Crystal form: Isometric, with octahedral habit, but crystals rare. Commonly massive, granular to compact.

H: $5\frac{1}{2}$

G: 4.6

Color: Iron black to brownish black. Luster metallic to submetallic; commonly pitchy. Subtranslucent. Dark brown streak.

Occurrence: A common constituent of peridotites, of other ultrabasic rocks, and of serpentine derived from them. Associated with olivine, serpentine, and corundum.

Remarks: Lack of magnetism and brown streak distinguishes it from magnetite. Association with olivine and serpentine is highly characteristic.

Hydroxides

The hydroxides are distinguished from the oxides by the presence of $(OH)^-$ groups, or H_2O molecules in the structure. The presence of $(OH)^-$ groups causes the bond strengths in these structures to be generally weaker than in the oxides. As a result, the hardness of hydroxides is generally lower than that of oxides.

BRUCITE

Crystal form:	Commonly foliated, massive. When in crystals, they are hexagonal, usually tabular on {0001}.
Cleavage:	Basal {0001} perfect. Folia flexible but not elastic. Sectile.
H:	$2\frac{1}{2}$
G:	2.39
Color:	White, gray, light green. Luster on base pearly, elsewhere vitreous to waxy. Transparent to translucent.
Occurrence:	Found in association with serpentine, dolomite, magnetite, and chromite. As a reaction product of Mg-silicates, especially serpentine.
Remarks:	Recognized by its foliated nature, light color, and pearly luster on cleavage faces. Distinguished from talc by being considerably harder, and from mica by being inelastic.

MANGANITE

Crystal form:	Monoclinic crystals that are prismatic parallel to the c axis and vertically striated. Often columnar to coarse–fibrous. Contact and penetration twins common.
Cleavage:	Pinacoidal {010} perfect, {110} and {001} good
H:	4
G:	4.3
Color:	Steel gray to iron black. Luster metallic. Opaque. Streak dark brown.
Occurrence:	In manganese deposits associated with other Mn-oxides and hydroxides. Also in low-temperature hydrothermal veins associated with barite, siderite, and calcite. It commonly alters to pyrolusite.
Remarks:	Its black color and prismatic crystals are most diagnostic. Hardness (4) and brown streak distinguish it from pyrolusite.

ROMANECHITE

Crystal form:	Massive, botryoidal, stalactitic; appears amorphous.
H:	5–6
G:	3.7–4.7
Color:	Black. Opaque. Luster submetallic. Streak brownish black.
Occurrence:	A product of secondary alteration of original Mn-carbonates, Mn-silicates, or both. Generally in association with pyrolusite in manganese ore deposits, in nodular deposits in bogs and on lake bottoms, and on see and ocean floors. Many of the hard botryoidal masses formerly called *psilomelane* are mixtures of manganese oxides of which romanechite is a major constituent.
Remarks:	Distinguished from other manganese oxides by its greater hardness and botryoidal form and from limonite by its black streak.

GOETHITE

Crystal form:	Rarely in distinct prismatic (orthorhombic), vertically striated crystals. Generally massive, reniform, stalactitic, and in radiating fibrous aggregation. Foliated. So-called *bog ore* is generally loose and porous.
Cleavage:	Pinacoidal {010} perfect
H:	$5–5\frac{1}{2}$
G:	4.37, but may be as low as 3.3 for impure material
Color:	Yellowish brown to dark brown. Luster adamantine to dull, silky in some fine, scaly, or fibrous varieties. Streak yellowish brown. Subtranslucent.
Occurrence:	A common mineral resulting from oxidation and weathering of iron-bearing minerals. It is a widespread deposit in bogs and around springs. It constitutes the *gossan* or "iron hat" over metalliferous vein deposits.
Remarks:	Distinguished from colloform hematite and romanechite by color and streak.

LIMONITE

Remarks:	Limonite is a field term used for very fine-grained to amorphous mixtures of brown ferric hydroxides whose real identities are unknown. Limonite, therefore, has variable compositions (and variable chemical and physical properties) and consists of a mixture of several iron hydroxides (commonly with goethite a major constituent) or of a mixture of several minerals such as hematite, goethite, or lepidochrocite, with or without additional absorbed water.

BAUXITE

The term bauxite is a rock term, not the name of a mineral species, because bauxite consists of a very fine-grained intergrowth of three aluminum hydroxides, diaspore, gibbsite, and boehmite. Bauxite is the major ore of aluminum.

Crystal form:	A mixture. Pisolitic, in round concretionary grains. Also massive, earthy, claylike.
H:	1–3
G:	2–2.55
Color:	White, gray, yellow, red. Yellow color caused by limonite staining. Luster dull to earthy.
Occurrence:	Bauxite is of supergene origin, commonly produced under subtropical to tropical conditions by prolonged weathering and leaching of silica from aluminum-bearing rocks. *Laterite,* formed in the tropics, is a soil consisting mainly of hydrous aluminum and ferric oxides.

Remarks: Recognized by its light, commonly yellowish-brown color, pisolitic and earthy texture, and low degree of hardness.

Halides

The few common halides that will be considered here are all isometric and have relatively low hardness.

HALITE

Crystal form: Isometric, with cubic habit; other forms rare. Some crystals hopper-shaped, meaning that the faces of the cube have grown more at the edges than in the center, giving the cube face a centrally depressed or hopper-shaped form. When in crystals or granular crystalline masses showing cubic cleavage, known as *rock salt.* Also massive, granular to compact.

Cleavage: Cubic {001} perfect

H: $2\frac{1}{2}$

G: 2.16

Color: Colorless or white, or when impure may show shades of yellow, blue, purple, and red. Transparent to translucent.

Taste: Salty

Occurrence: Occurs most commonly in evaporite deposits, associated with gypsum, anhydrite, dolomite, and shale, in rock salt beds ranging in thickness from a few feet to several hundred feet. Valuable deposits are also found in intrusive masses known as *salt domes,* which have their roots in thick, bedded deposits. Because salt is readily dissolved in surface waters, it crops out only in the driest of climates, and commercial deposits are confined almost entirely to the subsurface.

Remarks: Characterized by its cubic cleavage and taste, and distinguished from sylvite by a less bitter taste and yellow flame color,* caused by the presence of sodium.

SYLVITE

Crystal form: Isometric crystals with cube and octahedron frequently in combination. Usually in granular crystalline masses with good cubic cleavage; also compact.

Cleavage: Cubic {001} perfect

H: 2

G: 1.99

Cleavage: Colorless to white; also shades of blue, yellow, or red caused by impurities. Transparent when pure.

Taste: Salty, and more bitter than halite.

Occurrence: Sylvite has the same origin, mode of occurrence, and associations as halite but is much rarer.

Remarks: Similar in occurrence to halite and distinguished from halite by its bitter taste and sectility. On scratching a smooth surface, a knife produces a powder with halite, but very little powder with sylvite.

FLUORITE

Crystal form: Isometric, usually in cubes, and often in penetration twins. Other crystal forms are rare. Usually in crystals or in cleavable masses. Also massive; coarse- or fine-granular; columnar.

Cleavage: Octahedral {111} perfect

H: 4

G: 3.18

Color: Varies widely. Most commonly light green, yellow, bluish green, or purple; also colorless, white, rose, blue, or brown. Color is often distributed in bands in crystals or in the massive variety. Transparent to translucent. Luster vitreous. May show fluorescence under ultraviolet radiation.

Occurrence: A common, widely distributed mineral. In hydrothermal veins it may be the chief mineral, or it may be the gangue mineral in metallic ores, especially those of lead and silver. Also common in dolomites and limestones. May be associated with many different minerals, among them calcite, dolomite, gypsum, celestite, barite, quartz, galena, sphalerite, cassiterite, topaz, tourmaline, and apatite.

Remarks: Distinguished by its cubic crystals and perfect octahedral cleavage. Also by its vitreous luster and usually fine coloring, and by the fact that it can be scratched by a knife. It is harder than calcite and does not effervesce with cold HCl.

---

*A flame test is easily performed in the laboratory if a Bunsen burner, platinum wire loop on a glass rod, and a mortar and pestle (for grinding the mineral to a fine powder) are available. The flame test illustrates the volatilization of some elements in the flame and the subsequent coloration of the flame. It is best executed by introducing the fine powder of the mineral into the Bunsen burner on a piece of platinum wire. Sodium gives an intensely yellow flame, and indeed sodium is so pervasive in laboratory dust that the flame may show a yellow coloration even for a mineral that contains no sodium. Flame tests results are also noted for several carbonates and sulfates in exercise 29.

Student Name

TABLE 28.1. Tabulation of physical properties and subsequent mineral identification for three unknowns.

Specimen number: . . . .

Crystal Morphology/Symmetry/Twinning/Striations:

Cleavage/Parting/Fracture:

Color(s)

Hardness:

Tenacity:

Luster:

Specific gravity (heft):

Streak:

Other distinguishing features
(effervescence in HCl,
magnetism, smell, exsolution
lamellae, etc.):

Name:

Mineral classification; Formula:

Association:

Specimen number: . . . .

Crystal Morphology/Symmetry/Twinning/Striations:

Cleavage/Parting/Fracture:

Color(s)

Hardness:

Tenacity:

Luster:

Specific gravity (heft):

Streak:

Other distinguishing features
(effervescence in HCl,
magnetism, smell, exsolution
lamellae, etc.):

Name:

Mineral classification; Formula:

Association:

Specimen number: . . . .

Crystal Morphology/Symmetry/Twinning/Striations:

Cleavage/Parting/Fracture:

Color(s)

Hardness:

Tenacity:

Luster:

Specific gravity (heft):

Streak:

Other distinguishing features
(effervescence in HCl,
magnetism, smell, exsolution
lamellae, etc.):

Name:

Mineral classification; Formula:

Association:

TABLE 28.2. Tabulation of physical properties and subsequent mineral identification for ten unknowns.

Specimen number	Luster (metallic versus non-metallic)	Hardness	Cleavage or fracture	Crystal form	Color	Other (streak, texture, striations)	Mineral Name

Please make photocopies of this form if more copies are needed.

Common Carbonates, Sulfates, Tungstate, and Phosphates, and their Physical Properties

PURPOSE OF EXERCISE

An overview of some of the most diagnostic physical properties and common occurrences of carbonates, sulfates, and phosphates.

FURTHER READING AND CD-ROM INSTRUCTIONS

Klein, C. and Dutrow, B., 2008, *Manual of Mineral Science*, 23rd ed., Wiley, Hoboken, New Jersey, pp. 399–433.

CD-ROM, entitled *Mineralogy Tutorials*, version 3.0, that accompanies *Manual of Mineral Science*, 23rd ed., click on "Module IV", and subsequently on the button "Sulfates, Phosphates, and Carbonates".

Nesse, W. D., 2000, *Introduction to Mineralogy*, Oxford University Press, New York, pp. 326–355.

Perkins, D., 2002, *Mineralogy*, 2nd ed., Prentice Hall, Upper Saddle River, New Jersey, pp. 376–399.

Wenk, H. R. and Bulakh, A., 2004, *Minerals: Their Constitution and Origin*, Cambridge University Press, New York, New York, pp. 359–387.

Deer, W. A., Howie, R. A. and Zussman, J., 1992, *An Introduction to The Rock-Forming Minerals*, 2nd ed., John Wiley and Sons, Hoboken, New Jersey, pp. 606–672.

Background Information: The study of hand specimens continues with carbonates, which includes the common members of the *calcite, aragonite,* and *dolomite groups;* with sulfates and the common minerals of the *barite group* as well as *gypsum;* with the tungstate *scheelite;* and with several phosphates, namely *apatite, amblygonite,* and *turquoise.*

MATERIALS

Representative collections of members of the carbonates, including calcite, magnesite, siderite, rhodochrosite, smithsonite, aragonite, witherite, strontianite, cerussite, dolomite, malachite, and azurite; of the sulfates, including barite, celestite, anglesite, anhydrite, and gypsum; of the tungstate scheelite; and of the phosphates apatite, amblygonite, and turquoise. Basic tools for hardness, specific gravity, and streak testing as well as a hand lens, a binocular microscope, or both are needed. Index cards or a notebook should be available for abstracting the most important data for each mineral (see Fig. 24.1). Tables 29.1 and 29.2 can also be used.

ASSIGNMENT

Hand specimen study of the various specimens available of the carbonates, sulfates, tungstates, and phosphates. Consult the listing of highly diagnostic physical properties that follows, as well as the determinative tables and mineral descriptions in the above listed textbooks.

Carbonates

The anhydrous carbonates fall into three structurally different groups: the *calcite group,* the *aragonite* group, and the *dolomite* group. In addition to the carbonates in these groups, the hydrous copper carbonates, azurite and malachite, are the only important carbonates.

Calcite Group

Five common carbonates are members of the calcite group. These are calcite, magnesite, siderite, rhodochrosite, and smithsonite. These minerals are all rhombohedral and isostructural with calcite. All show perfect rhombohedral $\{10\bar{1}1\}$ cleavage.

CALCITE

Crystal form:	Hexagonal. Crystals extremely varied in habit and often complex. Three important habits are (1) prismatic, in long or short prisms, with a base or rhombohedral terminations; (2) rhombohedral, in which rhombohedral forms predominate; and (3) scalenohedral, with dominant scalenohedral forms in addition to prism faces. Calcite is usually in crystals or in coarse- to fine-grained aggregates. Also compact, earthy (as in *chalk*), and stalactitic (as in cave deposits).
Cleavage:	Rhombohedral $\{10\bar{1}1\}$ perfect; cleavage angle = 74°55′. Parting along twin lamellae of $\{01\bar{1}2\}$.

H: 3 on cleavage, $2\frac{1}{2}$ on base

G: 2.71

Color: Usually white to colorless, but may be variously tinted gray, red, green, blue, yellow; also, when impure, brown to black. Luster vitreous to earthy. Transparent to translucent. *Iceland spar* is a chemically pure and optically colorless variety of calcite.

Occurrence: One of the most common rock-forming minerals. The main constituent of *limestone* and of *marble,* metamorphosed limestone. Also the main mineral in *chalk,* which is fine-grained and pulverulent (meaning, easily powdered). Common as stalagmites, stalactites, and incrustations in cave deposits. Cellular deposits formed around hot or cold calcareous springs are known as *travertine* or *tufa.* Calcite may also be a primary mineral in some igneous rocks such as carbonatites and nepheline syenite. It is a late crystallization product in the cavities of lavas. It is also a common mineral in hydrothermal veins associated with sulfide ores.

Remarks: Characterized by its hardness (3), rhombohedral cleavage, vitreous luster, light color, and its ready effervescence in cold dilute HCl. Distinguished from dolomite by the fact that coarse fragments of calcite effervesce freely in cold dilute HCl, and distinguished from aragonite by lower specific gravity and rhombohedral cleavage.

MAGNESITE

Crystal form: Rarely in rhombohedral crystals. Usually cryptocrystalline in white, compact, earthy masses; also in cleavable granular masses, coarse to fine.

Cleavage: Rhombohedral {10$\bar{1}$1} perfect

H: $3\frac{1}{2}$–5

G: 3.0–3.2

Color: White, gray, yellow, brown. Luster vitreous. Transparent to translucent.

Occurrence: Common in veins and irregular masses derived from the alteration of Mg-rich metamorphic and igneous rocks; as such, a constituent of serpentinites and altered peridotites. Such magnesite is compact and cryptocrystalline and may be associated with opaline silica. Cleavable magnesite is found in talc, chlorite, and mica schists and in dolomitic limestones.

Remarks: The white massive variety resembles chert and is distinguished from it by a lower degree of hardness. Cleavable varieties are distinguished from dolomite by higher specific gravity. Magnesite, almost nonreactive in cold HCl, dissolves with effervescence in hot HCl.

SIDERITE

Crystal form: Commonly in rhombohedral crystals, which may show curved faces. Also in globular concretions and cleavable masses. May be botryoidal, compact, and earthy.

Cleavage: Rhombohedral {10$\bar{1}$1} perfect

H: $3\frac{1}{2}$–4

G: 3.96, for pure $FeCO_3$

Color: Usually light to dark brown. Luster vitreous. Transparent to translucent.

Occurrence: As a vein mineral, in well-crystallized form, it is associated with metallic ores containing silver minerals, pyrite, chalcopyrite, and galena. It also occurs as *clay ironstone,* with admixed clay minerals, in concretions with concentric layers. Also found in shales and coal measures as *blackband ore,* contaminated with carbonaceous material. It is also a common constituent of sedimentary Precambrian iron-formations, in association with chert and magnetite.

Remarks: Its brownish color and high specific gravity distinguish it from other carbonates. Soluble in powdered form in cold HCl or as fragments in hot HCl with effervescence. Distinguished from sphalerite by its rhombohedral cleavage.

RHODOCHROSITE

Crystal form: Rarely in rhombohedral crystals, which may show curved faces. Usually cleavable, massive; granular to compact.

Cleavage: Rhombohedral {10$\bar{1}$1} perfect

H: $3\frac{1}{2}$–4

G: 3.5–3.7

Color: Usually some shade of rose red; may be light pink to dark brown. Luster vitreous. Transparent to translucent.

Occurrence: A constituent of hydrothermal veins with ore minerals of silver, lead, and copper, and of manganese deposits.

Remarks: The combination of pink color and rhombohedral cleavage is highly diagnostic. Its hardness (4) distinguishes it from rhodonite, which has a hardness of 6. Soluble in hot HCl with effervescence.

SMITHSONITE

Crystal form: Usually reniform, botryoidal, or stalactitic. Rarely in small rhombohedral or scalenohedral crystals. Also in crystalline incrustations, or granular to earthy.

Cleavage: Rhombohedral {10$\bar{1}$1} perfect

H: 4–$4\frac{1}{2}$

G: 4.30–4.45

Color: Often dirty brown; may also be colorless, white, green, blue, or pink. The yellow variety is known as *turkey-fat* ore. Luster vitreous. Translucent.

Occurrence: Smithsonite is a zinc ore of supergene origin, usually found with zinc deposits in limestones. Associated with sphalerite, galena, hemimorphite, cerussite, calcite, and limonite.

Remarks: Effervesces in powdered form and in hot HCl, and is further characterized by its usually reniform, botryoidal, or stalactitic habit.

Color may be highly variable. Further distinguished from other carbonates by its hardness of 4 and high specific gravity.

Aragonite Group

Four relatively common carbonates are members of the aragonite group. These are aragonite, witherite, strontianite, and cerussite. These members are orthorhombic and isostructural with aragonite.

ARAGONITE

Crystal form:	Orthorhombic crystals with three common habits: (1) acicular pyramidal, usually in radiating groups of crystals; (2) tabular with prominent vertical pinacoid and prisms; and (3) in cyclic hexagonal-like twins with a basal plane, which is commonly striated in three different directions. Also in reniform, columnar, and stalactitic aggregates.
Cleavage:	Pinacoidal {010} distinct, prismatic {110} poor
H:	$3\frac{1}{2}$–4 (harder than calcite)
G:	2.95 (higher than calcite)
Color:	Colorless, white, pale yellow, and variously tinted. Luster vitreous. Transparent to translucent.
Occurrence:	Much less common than calcite, of which it is the orthorhombic polymorph. Deposited from hot springs, and associated with gypsum. As fibrous crusts on serpentine and in amygdaloidal cavities in basalt. In metamorphic assemblages of the blueschist facies, as a result of crystallization at high pressure but relatively low temperature. The pearly layer of many shells and the pearl itself are aragonite.
Remarks:	Effervesces in cold dilute HCl. Distinguished from calcite by its higher specific gravity and lack of rhombohedral cleavage. Cleavage fragments of columnar calcite are terminated by rhombohedral cleavage transecting the columnar crystals, whereas columnar aragonite has cleavage parallel to elongation. Distinguished from witherite and strontianite by lower specific gravity and lack of distinctive flame coloration (see flame test, footnote in exercise 28).

WITHERITE

Crystal form:	Orthorhombic, with crystals always twinned on {110} forming pseudohexagonal dipyramids by the intergrowth of three individuals. Crystals often deeply striated horizontally. Also botryoidal to globular; columnar or granular.
Cleavage:	Pinacoidal {010} distinct, prismatic {110} poor
H:	$3\frac{1}{2}$
G:	4.3 (high for a nonmetallic mineral)
Color:	Colorless, white, gray. Luster vitreous. Translucent.
Occurrence:	In hydrothermal veins associated with galena.
Remarks:	Soluble in cold HCl with effervescence, and gives a yellowish-green flame test (for Ba; see footnote about flame tests on p. 367). Its high specific gravity is characteristic, and it is distinguished from barite by its effervescence in acid.

STRONTIANITE

Crystal form:	Orthorhombic, usually in radiating clusters of acicular crystals (like type 1 of aragonite). Commonly twinned on {110} giving a pseudohexagonal appearance. Also columnar, fibrous, and granular.
Cleavage:	Prismatic {110} good
H:	$3\frac{1}{2}$–4
G:	3.7 (high for a nonmetallic mineral)
Color:	White, gray, yellow, green. Luster vitreous. Transparent to translucent.
Occurrence:	Occurs in low-temperature hydrothermal veins and deposits associated with barite, celestite, and calcite. Also present in sulfide veins, and a rare constituent of igneous rocks.
Remarks:	Dissolves with effervescence in dilute HCl, and gives a crimson flame test (for Sr). Characterized by high specific gravity and effervescence in HCl. Can be distinguished from witherite and aragonite by flame test (see footnote in exercise 28). Distinguished from celestite by poorer cleavage and effervescence in acid.

CERUSSITE

Crystal form:	Orthorhombic crystals commonly with varied habit and many forms. Often tabular on {010}, and in reticulated groups of plates crossing each other at 60° angles. Occurs as pseudo-hexagonal twins with deep reentrant angles in the vertical zone. Also in granular crystalline aggregates; fibrous, compact, and earthy.
Cleavage:	Prismatic {110} good, {021} fair
H:	3–$3\frac{1}{2}$
G:	6.65 (very high for a nonmetallic mineral)
Color:	Colorless, white, or gray. Luster adamantine. Transparent to subtranslucent.
Occurrence:	An important and common supergene lead ore associated with primary minerals such as galena and sphalerite, and secondary minerals such as anglesite, smithsonite, and limonite.
Remarks:	Recognized by its high specific gravity (very high for a nonmetallic mineral), white color, and adamantine luster.

Dolomite Group

This carbonate group includes three end-member compositions, all of which are isostructural. These are dolomite, $CaMg(CO_3)_2$, ankerite, $CaFe(CO_3)_2$, and kutnahorite, $CaMn(CO_3)_2$. Of these three dolomite is by far the most common, but considerable solid solution occurs in dolomite toward the Fe-rich end-member ankerite. In the following description dolomite and ankerite are treated together.

DOLOMITE–ANKERITE

Crystal form: *Dolomite:* Usually as rhombohedra, often with curved faces, and when strongly curved known as "saddle-shaped" crystals. Other forms are rare. In coarse, cleavable masses to fine-grained to compact.

Ankerite: Generally not found in well-formed crystals. When in crystals, they resemble those of dolomite.

Cleavage: Rhombohedral $\{10\bar{1}1\}$ perfect

H: $3\frac{1}{2}$–4

G: 2.85 (for dolomite) and ~ 3.1 (for ankeritic compositions)

Color: *Dolomite:* Colorless, white, gray, green, or some shade of light pink, or flesh color.
Ankerite: Typically yellowish white to yellowish brown (owing to oxidation of some of the iron). Luster vitreous. Transparent to translucent.

Occurrence: *Dolomite:* Most common in sedimentary rocks known as *dolomite* or *dolostone,* and in *dolomitic marble.* Also as a hydrothermal vein mineral, especially in lead and zinc veins that traverse limestone, associated with fluorite, calcite, barite, and siderite.

Ankerite: a common carbonate in Precambrian iron-formations in association with chert, magnetite, and hematite.

Remarks: Dolomite and ankerite have similar properties except for the color differences already noted. In cold, dilute HCl large fragments of dolomite and ankerite are slowly attacked; they become soluble, with effervescence, only in hot HCl. Powdered mineral is readily soluble in cold acid. The crystallized variety is recognized by the curved rhombohedral crystals and flesh-pink color. The massive rock variety is distinguished from limestone by much less vigorous reaction with HCl.

MALACHITE

Crystal form: Slender monoclinic, prismatic crystals but seldom distinct. Crystals may be pseudomorphous after azurite. Usually in botryoidal or stalactitic masses with radiating fibers. Also granular and earthy.

Cleavage: Pinacoidal $\{\bar{2}01\}$ perfect but rarely seen

H: $3\frac{1}{2}$–4

G: 3.9–4.03

Color: Bright green. Luster adamantine to vitreous in crystals; often silky in fibrous varieties; dull in earthy type. Translucent. Streak pale green.

Occurrence: A common supergene copper mineral found in the oxidized portions of copper veins associated with azurite, cuprite, native copper, and iron oxides.

Remarks: Soluble in cold dilute HCl with effervescence, giving a green solution. Recognized by its bright green color and botryoidal forms, and distinguished from other green copper minerals by its effervescence in acid.

AZURITE

Crystal form: Monoclinic with crystals frequently complex and malformed. Also in radiating spherical groups.

Cleavage: Prismatic $\{011\}$ perfect, pinacoidal $\{100\}$ fair

H: $3\frac{1}{2}$–4

G: 3.77

Color: Intense azure blue. Luster vitreous. Transparent to translucent.

Occurrence: Less common than malachite but of the same origin and associations.

Remarks: Characterized chiefly by its effervescence in cold dilute HCl, its hardness, its azure-blue color, and common association with malachite.

Sulfates

The most important and common anhydrous sulfates are those of the barite group, which includes barite, celestite, and anglesite. Another fairly common anhydrous sulfate is anhydrite. The most common hydrous sulfate to be considered is gypsum.

BARITE

Crystal form: Orthorhombic crystals usually tabular on $\{001\}$. Crystals may be very complex. Frequently in divergent groups or tabular crystals forming "crested barite" or "barite roses." Also coarsely laminated, granular, or earthy.

Cleavage: Pinacoidal $\{001\}$ perfect, with prismatic $\{210\}$ less perfect

H: 3–$3\frac{1}{2}$

G: 4.5 (heavy for a nonmetallic mineral)

Color: Colorless, white, and light shades of blue, yellow, red. Luster vitreous; on some specimens pearly on the base. Transparent to translucent.

Occurrence: Occurs most commonly in hydrothermal veins associated with ores of silver, lead, copper, cobalt, and manganese.

Remarks: Recognized by its high specific gravity, characteristic cleavage, and crystal form. Gives a yellowish-green flame test (for Ba; see footnote about flame tests on p. 367).

CELESTITE

Crystal form: Orthorhombic crystals closely resemble those of barite. Commonly tabular on $\{001\}$ or prismatic parallel to the *a* or *b* axis. Also radiating fibrous; granular.

Cleavage: Basal $\{001\}$ perfect, prismatic $\{210\}$ good

H: 3–$3\frac{1}{2}$

G: 3.95–3.97

Color: Colorless, white, often faintly blue or red. Luster vitreous to pearly. Transparent to translucent.

Occurrence: Most commonly as disseminations through limestone or sandstone, or in nests and lining cavities in such rocks. Associated with calcite, dolomite, gypsum, halite, sulfur, and fluorite. Common in lead veins with galena.

Remarks:	Closely resembles barite but is differentiated by lower specific gravity and a crimson flame test (for Sr). Fine radiating or fibrous habit of some celestite is also characteristic, as may be its usual pale blue color.

ANGLESITE

Crystal form:	Orthorhombic crystals with habit similar to that of barite but more varied. Also massive, granular to compact. Frequently earthy, in concentric layers that may have an unaltered core of galena.
Cleavage:	Basal {001} good, prismatic {210} imperfect. Fracture conchoidal.
H:	3.0
G:	6.2–6.4 (very high for a nonmetallic mineral)
Color:	Colorless, white, gray, pale shades of yellow. May be colored dark gray by impurities. Luster adamantine when crystalline, dull when earthy. Transparent to translucent.
Occurrence:	A common supergene mineral found in the oxidized portions of lead deposits. Associated with galena, cerussite, sphalerite, smithsonite, hemimorphite, and iron oxides.
Remarks:	Recognized by its high specific gravity, its adamantine luster, and its common association with galena.

ANHYDRITE

Crystal form:	Orthorhombic but rarely in crystals. Usually massive or in crystalline masses resembling an isometric mineral with cubic cleavage. Also fibrous, granular, massive.
Cleavage:	Three pinacoidal cleavages: {010} perfect, {100} nearly perfect, and {001} good.
H:	$3-3\frac{1}{2}$
G:	2.89–2.98
Color:	Colorless to bluish or violet. May also be white or tinged with rose, red, or brown. Luster vitreous to pearly on cleavage.
Occurrence:	Is found in much the same manner as gypsum, with which it is commonly associated, but is less common than gypsum. Occurs in beds associated with salt deposits in the cap rock of salt domes, and in limestones. Also in amygdaloidal cavities in basalts.
Remarks:	Characterized by its three cleavages at right angles. It is distinguished from calcite by its higher specific gravity and from gypsum by its greater hardness. The cleavages are markedly different from those of barite. Some massive anhydrite varieties may be difficult to recognize as such.

GYPSUM

Crystal form:	Monoclinic crystals of simple habit. Commonly tabular on {010}; diamond-shaped with beveled edges. Twinning on {100} common, resulting in swallowtail twins.
Cleavage:	Pinacoidal {010} perfect, yielding thin folia; {100} with conchoidal surface; {110} with fibrous fracture.
H:	2 (can be scratched by the fingernail)
G:	2.32
Color:	Colorless, white, gray; various shades of yellow, red, brown, from impurities. Luster usually vitreous; also pearly and silky. Transparent to translucent. *Satin spar* is a fibrous gypsum with silky luster. *Alabaster* is a fine-grained massive variety. *Selenite* is a variety that yields broad, colorless, and transparent cleavage folia.
Occurrence:	Gypsum is commonly distributed in sedimentary rocks, often as thick beds. Occurs interstratified with limestone and shale and as a layer underlying beds of rock salt. Frequently formed by the alternation of anhydrite. Also found in volcanic regions, and as a gangue mineral in metallic veins. Common associations are halite, anhydrite, dolomite, calcite, sulfur, pyrite, and quartz.
Remarks:	Characterized by its softness and its three unequal cleavages.

Tungstate

Of the various tungstates, only one relatively common species will be described, namely scheelite, $CaWO_4$.

SCHEELITE

Crystal form:	Tetragonal dipyramidal crystals of which the dipyramid {112} closely resembles the octahedron in angles. Also massive granular.
Cleavage:	Prismatic {101} distinct
H:	$4\frac{1}{2}-5$
G:	5.9–6.1 (very high for a nonmetallic mineral)
Color:	White, yellow, green, brown. Luster vitreous to adamantine. Translucent, some specimens transparent. Most scheelite will fluoresce with bluish-white color in short ultraviolet radiation.
Occurrence:	Found in granite pegmatites, contact metamorphic deposits, and high-temperature hydrothermal veins associated with granitic rocks. Common associations are cassiterite, topaz, fluorite, apatite, and molybdenite.
Remarks:	Recognized by its high specific gravity, crystal form, fluorescence in shortwave ultraviolet light, and generally light color.

Phosphates

Of the many phosphate minerals, only a few are common. Here descriptions will be given for apatite, amblygonite, and turquoise.

APATITE

Crystal form:	Hexagonal, commonly in crystals of long prismatic habit; some short prismatic or

tabular. Also massive granular to compact masses.

Cleavage: Basal {0001} poor

H: 5 (can just be scratched by a knife)

G: 3.15–3.20

Color: Usually some shade of green or brown; also blue, violet, colorless. Luster vitreous to subresinous. Transparent to translucent.

Occurrence: Apatite is widely distributed in igneous, metamorphic, and sedimentary rocks. It is also found in pegmatites and other veins, probably of hydrothermal origin. Phosphate materials of bones and teeth are members of the apatite group.

The apatite in igneous rocks is commonly in very fine-grained to microscopic crystals in accessory amounts. It can occur in very large crystals in coarsely crystalline limestones and marbles. The variety *collophane* (a massive, cryptocrystalline type) constitutes the bulk of *phosphorite* or *phosphate rock*.

Remarks: Usually recognized by a combination of its color, hexagonal crystal form, and hardness. It is distinguished from beryl by inferior hardness and from quartz by color and hardness. Massive, granular varieties may resemble diopside, but apatite is of inferior hardness.

AMBLYGONITE

Crystal form: Triclinic but usually occurs in coarse, cleavable masses.

Cleavage: {100} perfect, {110} good, and {0$\bar{1}$1} distinct

H: 6

G: 3.0–3.1

Color: White to pale green or blue, rarely yellow. Luster vitreous, pearly on {100} cleavage. Translucent.

Occurrence: Found mainly in Li-bearing granite pegmatites in association with spodumene, tourmaline, lepidolite, and apatite.

Remarks: Resembles albite, with which it is commonly associated. However, different cleavage angles and the less perfect cleavage of amblygonite, as well as the higher specific gravity, distinguish it from albite. A flame test gives a diagnostic red coloring for Li. See footnote about flame tests on p. 367.

TURQUOISE

Crystal form: Rarely in minute triclinic crystals, usually cryptocrystalline. Massive compact, reniform, stalactitic. In thin seams, incrustations, and disseminated grains.

Cleavage: Basal {001} perfect

H: 6

G: 2.6–2.8

Color: Blue, bluish green, green. Luster waxlike. Transmits light on thin edges.

Occurrence: A secondary mineral found in small veins and stringers traversing more or less decomposed volcanic rocks in arid regions.

Remarks: Easily recognized by its color. It is harder than chrysocolla, the only common mineral that it resembles.

TABLE 29.1. Tabulation of physical properties and subsequent mineral identification for three unknowns.

Specimen number: . . . .

Crystal Morphology/Symmetry/Twinning/Striations:

Cleavage/Parting/Fracture:

Hardness:

Luster:

Streak:

Color(s)

Tenacity:

Specific gravity (heft):

Other distinguishing features (effervescence in HCl, magnetism, smell, exsolution lamellae, etc.):

Name:

Mineral classification; Formula:

Association:

Specimen number: . . . .

Crystal Morphology/Symmetry/Twinning/Striations:

Cleavage/Parting/Fracture:

Hardness:

Luster:

Streak:

Color(s)

Tenacity:

Specific gravity (heft):

Other distinguishing features (effervescence in HCl, magnetism, smell, exsolution lamellae, etc.):

Name:

Mineral classification; Formula:

Association:

Specimen number: . . . .

Crystal Morphology/Symmetry/Twinning/Striations:

Cleavage/Parting/Fracture:

Hardness:

Luster:

Streak:

Color(s)

Tenacity:

Specific gravity (heft):

Other distinguishing features (effervescence in HCl, magnetism, smell, exsolution lamellae, etc.):

Name:

Mineral classification; Formula:

Association:

TABLE 29.2. Tabulation of physical properties and subsequent mineral identification for ten unknowns.

Specimen number	Luster (metallic versus non-metallic)	Hardness	Cleavage or fracture	Crystal form	Color	Other (streak, texture, striations)	Mineral Name

Please make photocopies of this form if more copies are needed.

Common Native Elements, Sulfides, and a Sulfosalt and their Physical Properties

PURPOSE OF EXERCISE

An overview of some of the most diagnostic physical properties and common occurrences of native elements and sulfides.

FURTHER READING AND CD-ROM INSTRUCTIONS

Klein, C. and Dutrow, B., 2008, *Manual of Mineral Science*, 23rd ed., Wiley, Hoboken, New Jersey, pp. 331–367.

CD-ROM, entitled *Mineralogy Tutorials*, version 3.0, that accompanies *Manual of Mineral Science*, 23rd ed., click on "Module IV", and subsequently on the button "Sulfides, Halides, and Native Elements".

Nesse, W. D., 2000, *Introduction to Mineralogy*, Oxford University Press, New York, pp. 378–404.

Perkins, D., 2002, *Mineralogy*, 2nd ed., Prentice Hall, Upper Saddle River, New Jersey, pp. 346–361.

Wenk, H. R. and Bulakh, A., 2004, *Minerals: Their Constitution and Origin*, Cambridge University Press, New York, New York, pp. 337–346; 388–405.

Deer, W. A., Howie, R. A. and Zussman, J., 1992, *An Introduction to The Rock-Forming Minerals*, 2nd ed., John Wiley and Sons, Hoboken, New Jersey, pp. 583–605.

Background Information: The study of hand specimens of individual minerals concludes with native elements, which include *native metals* as well as *nonmetals*, the diverse and large group of *sulfides*, and a *sulfosalt*.

MATERIALS

Representative collections of members of the native element group: among the native metals, gold, silver, and copper; among the native nonmetals, sulfur, diamond, and graphite; and of the sulfide group, chalcocite, bornite, galena, sphalerite, chalcopyrite, pyrrhotite, pentlandite, covellite, cinnabar, realgar, orpiment, stibnite, pyrite, marcasite, molybdenite, and arsenopyrite; and among the sulfosalts, enargite. Basic testing tools for hardness, specific gravity, and streak as well as a hand lens, a binocular microscope, or both are needed. Index cards or a notebook should be available for abstracting the most important data for each mineral (see Fig. 24.1). Tables 30.1 and 30.2 can also be used.

ASSIGNMENT

Hand specimen study of the various specimens available of the native elements, sulfides, and sulfosalts. Consult the listing of highly diagnostic physical properties that follows, as well as the determinative tables and mineral descriptions in the above listed textbooks.

Native Elements

Among the native elements, the most common and important *native metals* are members of the *gold group:* gold, silver, and copper, all of which are isometric and isostructural and have metallic luster and high specific gravity. The important *nonmetals* are sulfur and carbon, in the form of diamond and graphite.

GOLD

Crystal form: Isometric, with crystals commonly octahedral. Often in arborescent crystal groups. Crystals may be irregularly shaped, passing into filliform, reticulated, and dendritic shapes. Most commonly in irregular plates, scales, or masses. Also in rounded or flattened grains, "nuggets."

H: $2\frac{1}{2}$–3; fracture hackly. Very malleable and sectile.

G: 19.3 when pure. The presence of other metals (usually silver) decreases the specific gravity, which may be as low as 15.

Color: Various shades of yellow depending on the purity, becoming paler with increasing silver content. Metallic luster.

Occurrence: The chief sources of gold are (1) hydrothermal gold–quartz veins in which it is commonly so fine-grained that its presence cannot be detected with the eye, and (2) placer deposits, and their lithified equivalents known as conglomerates, in which the gold occurs as fine detrital grains and nuggets.

Remarks: The chances are few that you will be handling gold specimens in the laboratory! However, highly secured exhibits are commonly present in natural history museums. Gold is characterized by its color, malleability, and sectility.

In small grains it can be easily confused with pyrite and chalcopyrite, hence the common designation *fool's gold* for these minerals. Pyrite has a much greater hardness, and chalcopyrite is brittle instead of malleable and sectile.

SILVER

Crystal form:	Isometric crystals are commonly malformed and in branching, arborescent, or reticulated groups. Also in coarse or fine wire. Usually in irregular masses, plates, and scales.
H:	$2\frac{1}{2}$–3. Fracture hackly. Malleable and ductile.
G:	10.5 when pure, 10–12 when impure.
Color:	Silver white, often tarnished to brown or gray-black. Streak silver white. Luster metallic. Opaque.
Occurrence:	Native silver in large deposits has been precipitated from primary hydrothermal solutions. It is also distributed, in smaller amounts, in the oxidized zones of ore deposits.
Remarks:	As is true of gold, you will probably have no chance to handle specimens of native silver in the laboratory. However, good silver specimens are commonly seen in the exhibits of natural history museums. Native silver is distinguished by its malleability, color on the fresh surface, and high specific gravity.

COPPER

Crystal form:	Isometric crystals with tetrahexahedral, cube, dodecahedral, and octahedral faces are common. Crystals are usually malformed and in branching and arborescent groups. Usually in irregular masses, plates, and scales, and twisted and wirelike forms.
H:	$2\frac{1}{2}$–3. Highly ductile and malleable. Fracture hackly.
G:	8.9
Color:	Copper red on fresh surface, usually dark with dull luster because of tarnish.
Occurrence:	Small amounts of native copper occur in the oxidized zones of copper deposits associated with cuprite, malachite, and azurite. Primary deposits of native copper are associated with basaltic lavas, where copper was deposited through the reaction of hydrothermal solutions with iron oxide minerals.
Remarks:	Native copper is recognized by its red color on fresh surfaces, hackly fracture, high specific gravity, and malleability.

SULFUR

Crystal form:	Orthorhombic crystals with pyramidal habit, commonly with two dipyramids, prism {011}, and a base in combination. Frequently in masses, imperfectly crystallized. Also massive, reniform, stalactitic, as incrustations, earthy.
Cleavage:	None. Fracture conchoidal to uneven. Brittle.
H:	$1\frac{1}{2}$–$2\frac{1}{2}$
G:	2.05–2.09
Color:	Sulfur yellow, varying with impurities to yellow shades of green, gray, and red. Luster resinous. Transparent to translucent.
Occurrence:	Is found commonly at or near the crater rims of active or extinct volcanoes, deposited from volcanic gases. It also occurs in veins with metallic sulfides formed by the oxidation of the sulfides. It is most common in Tertiary sedimentary rocks associated with anhydrite, gypsum, and limestone. Other associations include celestite, gypsum, calcite, and aragonite.
Remarks:	Sulfur is a poor conductor of heat. When a crystal is held in the hand close to the ear, it will be heard to crack. The surface layers expand because of the heat of the hand, but the interior, because heat conduction is slow, is unaffected, and thus the crystal cracks. Crystals of sulfur should therefore be handled with care.

Sulfur is identified by its yellow color, absence of cleavage (which distinguishes it from orpiment), low degree of hardness, and brittleness. |

DIAMOND

Crystal form:	Isometric crystals with octahedral form most common; cubes and dodecahedrons may occur as well. Octahedral faces are frequently curved, and twins on {111} (spinel law) are common. *Bort*, a variety of diamond, has rounded forms and a rough exterior resulting from a radial or cryptocrystalline aggregate. The term is also applied to badly colored or flawed diamonds without gem value.
Cleavage:	Octahedral {111} perfect
H:	10 (hardest known mineral)
G:	3.51
Color:	Usually pale yellow or colorless; also pale shades of red, orange, blue, green, and brown. Deeper shades are rare. Luster adamantine; uncut crystals have a characteristic greasy appearance. *Carbonado* or *carbon* is a black or grayish-black bort. It is noncleavable, opaque, and less brittle than crystals.
Occurrence:	The primary occurrence of diamonds is in altered peridotite called *kimberlite*, as intrusive bodies that are commonly circular with a pipelike shape and referred to as "diamond pipes." Kimberlite consists of rounded and corroded phenocrysts of olivine, phlogopite, ilmenite, and pyrope in a fine-grained matrix of olivine, phlogopite, serpentine, spinels, calcite, and/or dolomite. A large percentage of diamonds are recovered from alluvial deposits (known as *placers*), where they accumulate because of their inert nature, great hardness, and fairly high specific gravity.
Remarks:	Diamond is distinguished from minerals that resemble it by its great hardness, adamantine luster, and cleavage. Although you will undoubtedly not be handling cut or uncut diamonds in the laboratory, you might look at several diamond displays in jewelers' windows or cases. You may also look for

a synthetic diamond simulant known as cubic zirconia (abbreviated CZ), which has properties so much like those of diamond that it is a very successful and relatively inexpensive substitute for gem diamond. The size (and weight) of diamonds, and other gemstones, is expressed in *carats*, where 1 carat = 200 milligrams.

GRAPHITE

Crystal form: Tabular crystals with hexagonal outline and a prominent basal face are quite common. Usually in foliated or scaly masses; also radiated or granular.

Cleavage: Basal {0001} perfect. Folia flexible but not elastic.

H: 1–2 (readily marks paper and soils the fingers). Greasy feel.

G: 2.23

Color: Black, with black streak. Luster metallic, sometimes dull, earthy.

Occurrence: Graphite occurs commonly in metamorphic rocks such as crystalline limestones, schists, and gneisses. It is also present in hydrothermal veins associated with quartz, orthoclase, tourmaline, apatite, pyrite, and titanite. It may also occur as an original constituent of some igneous rocks.

Remarks: It is recognized by its black color, foliated nature, extreme softness, greasy feel, low specific gravity, and its easy marking on paper. It is distinguished from molybdenite by its black color (molybdenite has a blue tone), and black streak on porcelain (molydenite has a grayish-black streak).

SULFIDES

The sulfide group of minerals includes the majority of metallic ore minerals. With them are classified the sulfosalts, the most common species of which, enargite, will be discussed.

Most of the sulfides are opaque with distinctive colors and characteristically colored streaks and with relatively high specific gravities. Those that are nonopaque, such as cinnabar, realgar, and orpiment, transmit light only on their edges.

CHALCOCITE

Crystal form: Usually not in crystals, most commonly massive and very fine-grained. If in orthorhombic crystals, usually small and tabular with hexagonal outline; striated parallel to the *a* axis.

Cleavage: Prismatic {110} poor. Fracture conchoidal.

H: $2\frac{1}{2}$–3, imperfectly sectile

G: 5.5–5.8

Color: Shining lead gray, tarnished to dull black upon exposure. Streak grayish black. Luster metallic, although some chalcocite is soft and sooty.

Occurrence: This is one of the most important copper ore minerals. It occurs as a primary mineral in hydrothermal veins, with bornite, chalcopyrite,

enargite, and pyrite. Its most common occurrence is as a supergene mineral in enriched zones of sulfide deposits, often forming "chalcocite blankets" at the level of the water table. It is also a constituent of "porphyry copper" deposits, disseminated throughout porphyritic granodiorite intrusives. (Porphyritic is the term for the texture of an igneous rock in which larger crystals—*phenocrysts*—are set in a finer-grained groundmass.)

Remarks: It is distinguished by its lead-gray color, a low degree of hardness, and slight sectility.

BORNITE

Crystal form: Rarely in tetragonal crystals; usually massive.

H: 3

G: 5.06–5.08

Color: Brownish bronze on fresh surface but quickly tarnished to variegated purple and blue (hence called *peacock ore*) and finally to almost black on exposure. Luster metallic. Streak grayish black.

Occurrence: Commonly associated with other sulfides such as chalcocite, chalcopyrite, covellite, pyrrhotite, and pyrite in hypogene deposits. It occurs less frequently as a supergene mineral in the upper enriched parts of copper veins. Also found as disseminations in basic rocks, in contact metamorphic deposits, and in pegmatites.

Remarks: Distinguished by its characteristic brown color on the fresh fracture and the purplish tarnish.

GALENA

Crystal form: Isometric crystals with the most common form the cube, sometimes truncated by the octahedron. Commonly in coarse- to fine-grained cleavable and granular masses.

Cleavage: Cubic {001} perfect

H: $2\frac{1}{2}$

G: 7.4–7.6

Color: Lead-gray with lead-gray streak. Luster bright metallic.

Occurrence: Galena is a very common metallic sulfide, found in veins associated with sphalerite, marcasite, chalcopyrite, cerussite, anglesite, calcite, quartz, barite, and fluorite. It also occurs in contact metamorphic deposits and in pegmatites.

Remarks: Easily recognized by its good cubic cleavage, high specific gravity, softness, and lead-gray color and streak. It is distinguished from stibnite by its cubic cleavage, specific gravity, and darker color.

SPHALERITE

Crystal form: Isometric with the tetrahedron, dodecahedron, and cube as common forms. Crystals are frequently highly complex and usually malformed or in rounded aggregates. Usually in cleavable masses, coarse- to fine-granular. Also compact, botryoidal, cryptocrystalline.

Cleavage: Dodecahedral {011} perfect but some sphalerite is too fine-grained to show cleavage.

H: $3\frac{1}{2}$–4

G: 3.9–4.1

Color: Colorless when pure ZnS and green when nearly so. Commonly yellow, brown to black, darkening with increase in iron. Also red (*ruby zinc*). Luster nonmetallic and resinous to submetallic; also adamantine. Transparent to translucent. Streak white to yellow and brown.

Occurrence: Sphalerite, the most common ore mineral of zinc, is widely distributed. Its occurrence and mode of origin are similar to those of galena, with which it is commonly associated.

Remarks: Recognized by its striking resinous luster, perfect cleavage, and hardness ($3\frac{1}{2}$–4). Sphalerite may be difficult to recognize because of its wide variability in color and luster, with yellow, yellow-brown, brown, and dark brown the most common colors. The dark varieties (*black jack*) can be told by the reddish-brown streak, always lighter than that of the massive mineral.

CHALCOPYRITE

Crystal form: Tetragonal, commonly appearing tetrahedral on account of the disphenoid {112}. Usually massive.

H: $3\frac{1}{2}$–4; brittle

G: 3.1–3.4

Color: Brass yellow; often tarnished to bronze or irridescent. Luster metallic. Streak greenish black.

Occurrence: Very common in occurrence and one of the most important ore minerals of copper. In low-temperature hydrothermal vein deposits it is found with galena, sphalerite, and dolomite. In higher-temperature deposits it may occur with pyrrhotite and pentlandite. It is a primary mineral of porphyry copper deposits (see also chalcocite). It is also found as a primary constituent of igneous rocks; in pegmatites; in contact metamorphic deposits; and disseminated in metamorphic schists.

Remarks: Recognized by its brass-yellow color and greenish-black streak. Distinguished from pyrite by being softer than a steel knife and from gold by being brittle. Known as "fool's gold," a term also applied to pyrite.

PYRRHOTITE

Crystal form: Hexagonal (for the high-temperature form), but crystals uncommon. Usually massive, granular.

H: 4

G: 4.58–4.65

Color: Brownish bronze. Luster metallic. Streak black. Opaque.

Magnetism: Intensity may be variable. Generally easily attracted to a magnet, but much less magnetic than magnetite.

Occurrence: Occurs as disseminated grains in basic igneous rocks such as norites. In sulfide ore deposits commonly associated with pentlandite, chalcopyrite, and other sulfides.

Remarks: It is distinguished from chalcopyrite by its color and magnetism, and from pyrite by color and hardness. The massive variety may be associated with pentlandite, from which it is distinguished only with difficulty (see pentlandite, and Ni test).

PENTLANDITE

Crystal form: Isometric, but most commonly massive, in granular aggregates with octahedral parting.

Cleavage: None; octahedral parting on {111}.

H: $3\frac{1}{2}$–4

G: 4.6–5.0

Color: Yellowish bronze. Luster metallic. Streak light bronze brown. Opaque.

Occurrence: The major ore mineral for nickel, found in basic igneous rocks associated with nickel sulfides, pyrrhotite, and chalcopyrite; probably formed by processes of magmatic segregation.

Remarks: Pentlandite closely resembles pyrrhotite, with which it is commonly associated. It is best distinguished from pyrrhotite by its octahedral parting and lack of magnetism. The presence of nickel can be quickly established by a test with dimethylglioxime. (When a little dimethylglioxime powder is dissolved in some water on a pentlandite specimen, a scarlet coloration forms, indicating the presence of nickel.)

COVELLITE

Crystal form: Rarely in tabular, hexagonal crystals. Usually massive as coatings or disseminations through other copper minerals.

Cleavage: Basal {0001} perfect

H: $1\frac{1}{2}$–2

G: 4.6–4.76

Color: Indigo blue or darker. Often irridescent. Luster metallic. Streak lead gray to black. Opaque.

Occurrence: It is not an abundant mineral but is commonly found in most copper ore deposits as a supergene mineral. It is found with other copper minerals, mainly chalcocite, chalcopyrite, bornite, and enargite, and derived from them by alteration.

Remarks: Characterized by the indigo-blue color, micaceous cleavage yielding flexible plates, and association with other copper minerals.

CINNABAR

Crystal form: Hexagonal (for the low-temperature form), with rhombohedral or thick-tabular crystals (with well-developed base, {0001}) most common. Generally fine-granular, massive; also earthy, as incrustations and disseminations through the host rock.

Cleavage: Rhombohedral {10$\overline{1}$0} perfect

H: $2\frac{1}{2}$

G: 8.10 (when pure)

Color: Vermillion red when pure to brownish red when impure. Streak scarlet. Transparent to translucent.

Occurrence: Cinnabar is the most important ore of mercury but is found at only a few localities. It occurs as impregnations and vein fillings near recent volcanic rocks and hot springs. Associated with pyrite, marcasite, stibnite, and sulfides of copper. Gangue minerals may include opal, chalcedony, quartz, barite, calcite, and fluorite.

Remarks: Recognized by its red color, scarlet streak, high specific gravity, and cleavage.

REALGAR

Crystal form: Monoclinic, but crystals uncommon. Generally coarse- to fine-granular and often earthy and as encrustations.

Cleavage: Pinacoidal {010} good

H: $1\frac{1}{2}$–2, sectile

G: 3.48

Color: Red to orange. Streak also red to orange. Luster resinous. Translucent to transparent.

Occurrence: Found in hydrothermal veins with orpiment, associated with lead and silver minerals, as well as with stibnite. It also occurs as a volcanic sublimation product and as a deposit from hot springs.

Remarks: Distinguished by its red color, orange-red streak, resinous luster, and almost invariable association with orpiment.

ORPIMENT

Crystal form: Monoclinic, but crystals rare. Usually in foliated, columnar, or fibrous masses.

Cleavage: Pinacoidal {010} perfect. Cleavage laminae flexible but not elastic.

H: $1\frac{1}{2}$–2, sectile

G: 3.49

Color: Lemon yellow. Streak pale yellow. Luster resinous. Translucent.

Occurrence: A rare mineral, associated usually with realgar and formed under similar conditions.

Remarks: Characterized by its yellow color, foliated structure, and excellent cleavage. Distinguished from sulfur by its cleavage and sectility.

STIBNITE

Crystal form: Orthorhombic. Crystals common. Slender prismatic habit with the prism zone vertically striated. Most prisms appear bent or twisted as a result of translation gliding. Often in radiating crystal groups or bladed forms with prominent cleavage. Also massive, coarse- to fine-granular.

Cleavage: Pinacoidal {010} perfect, with striations parallel to [100]

H: 2

G: 4.52–4.62

Color: Lead-gray to black. Streak also lead-gray to black. Metallic luster, splendent on cleavage surfaces. Opaque.

Occurrence: Found in low-temperature hydrothermal veins or replacement deposits and in hotsprings deposits. It is associated with other antimony minerals that have formed as the product of its decomposition, and with galena, cinnabar, sphalerite, barite, realgar, orpiment, and gold.

Remarks: Characterized by its bladed habit, perfect cleavage in one direction, lead-gray color, and soft black streak. It is distinguished from galena by its cleavage, habit, and lower specific gravity.

PYRITE

Crystal form: Isometric. Frequently in crystals. The most common forms are the cube (the faces of which are usually striated), the pyritohedron, and the octahedron. Also massive, granular, reniform, globular, and stalactitic.

Cleavage: None. Fracture conchoidal. Brittle.

H: 6–$6\frac{1}{2}$ (unusually hard for a sulfide)

G: 5.02

Color: Pale brass yellow; may be darker because of tarnish. Luster metallic, splendent. Streak greenish or brownish black. Opaque.

Occurrence: One of the most common and widespread sulfide minerals, it is formed at both high and low temperatures. It occurs as magmatic segregations, as an accessory mineral in igneous rocks, and in contact metamorphic deposits and hydrothermal veins. It is also common in sedimentary rocks, especially shales and limestones. In vein deposits it is associated with many minerals but most frequently with chalcopyrite, sphalerite, and galena.

Remarks: Distinguished from chalcopyrite by its paler color and greater hardness, from gold by its brittleness and hardness, and from marcasite by its deeper color and crystal form. Pyrite is easily altered to limonite. Pseudomorphic crystals of limonite after pyrite are common. Pyrite veins are usually capped by a cellular deposit of limonite, termed *gossan*.

MARCASITE

Crystal form: Orthorhombic. Polymorphous with pyrite. Crystals commonly tabular on {010}; less commonly prismatic parallel to [001]. Often twinned, giving cockscomb and spear-headed groups. Usually in radiating forms. Often stalactitic, having an inner core with radiating structure and covered with irregular crystal groups. Also globular and reniform.

H: 6–$6\frac{1}{2}$

G: 4.89

Color: Pale bronze-yellow to almost white on fresh fracture. Yellow to brown tarnish. Streak grayish black. Luster metallic. Opaque.

Occurrence: It is found in metalliferous veins, frequently with lead and zinc ores. It is less stable than pyrite (being easily decomposed) and much less common. It is commonly found as a supergene mineral deposited at low temperatures, at near-surface conditions. It is also found as replacement deposits in limestone and in concretions in shales.

Remarks: Usually recognized and distinguished from pyrite by its pale yellow color, its crystals, or its fibrous habit.

MOLYBDENITE

Crystal form: Hexagonal. Crystals in hexagonal plates or short, slightly tapering prisms. Commonly foliated, massive, or in scales.

Cleavage: Basal {0001} perfect, with laminae flexible but not elastic.

H: $1-1\frac{1}{2}$, greasy feel

G: 4.62–4.73

Color: Lead-gray. Streak grayish black. Luster metallic. Opaque.

Occurrence: Found as an accessory mineral in some granites; in pegmatites and aplites. Common in high-temperature vein deposits associated with cassiterite, scheelite, and fluorite. Also in contact metamorphic deposits with Ca-silicates, scheelite, and chalcopyrite.

Remarks: Although molybdenite resembles graphite, it can be distinguished from it by higher specific gravity and by a blue tone to its color (graphite has a brown tinge). On glazed poreclain it gives a greenish streak, graphite a black streak.

ARSENOPYRITE

Crystal form: Monoclinic. Crystals are commonly prismatic and elongated along the c axis and less

commonly along the b. Sometimes with faces striated parallel to [101]. Also granular and compact.

Cleavage: Prismatic {101} poor

H: $5\frac{1}{2}-6$

G: 6.07

Color: Silver-white. Streak black. Luster metallic. Opaque.

Occurrence: Arsenopyrite is the most common mineral containing arsenic. It occurs with tin and tungsten ores in high-temperature hydrothermal deposits. Also associated with silver and copper ores, galena, sphalerite, pyrite, and chalcopyrite. Also found as an accessory in pegmatites, in contact metamorphic deposits, and disseminated in crystalline limestones.

Remarks: Distinguished from marcasite by its silver-white color. Unambiguous identification may require some chemical tests for As and S.

ENARGITE

Crystal form: Orthorhombic. Crystals elongated parallel to the c axis and vertically striated. Also tabular parallel to {001}. Columnar, bladed, massive.

Cleavage: Prismatic {110} perfect, {100} and {010} distinct

H: 3

G: 4.45

Color: Grayish black to iron-black. Streak also grayish black to iron-black. Luster metallic. Opaque.

Occurrence: Found in vein and replacement deposits formed at moderate temperatures associated with pyrite, sphalerite, bornite, galena, and chalcocite.

Remarks: Characterized by its color and cleavage. Striated crystals are also diagnostic.

TABLE 30.1. Tabulation of physical properties and subsequent mineral identification for three unknowns.

Specimen number: . . . .

Crystal Morphology/Symmetry/Twinning/Striations:

Cleavage/Parting/Fracture: Color(s)

Hardness: Tenacity:

Luster: Specific gravity (heft):

Streak: Other distinguishing features
                                                             (effervescence in HCl,
                                                             magnetism, smell, exsolution
                                                             lamellae, etc.):

_____

Name:

Mineral classification; Formula:

Association:

Specimen number: . . . .

Crystal Morphology/Symmetry/Twinning/Striations:

Cleavage/Parting/Fracture: Color(s)

Hardness: Tenacity:

Luster: Specific gravity (heft):

Streak: Other distinguishing features
                                                             (effervescence in HCl,
                                                             magnetism, smell, exsolution
                                                             lamellae, etc.):

_____

Name:

Mineral classification; Formula:

Association:

Specimen number: . . . .

Crystal Morphology/Symmetry/Twinning/Striations:

Cleavage/Parting/Fracture: Color(s)

Hardness: Tenacity:

Luster: Specific gravity (heft):

Streak: Other distinguishing features
                                                             (effervescence in HCl,
                                                             magnetism, smell, exsolution
                                                             lamellae, etc.):

_____

Name:

Mineral classification; Formula:

Association:

Please make photocopies of this form if more copies are needed.

TABLE 30.2. Tabulation of physical properties and subsequent mineral identification for ten unknowns.

Specimen number	Luster (metallic versus non-metallic)	Hardness	Cleavage or fracture	Crystal form	Color	Other (streak, texture, striations)	Mineral Name

Please make photocopies of this form if more copies are needed.

Igneous Rocks in Hand Specimens and their Classification

PURPOSE OF EXERCISE

Identification of minerals that together compose igneous rocks, and subsequent classification of igneous rocks on the basis of their mineralogy and textures.

FURTHER READING

Klein, C. and Dutrow, B., 2008, *Manual of Mineral Science*, 23rd ed., Wiley, Hoboken, New Jersey, pp. 574–585.

Nesse, W. D., 2000, *Introduction to Mineralogy*, Oxford University Press, New York, pp. 186–190.

Perkins, D., 2002, *Mineralogy*, 2nd ed., Prentice Hall, Upper Saddle River, New Jersey, pp. 88–93, 110–116.

Wenk, H. R. and Bulakh, A., 2004, *Minerals: Their Constitution and Origin*, Cambridge University Press, New York, New York, pp. 331–335, 483–490, 587–592.

Best, M. G., 2003, *Igneous and Metamorphic Petrology*, 2nd ed., Blackwell Publishing, Malden, Massachusetts, 729p.

Blatt, H., Tracey, R. J. and Owens, B. E., 2006, *Petrology: Igneous, Sedimentary and Metamorphic*, 3rd ed., W. H. Freeman, New York, New York, 540p.

Winter, J. D., 2001, *An Introduction to Igneous and Metamorphic Petrology*, Prentice Hall, Upper Saddle River, New Jersey, 697p.

Background Information:

a. *General occurrence and texture.* There are two types of igneous rocks, *extrusive* and *intrusive*. The first group includes igneous rocks that reached the earth's surface in a molten or partly molten state. Modern volcanoes produce lava flows that pour from a vent or fracture in the earth's crust. Such extrusive or *volcanic* rocks tend to cool and crystallize rapidly, with the result that their grain size is generally small. If the grain size of the rock is so fine that the mineral constituents cannot be distinguished by the unaided eye (or with a 10× magnification hand lens), it is referred to as *aphanitic*, from the Greek root *aphano*, meaning invisible. If the cooling has been so rapid as to prevent the formation of even small crystals of the mineral constituents, the resulting rock may be a *glass*. Ordinarily the mineral constituents of fine-grained extrusive rocks can be determined only by microscopic examination of thin sections of the rocks.

Intrusive or *plutonic* rocks are the result of crystallization from a magma that did not reach the earth's surface. Large magmatic intrusions that are irregularly shaped and discordant with the surrounding country rock are referred to as *batholiths* or *stocks*, depending on their size. A magma that is deeply buried in the earth's crust generally cools slowly, and the mineral constituents crystallizing from it have time to grow to considerable size, giving the rock a medium- to coarse-grained texture. The mineral grains in these rocks can generally be identified with the naked eye, and such rocks are referred to as *phaneritic*, from the Greek word *phaneros*, meaning visible. When a magma intrudes as a tabular body concordant with the country rock, it is known as a *sill*; if discordant, it is called a *dike*. The textures of sills and dikes are usually finer-grained than those of other plutonic rocks but coarser than those of volcanic rocks; these rocks of intermediate grain size are known as *hypabyssal*, from the Greek words *hypo*, meaning less than, and *abyssos*, meaning deep.

Some igneous rocks show distinct crystals of some minerals embedded in a much finer-grained or glassy matrix. The larger crystals are *phenocrysts*, and the finer-grained material is the *groundmass*. Such rocks are known as *porphyries*. The phenocrysts may vary in size from crystals an inch or more across down to very small individuals. The groundmass may be composed of fairly coarse-grained material, or its grains may be microscopic. The difference in size between the phenocrysts and the particles of the groundmass is the distinguishing feature of a porphyry. The porphyritic texture develops when some of the crystals grow to considerable size before the main mass of the magma consolidates into the finer-grained and uniformly grained material. Many types of igneous rocks may have a porphyritic variety, such as *granite porphyry*, *diorite porphyry*, *rhyolite porphyry*. Porphyritic varieties are most common in volcanic rocks.

b. *Chemical composition.* The bulk chemical compositions of igneous rocks exhibit a fairly limited range. The most abundant oxide component, SiO_2, ranges from about 40 to 75 weight percent in common igneous rock types (see Table 31.1 and Fig. 31.1). Al_2O_3 ranges generally from about 10 to 20 weight percent (except for peridotite and dunite; see analyses in Table 31.1) and each of the other major components generally does not exceed 10 weight percent (except MgO in peridotite and dunite; see Table 31.1).

When the magma is fairly low in SiO_2, the resulting rocks will contain mainly relatively silica-poor but Fe–Mg-rich minerals such as olivine, pyroxene, hornblende, or biotite, calcic plagioclase, and little or no free SiO_2 (i.e., quartz, cristobalite, tridymite; see Fig. 31.1). These rocks, which tend to be dark because of their high

TABLE 31.1   Average Chemical Composition of Some Igneous Rocks

Oxide	Nepheline Syenite	Syenite	Granite	Tonalite	Diorite	Gabbro	Peridotite	Dunite
SiO_2	54.83	59.41	72.08	66.15	51.86	48.36	43.54	40.16
TiO_2	0.39	0.83	0.37	0.62	1.50	1.32	0.81	0.20
Al_2O_3	22.63	17.12	13.86	15.56	16.40	16.84	3.99	0.84
Fe_2O_3	1.56	2.19	0.86	1.36	2.73	2.55	2.51	1.88
FeO	3.45	2.83	1.67	3.42	6.97	7.92	9.84	11.87
MnO	trace	0.08	0.06	0.08	0.18	0.18	0.21	0.21
MgO	trace	2.02	0.52	1.94	6.12	8.06	34.02	43.16
CaO	1.94	4.06	1.33	4.65	8.40	11.07	3.46	0.75
Na_2O	10.63	3.92	3.08	3.90	3.36	2.26	0.56	0.31
K_2O	4.16	6.53	5.46	1.42	1.33	0.56	0.25	0.14
H_2O	0.18	0.63	0.53	0.69	0.80	0.64	0.76	0.44
P_2O_5	—	0.38	0.18	0.21	0.35	0.24	0.05	0.04
Total	99.77	100.00	100.00	100.00	100.00	100.00	100.00	100.00

SOURCE: All analyses except that for nepheline syenite from S. R. Nockolds, 1954, *Geological Society of American Bulletin*, vol. 65, pp. 1007–1032.

TABLE 31.2   Simplified Classification of the Igneous Rocks

Feldspar	Quartz > 5%		No Quartz, No Feldspathoids		Nepheline or Leucite > 5%	
	Coarse	Fine	Coarse	Fine	Coarse	Fine
K-feldspar[a] > plagioclase	Granite	Rhyolite	Syenite	Trachyte	Foid syenite	Phonolite
					Foid monzosyenite	Tephritic phonolite
Plagioclase > K-feldspar	Granodiorite	Dacite	Monzonite	Latite	Foid monzodiorite	Phonolitic basanite (olivine > 10%)
Plagioclase (oligoclase or andesine)	Tonalite	Quartz andesite	Monzodiorite	Andesite basalt	Foid monzogabbro	Phonolitic tephrite (olivine < 10%)
Plagioclase (labradorite to anothite)	Quartz diorite	Andesite	Gabbro	Basalt	Foid diorite	Basanite (olivine > 10%)
					Foid gabbro	Tephrite (olivine < 10%)
No feldspar			Peridotite (olivine dominant) Pyroxenite (pyroxene dominant) Hornblendite (hornblende dominant)		Ijolite	Nephelinite (− olivine) Nepheline basalt (+ olivine)

[a]K-feldspar includes orthoclase, microcline, and microperthite; in high-T volcanic rocks it can be sanidine or anorthoclase.

SOURCE: From *Manual of Mineral Science*, 23rd ed., p. 580.

FIGURE 31.1   Relationships of variation in chemical and mineral compositions of igneous rocks. (From *Manual of Mineral Science*, 23rd ed., p. 577)

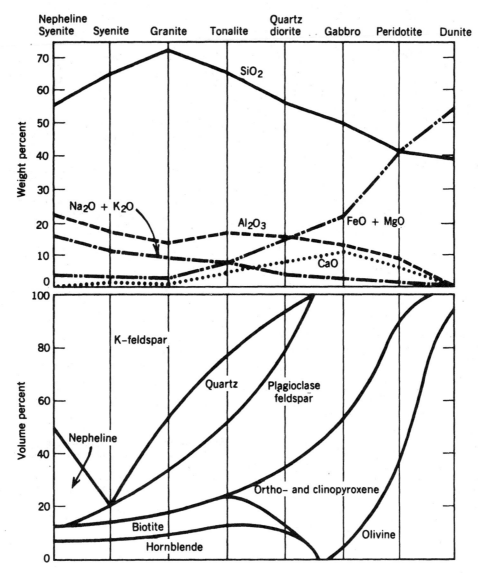

percentage of ferromagnesian minerals, are known as *mafic* rock types. Rocks rich in ferromagnesian minerals without plagioclase are known as *ultramafic*, that is, peridotite (see Table 31.2). When the melt is poor in SiO_2, *subsiliceous*, or silica-undersaturated, and high in alkalies and Al_2O_3 (as in nepheline syenite; see Table 31.1), the resultant crystallization products will contain SiO_2-poor minerals such as feldspathoids and will lack free SiO_2 as quartz (see Fig. 31.1). Crystallization of a melt high in SiO_2 (silica-over-saturated) forms rocks with abundant quartz and alkali feldspars, with or without muscovite, and only minor amounts of ferromagnesian minerals. Such rock types are referred to as *felsic* (alkali feldspar-rich) or *silicic*, and are lighter in overall color than mafic rocks. In general, the darker the rock the greater the abundance of ferromagnesian minerals and the lighter the rock the greater the abundance of quartz, feldspars, or feldspathoids.

c. *Classification*. Because of considerable variation in magmas in both chemistry and conditions of crystallization, igneous rock types show a wide variation in mineralogy and texture. There is a complete gradation from one rock type into another, so the names of igneous rocks and the boundaries between types are largely arbitrary (see Figs. 31.1 and 31.2).

Many schemes have been proposed for the classification of igneous rocks, but the most practical for the introductory student is based on mineralogy and texture. In general, four criteria are to be considered in classifying an igneous rock. (1) The relative amount of silica; quartz (or tridymite or cristobalite) indicates an excess of silica; feldspathoids indicate a deficiency of silica. (2) The kinds of feldspar (alkali feldspar versus plagioclase) and the relative amount of each kind. (3) The relative amounts and types of dark minerals. (4) The texture or size of the grains. Is the rock coarse- or fine-grained; that is, is it plutonic or volcanic?

It is clear that the exact determination of the kind of feldspar is impossible in the hand specimen. In many fine-grained rocks it is also impossible to recognize individual minerals. Such precise work must be carried out by the microscopic examination of thin sections of rocks. Nevertheless, it is important that the basis for the general classification be understood in order that a simplified hand specimen classification may have more meaning.

Three major divisions may be made on the basis of the silica content. (1) Quartz present in amounts greater than 5% (silica-oversaturated). (2) No quartz and no feldspathoids present (silica-saturated). (3) Feldspathoids in amounts greater than 5% (silica-undersaturated). These three divisions made on the basis of silica content are then subdivided according to the kind and amount (or the absence) of feldspar. Most of the rocks thus classified have coarse-grained and fine-grained varieties, which receive different names. Figure 31.2 illustrates the classification of the principal plutonic and volcanic rock types.

Table 31.2 gives examples of the principal rock types according to such a classification. Although these rock names are the most important, more than 600 have been proposed to indicate specific types.

d. *Mineralogical composition*. Many minerals are found in igneous rocks, but those that can be called rock-forming minerals are comparatively few. Table 31.3, which lists the major mineral constituents of igneous rocks, is divided into two parts: one part gives the common rock-forming minerals of igneous rocks and the second the accessory minerals of igneous rocks. Table 31.4 gives a volume percentage listing of mineral constituents in some common plutonic rock types. Figure 31.3 should be used as a guide for estimating the volume percentage of mineral constituents in the medium- to coarse-grained rocks. Such volume percentage estimates become impossible in very fine-grained rocks (such as many extrusive igneous rock types) without the aid of a petrographic microscope.

MATERIALS

A representative suite of diverse igneous rock types must be available in the laboratory. Most of the specimens should be intrusive because their relatively coarse grain size allows for mineral identification in hand specimens. A hand lens, a binocular microscope, or both should be available for this assignment.

ASSIGNMENT

1. Identify all the minerals present in each assigned specimen, and record their names in Table 31.5.

2. Estimate the volume percentage of each of the identified minerals, with the aid of Fig. 31.3, and record the estimated percentages next to the appropriate mineral name in Table 31.5. Record any mineral present in less than about 2 volume percent as a *trace*. The combination of the mineralogy and the volume percentage estimate of each of the minerals gives you the *mode* of the rock.

3. Carefully observe the texture of each of the rock types and record your observations in Table 31.5. The types of textural observations include grain size such as *fine-*, or *medium-*, or *coarse-grained; aphanitic, porphyritic*, in which larger crystals, *phenocrysts*, are set in a finer-grained *groundmass; pegmatitic*, which describes the texture of an exceptionally coarse-grained igneous rock; and so on.

4. On the basis of your findings on mineralogy, volume percentage of minerals, and texture, and with the aid of Fig. 31.2 and Tables 31.2 and 31.4, determine the appropriate rock name for each of the specimens. Record these in Table 31.5.

5. Last, make a brief entry in Table 31.5 on the origin of the rock type, such as volcanic versus intrusive, or high-temperature versus low-temperature, or fast-cooled versus slowly cooled, and so on.

FIGURE 31.2 General classification and nomenclature of (*a*) some common plutonic rock types and (*b*) some common volcanic rock types. This classification is based on the relative percentages of quartz, alkali feldspar, and plagioclase, measured in volume percent. (Adapted from Subcommission on the Systematics of Igneous Rocks, *Geotimes*, 1973, vol. 18, no. 10, pp. 26–30, and D. W. Hyndman, 1972, *Petrology of Igneous and Metamorphic Rocks*, McGraw-Hill, New York, p. 35.)

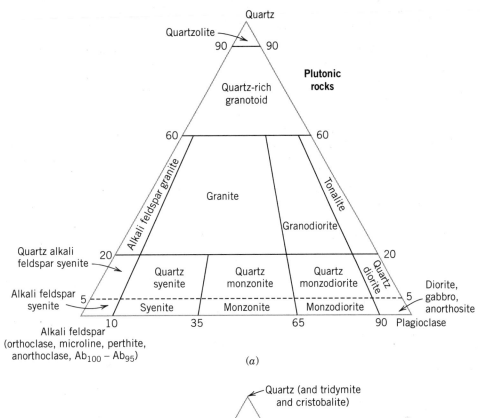

(*a*)

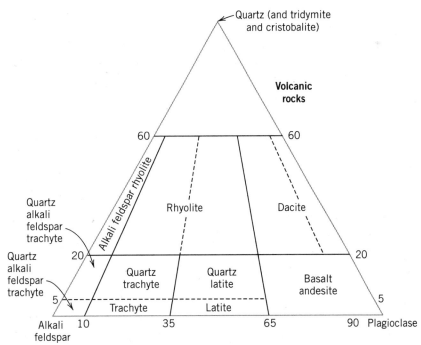

(*b*)

TABLE 31.3 *Mineralogy of the Igneous Rocks*

Common Rock-Forming Minerals	*Common Accessory Minerals*
1. Quartz, tridymite, cristobalite	1. Zircon

2. Feldspars
   Orthoclase
   Microcline
   Sanidine
   Plagioclase
	2. Titanite
	3. Magnetite
	4. Ilmenite
	5. Hematite

3. Nepheline	6. Apatite
4. Sodalite	7. Pyrite
5. Leucite	8. Rutile

6. Micas
   Muscovite
   Biotite
   Phlogopite
	9. Tourmaline
	10. Monazite
	11. Garnet

7. Pyroxenes
   Augite
   Orthopyroxene
   Aegirine	

8. Amphiboles
   Hornblende
   Arfvedsonite
   Riebeckite	

9. Olivine	

SOURCE: From *Manual of Mineral Science*, 23rd ed., Table 21.3, p. 581.

TABLE 31.4 Approximate Mineral Compositions of Some Plutonic Rock Types (in Volume Percent[a])

Plutonic Rock	Granite	Syenite	Grano-diorite	Quartz Diorite	Diorite	Gabbro	Olivine Diabase	Diabase	Dunite
Quartz	25		21	20	2				
Orthoclase and microperthite	40	72	15	6	3				
Oligoclase	26	12							
Andesine			46	56	64				
Labradorite						65	63	62	
Biotite	5	2	3	4	5	1		1	
Amphibole	1	7	13	8	12	3		1	
Orthopyroxene				1	3	6			2
Clinopyroxene		4		3	8	14	21	29	
Olivine						7	12	3	95
Magnetite	2	2	1	2	2	2	2	2	3
Ilmenite	1	1				2	2	2	
Apatite	Trace	Trace	Trace	Trace	Trace				
Titanite	Trace	Trace	1	Trace	Trace				
Color Index[b]	9	16	18	18	30	35	37	38	98–100

[a]The percentage values are based on grain counts of minerals in a thin section using a polarizing microscope. This is known as *modal analysis*.

[b]Color index—a number that represents the percentage, by volume, of dark-colored (i.e., mafic) minerals in a rock.

SOURCE: After E. S. Larsen, 1942, *Handbook of Physical Constants.*

FIGURE 31.3 Chart for determining the approximate volume percentage (*mode*) of minerals in rocks. Grain shapes are similar to those commonly observed. (From R. V. Dietrich and B. J. Skinner, 1979, *Rocks and Rock Minerals*, Wiley, New York, p. 118.)

5 volume percent

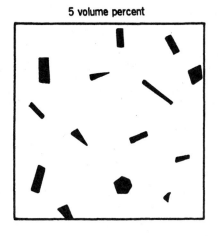

15 volume percent

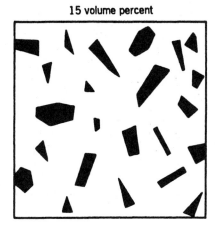

40 volume percent

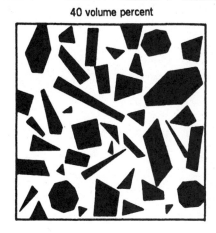

10 volume percent

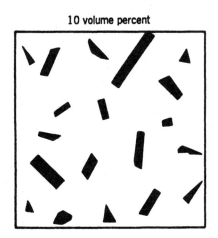

25 volume percent

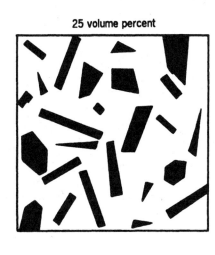

50 volume percent

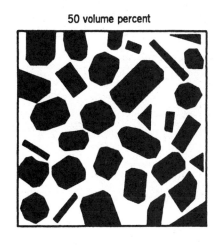

TABLE 31.5 Record of Mineralogical and Textural Findings on a Suite of Igneous Rock Types, and Your Brief Conclusions about Their Origin

Rock Specimen Number	Listing of Each of the Minerals	Volume Percent of Each	Texture	Rock Name	Brief Comments on Origin

Please make photocopies of this form if more copies are needed.

Sedimentary Rocks in Hand Specimens and their Classification

PURPOSE OF EXERCISE

Identification of the minerals that compose sedimentary rocks, and subsequent classification of sedimentary rock types on the basis of their mineralogy and texture.

FURTHER READING

Klein, C. and Dutrow, B., 2008, *Manual of Mineral Science*, 23rd ed., Wiley, Hoboken, New Jersey, pp. 585–596.

Nesse, W. D., 2000, *Introduction to Mineralogy*, Oxford University Press, New York, pp. 190–194.

Perkins, D., 2002, *Mineralogy*, 2nd ed., Prentice Hall, Upper Saddle River, New Jersey, pp. 117–118, 120, 127–132.

Wenk, H. R. and Bulakh, A., 2004, *Minerals: Their Constitution and Origin*, Cambridge University Press, New York, New York, pp. 351–358, 367–374.

Dunham, R. J., 1962, Classification of carbonate rocks according to depositional texture. *American Association of Petroleum Geologists Memoir*, no. 1, pp. 108–121.

Folk, R. L., 1959, Practical Petrographic classification of limestones. *American Association of Petroleum Geologists Bulletin*, no. 43, pp. 1–38.

Blatt, H., Tracey, R. J. and Owens, B. E., 2006, *Petrology: Igneous, Sedimentary and Metamorphic*, 3rd ed., W. H. Freeman, New York, New York, 540p.

Blatt, H., 1992, *Sedimentary Petrology*, 2nd ed., W. H. Freeman, New York, New York, 514p.

Nichols, G., 1999, *Sedimentology and Stratigraphy*, Blackwell Science Ltd., London, United Kingdom, 355p.

Boggs, S., Jr., 2001, *Principles of Sedimentology and Stratigraphy*, 3rd ed., Prentice Hall, Upper Saddle River, New Jersey, 726p.

Background Information: The materials of which sedimentary rocks are composed have been derived from the weathering of previously existing rock masses that were elevated above sea level. *Chemical weathering* decomposes minerals in the rocks, and *mechanical weathering* is responsible for the physical destruction of the original rock. The decomposition and disintergration products are transported to and deposited in areas of accumulation by the action of water or, less frequently, by glacial or wind action. Such loose deposits are converted into rocks by the processes of *diagenesis* and *lithification*, which include compaction and cementation of the loose materials.

The products of *chemical* decomposition may be transported in solution by water into lakes and seas, where chemical changes (such as those attributable to evaporation) or organisms may cause precipitation. These chemical (or biochemical) precipitates become, upon induration, diagenesis, and lithification, *chemical sedimentary rocks* (see Fig. 32.1). Such chemically deposited sediments are represented by carbonate (such as some limestones) and evaporite sequences and finely banded sedimentary iron-formations. These truly chemical sediments are also known as *orthochemical* (from the Greek, meaning *correct* or *true*) sedimentary rocks. If organisms have caused the precipitation of the major sedimentary mineral components, or if the precipitated minerals have undergone substantial movement (and redeposition) after their crystallization, the resulting sediment is referred to as *allochemical* (from the Greek, meaning *different*). Examples of such allochemical sedimentary rocks are oolitic and fossiliferous limestone; their constituent carbonate particles are termed *allochems*.

Generally, weathering includes both mechanical disintegration and chemical decomposition, and thus the end products consist of sedimentary materials formed through mechanical as well as chemical action. Mechanical weathering, a disintegration of the rock, produces solid fragments and mineral particles known as *detritus* or *clastic* material. This material, deposited as loose sediment, includes gravel and sand, which, upon lithification (compaction and cementation), forms conglomerates and sandstones (see Fig. 32.1). These sediments, with clastic textures, are known as *terrigenous* sediments. Detrital materials consist most commonly of chemically inert minerals such as quartz, garnet, zircon, rutile, and magnetite, and of rock fragments made of these minerals.

The two categories *chemical* and *terrigenous* are not mutually exclusive because most chemical sediments contain some detrital material and most clastic rocks also carry some chemical sediment.

Mineralogy of Sedimentary Rocks

The minerals of sedimentary rocks can be divided into two major groups: minerals that are resistant to the mechanical and chemical breakdown of the weathering cycle and minerals that are newly formed from the products of chemical weathering. Terrigenous sedimentary rocks consist mainly of the most resistant rock-forming minerals: quartz, K-feldspar, mica and lesser plagioclase,

FIGURE 32.1 Schematic diagram for the sequence: source rock—weathering—sedimentary rock. The arrows represent processes, the boxes represent products. (Modified after L. J. Suttner and J. Meyers, 1991, Field study of petrology of sedimentary rocks, in *Manual of Geologic Field Study of Northern Rocky Mountains*, Indiana University, Bloomington.)

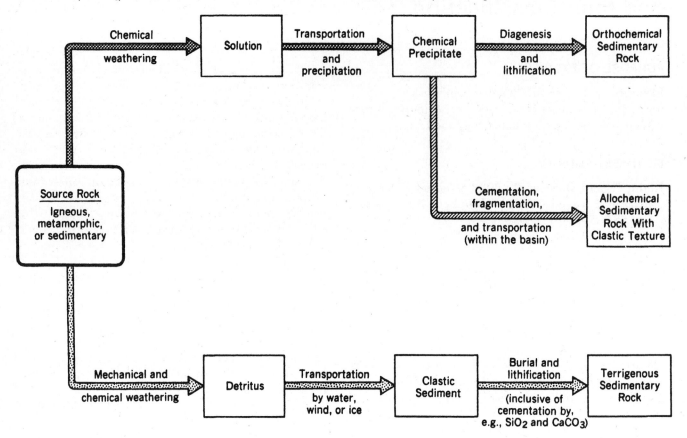

as well as small amounts of garnet, zircon, and spinel (magnetite). The terrigenous rock types may be regarded as accidental, mechanical mixtures of genetically unrelated resistant minerals. For example, a feldspar-rich sandstone may contain orthoclase and microcline, as well as several members of the plagioclase series. Such a random variety of feldspar compositions is not found in igneous assemblages because of physical–chemical controls in their crystallization sequence.

Sedimentary rocks that result from the inorganic or organic precipitation of minerals can be interpreted in large part in terms of chemical and physicochemical principles that apply at low temperatures (25°C) and low atmospheric pressure. The chemical sedimentary assemblages, therefore, are not random or accidental but reflect the concentrations of ions in solution, as well as conditions such as temperature, pressure, and salinity of the sedimentary basin. For example, the sequence of minerals in evaporite beds can be related to the concentration of ions in solution in the brine from which the sequence precipitated. Examples of common chemical precipitates are calcite, aragonite, gypsum, anhydrite, and halite. In sedimentary iron-formations hematite, magnetite, siderite, and ankerite as well as chert are considered products of chemical sedimentation.

Classification

Sedimentary rocks are, in general, *stratified*, that is, they have layers or beds distinguished from one another by differences in grain size, mineral composition, color, or internal structure. Other features that are uniquely diagnostic of a sedimentary origin are primary sedimentary structures; the presence of fossils; grains whose shape is the result of transportation (referred to as *clasts* or *detrital grains*); and the presence of a mineral that is invariably of sedimentary origin, such as glauconite.

All sedimentary rocks can be grouped in three broad categories, terrigenous, allochemical, and orthochemical, as shown in Table 32.1. A sedimentary rock type within each of these three broad categories is identified by its mineralogical composition and its texture (inclusive of grain size and grain shape).

Terrigenous sedimentary rocks. Terrigenous sedimentary rocks consist of detrital grains, which form the framework of the rock, and which are joined together by cement; these detrital grains (or clasts) are known as the framework grains. Variable amounts of matrix, which consists of fragmental material substantially smaller than the mean size of the framework grains, may also be present. Because of the normally very fine grain size of matrix

TABLE 32.1 Sedimentary Rocks Divided into Three Major Categories

I *Terrigenous Sedimentary Rocks (Clastic Texture)*	II *Allochemical Sedimentary Rocks (Biochemical/Biogenic, with Clastic Texture)*	III *Orthochemical Sedimentary Rocks*
Conglomerates, breccias, sandstones, and mudstones	Limestones, dolostones, phosphorites, chert, and coal	Evaporites, chert, travertine, and iron-formations
Agglomerates and volcaniclastic sandstones		

SOURCE: From *Manual of Mineral Science*, 23rd ed., p. 588.

material, it may be impossible to determine whether this is indeed of detrital rather than of diagenetic origin. The texture of rocks dominated by detrital material is a composite of the grain size, grain shapes, sorting, and angularity of the framework grains. A universally adopted grain size scale for detrital (or clastic) sediments is given in Table 32.2. Examples of various degrees of sorting are given in Fig. 32.2.

Conglomerates and *breccias* consist of large clasts (boulders, cobbles, pebbles, and granules) with or without a sandy matrix; conglomerates show pronounced rounding of the clasts whereas breccias contain more angular clasts. Volcanic debris may be a predominant component of some sedimentary rocks; such are commonly referred to as *pyroclastic rocks*. Coarse-grained pyroclastics with a grain size over 32 mm are known as *agglomerates* or *volcanic breccias*.

TABLE 32.2 Terms and Sizes for Clastic Sediments and Clastic Sedimentary Rock Types

Name		Millimeter	Micrometers	Φ^a
GRAVEL		4,096		−12
	Boulder	256		−8
	Cobble	64		−6
	Pebble	4		−2
	Granule	2		−1
SAND	Very coarse sand	1		0
	Coarse sand	0.5	500	1
	Medium Sand	0.25	250	2
	Fine Sand	0.125	125	3
	Very fine sand	0.062	62	4
MUD	Coarse silt	0.031	31	5
	Medium silt	0.016	16	6
	Fine silt	0.008	8	7
	Very fine silt	0.004	4	8
	Clay			

[a]The Φ scale, devised by W. C. Krumbein (1934), is based on a logarithmic transformation, $\Phi = -\log_2 S$, where S is grain size in millimeters. The Φ scale is commonly used in sedimentological studies because it is more convenient in presenting data than if values are given in millimeters

SOURCE: After J. A. Udden (1898) and C. K. Wentworth (1924). From *Manual of Mineral Science*, 23rd ed., p. 588.

FIGURE 32.2 Degrees of sorting in sandstones and conglomerates. (From R. R. Compton, 1962, *Manual of Field Geology*, Wiley, New York.)

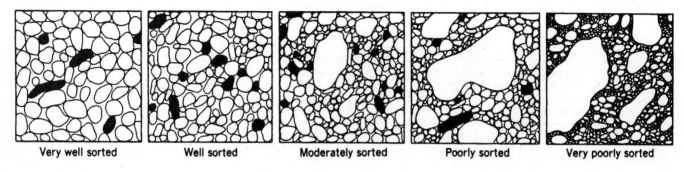

Very well sorted   Well sorted   Moderately sorted   Poorly sorted   Very poorly sorted

Sandstones are finer-grained (grain size between 2 and 0.062 mm) and most easily classified by determining the amounts of clastic (framework) grains composed of quartz (and chert), feldspar, and lithic fragments. A commonly accepted scheme of sandstone classification, as based on the population of these three components in the clastic grains, is given in Fig. 32.3. Such a classification scheme is most applicable to the study of sandstones in thin section under the microscope, because the name assignment is based on a modal (volume percent) analysis of the constituent clastic grains. However, by close inspection with a hand lens, a reasonably correct name can often be assigned. The matrix material of sandstones is commonly clay minerals and very fine-grained quartz. When the sandstone matrix constitutes more than about 10 volume percent of the rock, it is classified as a *wacke*. *Graywackes* are mostly dark gray sandstones with abundant matrix. Common cements in sandstones are quartz, calcite, and clay minerals. Diagenetic hematite may stain sandstones red.

Volcaniclastics are sandstones made up chiefly of volcanic fragments, volcanic glass, and crystals. Many tend to be green because there has been chlorite replacement.

Mudstone is a general term for sediments composed mainly of silt-sized (0.062 to 0.004 mm) and clay-sized (<0.004 mm) particles. Mudstones are essentially impossible to study in hand specimens because of their fine grain size. Modern laboratory methods for studying these very fine-grained rock types include X-ray diffraction, electron microprobe, and scanning electron microscope techniques. Siltstones and claystones are rock types made up mainly of silt and clay particle-size materials, respectively. Shale is characterized by its fissility, the ability to split into thin sheets, generally parallel to the bedding. Claystones are nonfissile, commonly with a massive or blocky texture. Slate is a mudstone with a well-developed cleavage, which may or may not be parallel to the bedding and is commonly the result of metamorphism.

A schematic classification of common terrigenous sedimentary rocks is given in Table 32.3.

Allochemical Carbonate Rocks.

Allochemical carbonate rocks show clastic (fragmental) textures analogous to those seen in terrigenous rocks, but the textural interpretation of these mineralogically simple rocks is not always straightforward. Calcite recrystallizes easily, and secondary dolomite, replacing the original calcium carbonate in a process known as *dolomitization*, often destroys the texture of the original carbonate. For these reasons textural interpretations of such "crystalline limestones" or "crystalline dolostones" may be impossible.

Limestones consist of two classes of constituents, orthochems and allochems. These constituents originate within the basin where limestone is deposited, and are therefore referred to as *intrabasinal*. Orthochem components are of two types: (1) *microcrystalline ooze*, which is a very fine-grained carbonate precipitate that has settled to the bottom of the basin, and (2) *sparry calcite cement*, which is coarser in grain size than the ooze and tends to be clear or translucent. This coarser-grained type of calcite is a pore-filling cement that was precipitated in place. Allochem components of limestone are of four types: *intraclasts, oolites, fossils,* and *pellets*. *Intraclasts* represent fragments of weakly consolidated carbonate sediment that have been torn up, transported, and redeposited by currents within the basin of deposition. They consist of various types of limestone and can range in size from very fine to pebble or boulder size. *Oolites*, in a size range of 0.1 to 1.0 mm in diameter, are spherical, show radial and concentric structures, and resemble fish roe. They are commonly formed around nuclei such as shell fragments, pellets, or quartz sand grains. They develop by chemical accretion under the rolling influence of waves in shallow marine environments. *Fossils* of many types are common constituents of limestones. *Pellets* are well-rounded, homogeneous aggregates of microcrystalline calcite in a size range of 0.03 to 0.2 mm in diameter. They consist for the most part of the feces of mollusks, worms, and crustaceans.

Because carbonate rocks tend to be mixtures of (1) allochems, (2) microcrystalline ooze, and (3) sparry calcite cement, Folk (1959) notes that, as a first-order approach, limestones can be classified on the basis of the volumetric abundance of these three types of materials. Such a classification of limestones, which ignores

FIGURE 32.3 Classification of common sandstones. (From M. E. Tucker, 1982, *Field Description of Sedimentary Rocks,* Wiley, New York, p. 21.)

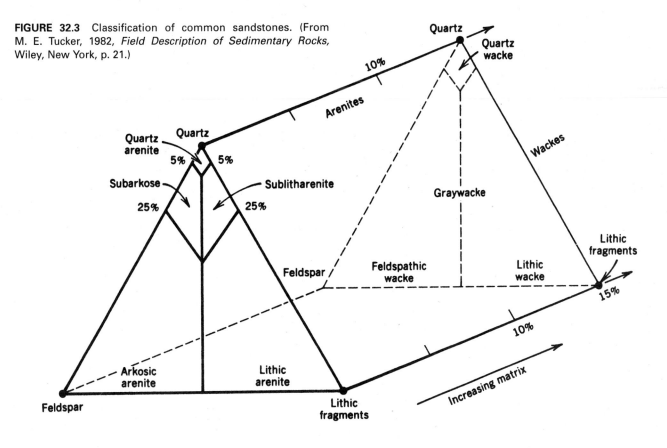

TABLE 32.3 Classification of Terrigenous Rocks

			Composition[a]		
			Lithic fragments (e.g., chert, limestone, volcanic, granite)	Quartz	Feldspar
Grain Size	Conglomerate (rounded), Breccia (angular)	Cobble	Cobble conglomerate (or breccia) (e.g., granite cobble conglomerate)	Quartz cobble conglomerate (or breccia)	
		Pebble	Pebble conglomerate (or breccia) (e.g., chert pebble conglomerate)	Quartz pebble conglomerate (or breccia)	
		Granule	Granule conglomerate (or breccia) (e.g., limestone granule conglomerate)	Quartz granule conglomerate (or breccia)	Feldspar granule conglomerate (or breccia)
	Sandstone	Sand	*Wacke (>10% matrix)*	Lithic[b] wacke	Quartz wacke	Feldspathic[c] wacke
			Arenite (<10% matrix)	Lithic arenite	Quartz arenite	Feldspathic[c] arenite
	Silt		(Composition cannot be evaluated because of fine grain size)		
			Siltstone		
	Mud		Mudstone—lacking fissility Shale—showing fissility		

[a]Compare this table with Fig. 32.3 for sandstone classification.

[b]Dark, highly indurated lithic wackes are also referred to as graywackes

[c]Red- or pink-colored feldspathic wackes and arenites can be referred to as arkoses.

SOURCE: Modified after L. J. Suttner and J. Meyers, 1991, Field study of the petrology of sedimentary rocks, in *Manual for Geologic Field Study of Northern Rocky Mountains,* Indiana University, Bloomington

any of the possibly present terrigenous components (e.g., detrital quartz sand grains), is given in Fig. 32.4. In this figure the *allochems* (intraclasts, oolites, fossils, and pellets) represent the framework material of the rock. The microcrystalline matrix is equivalent to a clay-rich matrix in a poorly washed sandstone. The sparry cement filling the pore spaces, as in quartz-rich sandstones, is a chemical precipitate. A rock composed only of microcrystalline limestone is referred to as a *micrite*. Folk (1959) recommends that after limestones have been divided into type I, type II, and type III (see the legend for Fig. 32.4), it is essential to note which allochems (intraclasts, oolites, fossils, or pellets) predominate. Once they are known, they can be incorporated into a scheme of nomenclature using as prefixes parts of the allochem names—("intra" from intraclasts; "oo" from oolite; "pel" from pellet; and "bio" for biogenic, in place of fossil.) Examples of such terms are biosparite and biomicrite, the names given to fossil fragments in two different matrix types. *Biosparite* is a limestone with more than 10% fossil allochems in a dominantly sparry calcite matrix; *biomicrite* is a limestone with more than 10% fossil allochems set in a micritic matrix. Other such terms are shown at the upper part of the triangle in Fig. 32.4.

Another commonly used classification of limestones, as outlined by R. J. Dunham (1962), emphasizes the depositional texture of limestones, rather than the micritic content of the rock used by Folk. Dunham's classification stresses the question "Were the framework grains (Folk's allochems) in close contact with each other, that is, were they well packed when they were deposited?" Because both classifications are almost equally popular, both are combined in a graphical representation in Fig. 32.5. The terms mudstone, wackestone, packstone, grainstone, and boundstone were introduced by Dunham (1962) to reflect, at one extreme, materials consisting of less than 10% grains (mudstone) and, at the other extreme, materials in which original components are predominant and are closely bound together (boundstone). The term *boundstone* is equivalent to *biolithite*, a limestone made up of organic structures that grew *in situ* (in place), forming a coherent rock mass during growth. *Wackestone* consists of more than 10% grains in a microcrystalline ooze (micrite), *packstone* is a grain-supported limestone with micrite matrix and sparry calcite cement, and *grainstone* is a grain-supported limestone with very little micrite, if any. Either of the two schemes of Fig. 32.5 can be used to classify limestones on the basis of careful observation by hand lens or binocular microscope in the laboratory.

Dolostones are Ca–Mg-rich carbonate rocks consisting mostly of the mineral dolomite. The term *dolomite* is commonly used for both the rock type and the mineral. Dolostones in limestone–dolostone sequences may show irregular or crosscutting relationships with the limestones, indicating that the dolostone has formed by the replacement of early calcite by later dolomite. Such observations form the basis for deciding that the rocks have undergone the process of *dolomitization*, which is the replacement of original calcium carbonate in limestone by Ca–Mg-carbonate (dolomite) at any time during or after deposition. Sometimes it is possible with a hand lens to

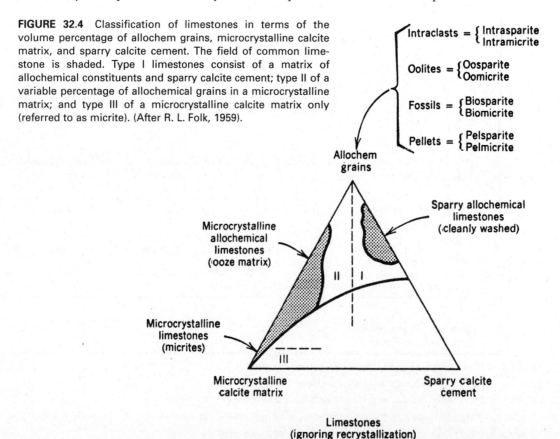

FIGURE 32.4 Classification of limestones in terms of the volume percentage of allochem grains, microcrystalline calcite matrix, and sparry calcite cement. The field of common limestone is shaded. Type I limestones consist of a matrix of allochemical constituents and sparry calcite cement; type II of a variable percentage of allochemical grains in a microcrystalline matrix; and type III of a microcrystalline calcite matrix only (referred to as micrite). (After R. L. Folk, 1959).

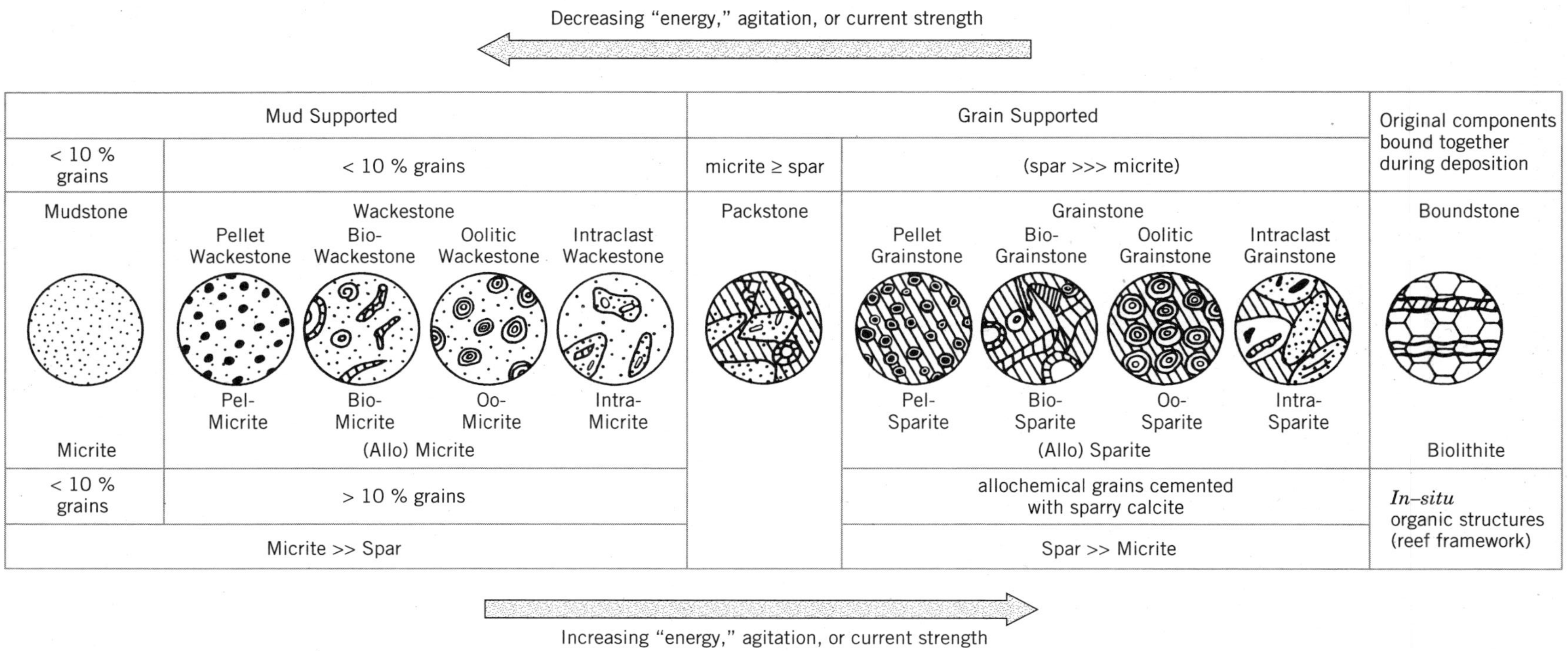

FIGURE 32.5 Classification of carbonate rocks after Folk (1959) and Dunham (1962). The headings at the top of the chart represent the classification of R. L. Folk, those at the top of the chart the classification of R. J. Dunham. (From L. J. Suttner and J. Meyers, 1991, Field study of the petrology of sedimentary rocks, in *Manual for Geologic Field Study of Northern Rocky Mountains*, Indiana University, Bloomington.)

see rhombic outlines of dolomite grains cutting across fossil fragments; however, such observations are best made in thin section with a petrographic microscope. Because of the secondary nature of dolostones and their commonly coarsely recrystallized grain size, little can be learned about their formation in hand specimens.

Orthochemical sedimentary rocks. Orthochemical sedimentary rocks have been formed by direct precipitation, through chemical action in a depositional basin as a result of environmental (e.g., climatic) changes. Sedimentary rocks that fall into this category are evaporites, banded iron-formation, some limestones, and travertine. Bedded cherts may also be of direct chemical origin.

MATERIALS

A representative suite of sedimentary rock types must be available in the laboratory. A hand lens, a binocular microscope, or both should be available for this assignment. A small dropper bottle with dilute hydrochloric acid is needed to identify and distinguish particular carbonate-containing sedimentary rocks.

ASSIGNMENT

1. Identify all the minerals present in each assigned specimen and record their names in Table 32.4.

2. Estimate the volume percentage of each of the identified minerals (with the aid of Fig. 31.3), and record the estimated percentages next to the appropriate mineral name in Table 32.4. Record any mineral present in less than about 2 volume percent as a *trace*.

3. Carefully evaluate the texture and grain size of each of the rock types and record your observations in Table 32.4.

4. On the basis of your findings on mineralogy, volume percentage of minerals, and texture, and with the aid of Table 32.1, decide whether a specific rock specimen is terrigenous (T), allochemical (A), or orthochemical (O). Enter T, A, or O in the appropriate column in Table 32.4.

5. Give the appropriate rock name for each of the specimens in Table 32.4 (consult the various figures and tables in this exercise).

6. Make a brief entry in Table 32.4 concerning the origin of each rock type.

Student Name _____

Table 32.4 Record of Mineralogical and Textural Findings on a Suite of Sedimentary Rock Types, and Your Classification of Each of the Specimens

Rock Specimen Number	Listing of Each of the Minerals	Volume Percent of Each	Comments on Texture and Grain Size	Terrigenous (T) Allochemical (A) or Ortochemical (O)	Rock Name	Brief Comments on Origin

Please make photocopies of this form if more copies are needed.

Metamorphic Rocks in Hand Specimens and Their Classification

PURPOSE OF EXERCISE

Identification of the minerals that make up metamorphic rocks, and subsequent classification of metamorphic rock types on the basis of their mineralogy and texture.

FURTHER READING

Klein, C. and Dutrow, B., 2008, Manual of Mineral Science, 23rd ed., Wiley, Hoboken, New Jersey, pp. 596–603.

Nesse, W. D., 2000, *Introduction to Mineralogy*, Oxford University Press, New York, pp. 194–200, 424.

Perkins, D., 2002, *Mineralogy*, 2nd ed., Prentice Hall, Upper Saddle River, New Jersey, pp. 133–153.

Wenk, H. R. and Bulakh, A., 2004, *Minerals: Their Constitution and Origin*, Cambridge University Press, New York, New York, pp. 440–447, 490–494.

Best, M-G., 2003, *Igneous and Metamorphic Petrology*, 2nd ed., Blackwell Publishing, Malden, Massachusetts, 729p.

Blatt, H., Tracey, R. J. and Owens, B. E., 2006, *Petrology: Igneous, Sedimentary and Metamorphic*, 3rd ed., W. H. Freeman, New York, New York, 540p.

Winter, J. D., 2001, *An Introduction to Igneous and Metamorphic Petrology*, Prentice Hall, Upper Saddle River, New Jersey, 697p.

Background Information: Metamorphic rocks are derived from preexisting rocks (igneous, sedimentary, or metamorphic) by mineralogical, textural, and structural changes. Such changes may be the result of marked variations in temperature, pressure, and shearing stress at considerable depth in the earth's crust. Weathering effects, at atmospheric conditions, are not considered part of metamorphism, nor are chemical reactions involving partial melting, for these are part of igneous processes. Metamorphic changes such as recrystallization and chemical reactions of mineral constituents take place essentially in the solid state, although the solids may exchange chemical species with a fluid phase consisting mainly of H_2O (as water, steam, or supercritical fluid, depending on the temperature and pressure at which the reactions took place). In addition to H_2O, CO_2 and CH_4 may be major compounds of the metamorphic fluid. The general conditions in which metamorphic rocks form lie between those of sedimentary rocks, which form at essentially atmospheric conditions of T and P, and those of igneous rocks,

which crystallize from a melt at high T. Metamorphic rocks may be the result of very large changes in pressure in conjunction with increasing metamorphic temperature; mineral assemblages in the earth's mantle have formed in response to very high confining pressures. Excluding gain or loss of H_2O and CO_2, many metamorphic reactions are generally considered to be essentially *isochemical*; this implies that during recrystallization and the process of chemical reactions, the bulk chemistry of the rocks has remained essentially constant. If this is not the case, and if additional elements have been introduced into the rock, by circulating fluids for example, it is said to have undergone *metasomatism*.

The most obvious textural feature of metamorphic rocks (except those of contact metamorphic origin; see below) is the alignment of minerals along planar surfaces. For example, a shale that has undergone only slight metamorphic changes may show well-developed cleavage along planar surfaces, producing a *slate*. With increased temperature of metamorphism, recrystallization of originally very fine grained minerals produces coarser-grained *schists*, in which minerals are aligned in parallel layers (known as *schistosity*). The coarsest-grained metamorphic rocks that show distinct mineralogic banding are known as *gneisses* (exhibiting *gneissosity*).

In general, metamorphic rocks can be divided into two groups: (1) those formed by contact metamorphism and (2) those formed by regional metamorphism. *Contact metamorphic rocks* form as concentric zones (*aureoles*) around hot igneous intrusive bodies. Such metamorphic rocks may lack schistosity, and along the relatively large temperature gradient from the intrusive contact to the unaffected country rock are zones that may differ greatly in mineral assemblages. (For an extensive discussion of mineral assemblage, see exercise 34.) Sandstones are converted to quartzites, shales are changed to *hornfels*, a fine-grained dense rock, and limestones recrystallize as marble. Metasomatism is commonly associated with contact metamorphism. Heated solutions emanating from the intruded igneous rock can bring about profound chemical changes in the adjacent country rock, causing new mineral assemblages as well as ore bodies to form.

In this exercise we will concern ourselves essentially with *regional* metamorphism, that is, with the metamorphic rocks resulting from increases in T or P, or both, on a regional scale (areas a few hundred to thousands of square miles in extent) in response to mountain building or to deep burial of rocks. Contact and regionally metamorphosed rocks generally reflect an increase in temperature (progressive metamorphism) in their assemblages. However, when assemblages that originate at high temperatures (e.g., igneous or high-temperature metamorphic rocks) fail to survive conditions of lower-temperature metamorphism, the process is referred to as *retrograde* (or retrogressive) metamorphism.

Mineralogic Composition

The mineralogic composition of a rock that has undergone metamorphic conditions will generally be very different from that of its unmetamorphosed equivalent. The extent of the differences is largely controlled by the marked changes of *T* and *P* during metamorphism; the relative degrees of metamorphism can be expressed in terms of grades, such as very low, low, medium, and high grade (see Fig. 33.1). *Changes in metamorphic grade are reflected in changes in the mineral assemblages of the rocks.*

The most common minerals in metamorphic rocks are listed in Table 33.1. Because of the generally very fine grain size of very low to low-grade metamorphic rocks, the determination of their mineralogy in hand specimens is not straightforward. For example, a *slate*, which is the low-grade metamorphic equivalent of an argillaceous shale (see Table 33.2), may appear under a hand lens to be composed mostly of ill-defined phyllosilicates such as muscovite and chlorite, but a microscope study would show it to consist of an association of quartz–feldspar–muscovite–chlorite. The mineralogy of somewhat higher-grade metamorphic rocks, which tend to be coarser-grained because they have undergone considerable recrystallization, is generally

easily determined in hand specimens. Examples of schistose and gneissic assemblages for medium- to high-grade metamorphic rocks that are carbonate-rich or argillaceous are given in Table 33.2. For such schistose and gneissic assemblages no new rock terms are introduced; these metamorphic rocks are named on the basis of their most common mineral constituents, listed in order of decreasing abundance. For example, a schist composed of 60 volume percent biotite, 30 volume percent quartz, and 10 volume percent kyanite would be known as a biotite–quartz–kyanite schist. Similarly a gneiss with 60 volume percent feldspar, 25 volume percent biotite, and 15 volume percent garnet would be named a feldspar–biotite–garnet gneiss. Several more-specialized rock terms based on the modal mineralogy of metamorphic rocks are *marble*, a metamorphosed limestone, generally with a well-defined recrystallized texture; *quartzite*, a rock composed essentially of quartz and derived through metamorphism of a quartz-rich sandstone; *serpentinite*, a rock composed essentially of the mineral serpentine; *soapstone*, a rock with a massive or schistose texture and composed mainly of fine-grained talc; *amphibolite*, an unfoliated or foliated rock made up mainly of amphibole and plagioclase; and *greenschist*, a well-foliated, generally fairly fine-grained, greenish rock composed mainly of chlorite, actinolite (or hornblende), and epidote.

Classification

On the basis mostly of the study of shales and their metamorphic equivalents, various metamorphic zones have been delineated; these are identified by the presence of certain *index* minerals. At successively higher grades of

FIGURE 33.1 Schematic *P–T* diagram outlining approximate fields for various metamorphic grades. The shaded area marked "diagenetic conditions" represents the general conditions of lithification of a sediment at low temperature. (Modified from *Manual of Mineral Science*, 23rd ed., p. 598.)

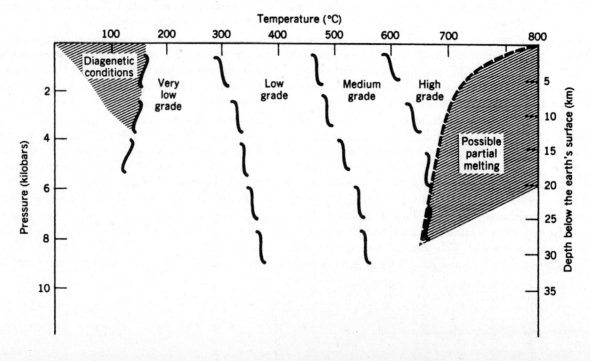

TABLE 33.1   Common Silicates of Metamorphic Rocks

Phyllosilicates	talc
	serpentine
	chlorite
	muscovite
	biotite
Inosilicates	anthophyllite
	cummingtonite–grunerite
	tremolite–actinolite
	hornblende
	glaucophane
	clinopyroxene with jadeite component (high pressure)
	diopside
	orthopyroxene
	wollastonite[a]
Tectosilicates	quartz
	plagioclase, except for very An-rich compositions
	microline and orthoclase
Nesosilicates	garnet (pyrope at high pressures)
	epidote
	kyanite, sillimanite, andalusite
	vesuvianite[a]
	forsterite
	staurolite
	chloritoid

[a]Especially common in contact metamorphic rocks.
SOURCE: From *Manual of Mineral Science*, 23rd ed., Table 21.2, p. 601.

metamorphism, argillaceous (Al-rich) rocks show the development of the following index minerals: first chlorite, then biotite, next almandine, subsequently staurolite, then kyanite, and at the highest temperatures sillimanite. When areas of specific index minerals are outlined on a geological map, such as a chlorite-rich region or a biotite-rich region, the line that marks the first appearance of an index mineral is known as an *isograd*. The isograds reflect positions of similar metamorphic grade in terms of T and P.

In regions where high pressures have been generated but where temperatures were low during metamorphism, the final mineral assemblages will be very different, for a specific bulk rock composition, from those formed in regions where temperatures have been high but pressures low. Argillaceous rocks metamorphosed at high tem-perature commonly contain sillimanite, whereas those subjected to high pressures (with some increase in temperature) contain kyanite. Similarly, high-temperature basalt assemblages occur as coarse-grained eclogites in the high-pressure regions of the lower crust of the earth.

In addition to the subdivision of P and T fields into very low, low, medium, and high grades of metamorphism (Fig. 33.1), metamorphic rocks are often classified in terms of *metamorphic facies*. A *metamorphic facies* is a set of metamorphic mineral assemblages, repeatedly

TABLE 33.2   Examples of Metamorphic Mineral Assemblages Produced During Prograde Metamorphism in Carbonate-Rich Rocks and Argillaceous Shale

Grade	Ca Carbonate-Rich Rock	Argillaceous (Al-rich) Shale
Very low grade	Calcite–dolomite–talc and calcite–quartz–talc	Muscovite–chlorite–quartz–feldspar
Low grade	Calcite–dolomite–tremolite and calcite–tremolite–quartz	Biotite–muscovite–chlorite–quartz–feldspar
Medium grade	Calcite–dolomite–diopside and calcite–diopside–quartz	Staurolite–garnet–biotite–muscovite–quartz–feldspar
High grade	Calcite–dolomite–forsterite and calcite–diopside–quartz	Sillimanite–garnet–biotite–muscovite–quartz–feldspar

SOURCE: From Manual of *Mineral Science*, 23rd ed., Table 21.13, p. 602.

associated in space and time, that have a constant and therefore predictable relation between mineral composition and chemical composition. A metamorphic facies, therefore, is defined not in terms of a single index mineral but by an association of mineral assemblages. All metamorphic facies are defined on the basis of the mineralogy in basalt and its various metamorphic products. The following are some examples of metamorphic facies. A *zeolite facies* represents the lowest grade of metamorphism. The mineral assemblages include zeolites, chlorite, muscovite, and quartz. A *greenschist facies* is the low-grade metamorphic facies of many regionally metamorphosed terranes. The mineral assemblages may include chlorite, epidote, muscovite, albite, and quartz. An *amphibolite facies* occurs in medium- to high-grade metamorphic terranes. The mineral constituents include hornblende, plagioclase, and almandine. Corresponding argillaceous rocks, metamorphosed to the same $T–P$ conditions, will contain sillimanite. The *glaucophane–lawsonite* schist (or *blueschist*) facies is represented by relatively low temperatures but elevated pressures of metamorphism in young orogenic zones, such as in California and Japan. Characteristic constituents are lawsonite, jadeite, albite, glaucophane, muscovite, and garnet. A *granulite facies* reflects the maximum temperature conditions of regional metamorphism such as have been commonly

attained in Precambrian terranes. Characteristic mineral constituents are plagioclase, hypersthene, orthopyroxene, garnet, and diopside. An *eclogite facies* represents the most deep-seated conditions of metamorphism. Characteristic mineral constituents are pyrope-rich garnet and omphacite-type pyroxene. Such assemblages are common in *kimberlite* pipes, many of which carry associated diamond. A diagram outlining the approximate fields of the various metamorphic facies in terms of P and T is given in Fig. 33.2.

MATERIALS

A representative suite of metamorphic rock types must be available in the laboratory. A hand lens, a binocular microscope, or both should be available for this assignment.

ASSIGNMENT

1. Identify all the minerals present in each assigned specimen, and record their names in Table 33.3.

2. Estimate the volume percentage of each of the identified minerals (with the aid of Fig. 31.3), and record the estimated percentages next to the appropriate mineral name in Table 33.3. Record any mineral present in less than about 2 volume percent as a *trace*.

FIGURE 33.2 Tentative scheme of metamorphic facies in relation to pressure ($P\,H_2O$) and temperature. All boundaries are gradational. Compare with Fig. 33.1. (From F. J. Turner, 1968, *Metamorphic Petrology*, McGraw-Hill, New York, P. 366, with some modifications. Copyright © 1968, by McGraw-Hill Inc. Used with permission of McGraw-Hill Book Company.)

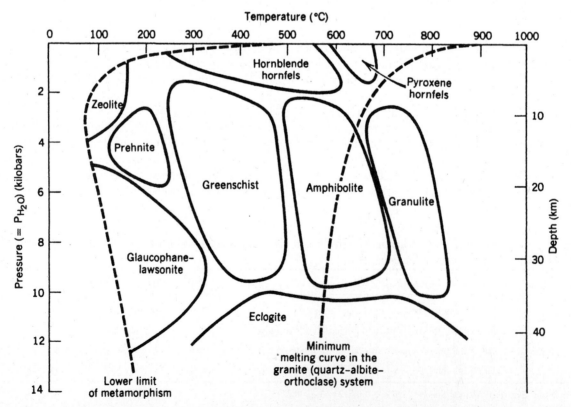

3. Carefully evaluate the texture and grain size of each of the rock types and record your observations in Table 33.3.

4. On the basis of your findings on mineralogy, volume percentage of minerals, and texture, give an appropriate rock name for each of the specimens and record this in Table 33.3.

5. Make a brief entry in Table 33.3 about the metamorphic conditions of origin, such as a specific metamorphic grade or metamorphic facies.

6. In the last column of Table 33.3, enter the name of the sedimentary, igneous, or metamorphic rock type that you think may have been the likely precursor to the metamorphic rock you have described.

TABLE 33.3 Record of Mineralogical and Textural Findings on a Suite of Metamorphic Rock Types, and Your Classification and Statement of Origin for Each of the Specimens.

Rock Specimen Number	Listing of All of the Minerals	Volume Percent of Each	Comments on Texture and Grain Size	Rock Name	Metamorphic Grade or Facies	Brief Comments on Origin and Probable Precursor Rock Type

Please make photocopies of this form if more copies are needed.

Plotting of Metamorphic Mineral Assemblages in the System SiO$_2$–MgO–Al$_2$O$_3$–H$_2$O: Low-, Medium-, and High-Temperature Assemblages

PURPOSE OF EXERCISE

Plotting of metamorphic mineral compositions on triangular diagrams as a basis for the evaluation of metamorphic assemblages in three approximate ranges of temperature: low (~200° to 400°C), medium (~400° to 650°C), and high (~650° to 850°C).

FURTHER READING AND CD-ROM INSTRUCTIONS

Klein, C. and Dutrow, B., 2008, *Manual of Mineral Science*, 23rd ed., Wiley, Hoboken, New Jersey, pp. 105–108.

CD-ROM, entitled *Mineralogy Tutorials*, version 3.0, that accompanies *Manual of Mineral Science*, 23rd ed., click on "Module I", and subsequently the button "Graphical Representation".

Nesse, W. D., 2000, *Introduction to Mineralogy*, Oxford University Press, New York, pp. 194–200, 424.

Perkins, D., 2002, *Mineralogy*, 2nd ed., Prentice Hall, Upper Saddle River, New Jersey, pp. 133–153.

Wenk, H. R. and Bulakh, A., 2004, *Minerals: Their Constitution and Origin*, Cambridge University Press, New York, New York, pp. 440–447, 490–494.

Blatt, H., Tracey, R. J. and Owens, B. E., 2006, *Petrology: Igneous, Sedimentary and Metamorphic*, 3rd ed., W. H. Freeman, New York, New York, 540p.

Best, M. G., 2003, *Igneous and Metamorphic Petrology*, 2nd ed., Blackwell Publishing, Malden, Massachusetts, 729p.

Winter, J. D., 2001, *An Introduction to Igneous and Metamorphic Petrology*, Prentice Hall, Upper Saddle River, New Jersey, 697p.

Deer, W. A., Howie, R. A. and Zussman, J., 1992, *An Introduction to The Rock-Forming Minerals*, 2nd ed., John Wiley and Sons, Hoboken, New Jersey, 696p.

Deer, W. A., Howie, R. A. and Zussman, 1962, *Rock-Forming Minerals*, 5 vols., John Wiley and Sons, New York. Eight completely revised volumes: 1A, *Orthosilicates*; 1B, *Disilicates and Ring Silicates*; 2A, *Single-Chain Silicates*; 2B, *Double-Chain Silicates*; 3A, *Micas*; 4A, *Framework Silicates (Feldspars)*; 4B, *Silica Minerals*; and 5B, *Non-Silicates* have been published since 1962 and are available from the Geological Society, London, Great Britain.

Palache, C., Berman, H., and Frondel, C., 1944, *The System of Mineralogy*, Vol. 1, Wiley, New York.

Zen, E-an, 1961, Mineralogy and petrology of the system Al$_2$O$_3$–SiO$_2$–H$_2$O in some pyrophyllite deposits of North Carolina, *American Mineralogist*, vol. 46, pp. 52–66.

Winkler, H. G. F., 1979, *Petrogenesis of Metamorphic Rocks*, 5th ed., Springer-Verlag, New York, 348p.

Turner, F. J., 1989, *Metamorphic Petrology*, 2nd ed., McGraw-Hill, New York, 512p.

Background Information: In exercise 7 you were introduced to the graphical representation of mineral compositions on triangular diagrams. Here we will develop this technique further so that you will be able to represent graphically mineral assemblages within a specific, and relatively simple, chemical system, SiO$_2$–MgO–Al$_2$O$_3$–H$_2$O.

As stated in exercise 33, *changes in metamorphic grade are reflected in changes in the mineral assemblages of the rocks.* In order to evaluate first hand all possible changes in mineral assemblages in the system SiO$_2$–MgO–Al$_2$O$_3$–H$_2$O, one would need a large collection of rock specimens that belong to this rather restricted chemical system. Furthermore, these rock specimens would have to represent a wide range of bulk chemistries within this chemical system. Acquiring such a collection of rock and mineral assemblage specimens within this chemical system, over a wide range of metamorphic temperatures, is a difficult task. In other words, the assessment of possible metamorphic assemblages within this chemical system will have to be done in some way other than the study of representative hand specimens. The only other option is to use published reports describing mineral coexistences (assemblages) over a range of metamorphic conditions in this chemical system. Because major reference works in mineralogy and petrology are not generally arranged in terms of chemical systems (such as SiO$_2$–MgO–Al$_2$O$_3$–H$_2$O), the only way in which the derived information on assemblages can be obtained (from the literature) is on the basis of specific minerals that belong to the system SiO$_2$–MgO–Al$_2$O$_3$–H$_2$O. The minerals that are to be considered are given in Table 34.1. In this listing specific (but also somewhat simplified) compositions are given for four minerals that are part of extensive solid solution series. The compositions of all the other minerals are much more specific, and if you do not already know them, they can be looked up in your assigned text. The minerals listed fall into several broad categories: anhydrous silicates, hydrous silicates, oxides, and hydroxides. With increasing metamorphic grade the assemblages become less hydrous, and finally completely anhydrous

TABLE 34.1 Listing of Minerals That Are Part of the Chemical System $SiO_2-MgO-Al_2O_3-H_2O$

Quartz

Corundum

Periclase

Spinel

Forsterite

Kyanite, sillimanite, andalusite

Enstatite

Anthophyllite

Gedrite — $Mg_5Al_2(Si_6Al_2O_{22})(OH)_2$ — an amphibole

Pyrope

Talc

Serpentine

Cordierite — $Mg_2Al_3(AlSi_5O_{18})$

Chlorite — $Mg_{12}Al_2Si_6O_{20}(OH)_{16}$

Staurolite — (Mg end-member) — $Mg_3Al_{18}Si_8O_{48}H_{2-4}$

Pyrophyllite

Kaolinite

Gibbsite

Brucite

through *progressive dehydration*. In other words, with increasing metamorphic temperatures, water is generally lost from the rock. This immediately suggests that the hydroxides and hydrous silicates will be part of the low- and medium-temperature assemblages, whereas the high-temperature assemblages would tend to be made up solely of anhydrous minerals.

As we have stated, the required information on mineral assemblages (as a function of changes in metamorphic temperature) will have to be obtained from references on mineralogy–petrology. Your mineralogy text can be used as a first but, for this exercise, generally not very satisfactory source. Let me illustrate here what sort of (commonly highly generalized) information you might obtain from standard mineralogy texts for two minerals from Table 34.1, namely talc and sillimanite. For both of these minerals you check under the heading *occurrence* (or *paragenesis*, a term used in some references). For talc you will learn "Talc characteristically occurs in low-grade metamorphic rocks. . ." (from *Manual of Mineral Science*, 23rd ed., p. 525) and for sillimanite you find "Sillimanite occurs as a constituent of high-temperature-metamorphosed Al-rich rocks" (also from *Manual of Mineral Science*, 23rd ed., p. 492). This type of information will allow you to separate most of the minerals given in Table 34.1 into broad and general categories such as low-temperature, medium-temperature, and high-temperature metamorphic. Notice, however, that some minerals are present and indeed stable over a wide range of metamorphic conditions, a prime example being quartz. Once you have broadly separated the minerals listed in terms of ranges of temperature of formation, you need more specific information about coexistences. For example, with what minerals does sillimanite occur (coexist) in the high-temperature metamorphic assemblages? Such specific information on coexistences is generally not available in mineralogy texts. For sillimanite you can look up Table 33.2 (p. 415) in this book, where some general examples of coexistences are listed. For the highest-temperature conditions you find "sillimanite–garnet–biotite–muscovite–quartz–feldspar" as an example in an Al-rich bulk composition. This would make you conclude that for this exercise the garnet (pyrope) and sillimanite may well be considered as an appropriate pair in an assemblage at high metamorphic grade. If indeed two minerals are found as a pair in a fairly well defined temperature range, they are considered an assemblage representative of that temperature range and are graphically shown by a tieline between their plotted composition points. The tieline between two mineral compositions denotes that they are found next to each other, that is, in an equilibrium assemblage. This information in the literature has generally been obtained through careful study of thin sections. In a triangular diagram, as is used in this exercise, the graphical expression of all possible tielines between mineral pairs will be as triangles of various sizes and shapes. A triangle is the graphical expression of three assemblage pairs forming the outer edges of the triangle (you may wish to refresh your memory on this by returning to exercise 7 in which tielines were plotted for various sulfide pairs in the system $Cu-Fe-S$).

Pertinent information on coexistences among minerals is easiest to locate in some of the references given above. These include many volumes of *Rock Forming Minerals* by Deer, Howie, and Zussman, one volume of *Dana's System* (for oxides and hydroxides), and texts on metamorphic petrology. The most detailed and pertinent information on assemblages is given in research publications for specific chemical systems. A good example of this is the article by E-an Zen (1961; reference given above) on mineral assemblages in the system Al_2O_3–SiO_2–H_2O.

MATERIALS

This assignment is essentially one of searching the pertinent literature for the assemblages and coexistences of the various minerals listed in Table 34.1 and relating these assemblages and coexistences to their metamorphic grade. The references should be available in the library. Hand specimens of any assemblages that represent rock compositions in the system SiO_2–MgO–Al_2O_3–H_2O are helpful for direct observation of some of the possible coexistences. Review the discussion of mineral assemblage in exercises 33 and 35.

ASSIGNMENT

1. Make four photocopies of the triangle in Fig. 34.1, one for each part of the assignment.

2. Using a copy of the triangular diagram, plot the compositions of all minerals listed in Table 34.1. Mark the top corner of your triangle as SiO_2, the bottom left corner as MgO, and the bottom right as Al_2O_3. The hydrous minerals (see also exercise 7) are plotted by ignoring the H_2O component. The theoretical justification for this is that H_2O (or OH) is part of the metamorphic fluid (which is involved in metamorphic reactions) as a *perfectly mobile component* (see, e.g., *Manual of Mineral Science*, 23rd ed., p. 598, for further discussion).

3. Mark the appropriate mineral name next to each compositional point.

4. On the basis of your literature search (and perhaps some representative assemblages observed in hand specimens in the laboratory), decide which of the minerals listed in Table 34.1 belong to (a) low-temperature, (b) medium-temperature, and (c) high-temperature metamorphic categories. Plot on three copies of the unmarked triangular diagram only the minerals that fall into category a, into category b, or into category c. Label each of these diagrams as either low *T*, medium *T*, or high *T*, and label all mineral compositions. As noted in the statement of the *purpose of exercise*, low-temperature metamorphism spans approximately the range of 200° to 400°C, medium-temperature metamorphism represents about 400° to 650°C, and high-temperature metamorphism reflects the approximate range of 650° to 800°C.

5. In each of the three triangular diagrams enter tielines between coexisting minerals as deduced from your careful reading of *occurrence* or *paragenesis* in the references noted. Notice that many of the minerals will occur as groups of three, in which case their occurrence is outlined by a triangle of tielines. Complete as many tielines as you can find information for. Because the temperature ranges for the evaluation of mineral associations in this exercise are large (200°C for the low- and high-*T* assemblages and 250°C for the medium-*T* assemblages), you will find a considerable number of *crossing tielines*. A crossing tieline is one that intersects the tieline between another mineral pair. If you have been successful in arranging the three mineral association categories (low *T*, medium *T*, and high *T*) quite separately from one another, you will tend to have the least number of crossing tielines.

6. The assignment is complete when you have four final triangular diagrams: (a) a plot of all mineral compositions listed in Table 35.1; (b) a low-temperature assemblage diagram; (c) a medium-temperature assemblage diagram; and (d) a high-temperature assemblage diagram. The three assemblage diagrams should show the maximum number of joins for mineral pairs, giving a multitude of triangles.

FIGURE 34.1 Triangular diagram with percentage lines at 2% intervals. Make at least four photocopies of this triangle for the exercise.

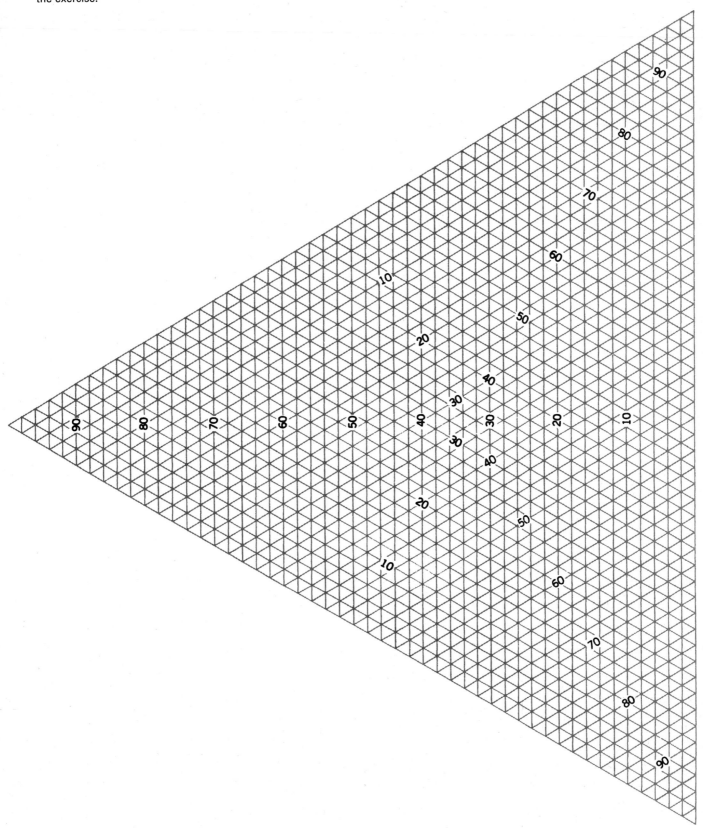

Metallic and Nonmetallic Ore Minerals and Ores: Their Classification, Associations, and Origin

PURPOSE OF EXERCISE

(1) To review the common ore minerals (metallic and nonmetallic), their chemistry, and their common uses; (2) to gain some understanding of metallic ore deposits, their geologic origin, and their classification; and (3) to evaluate, by laboratory study, specimens and assemblages from various ore deposits.

FURTHER READING

Klein, C. and Dutrow, B., 2008, *Manual of Mineral Science*, 23rd ed., Wiley, Hoboken, New Jersey, pp. 331–367.

Nesse, W. D., 2000, *Introduction to Mineralogy*, Oxford University Press, New York, pp. 378–382.

Perkins, D., 2002, *Mineralogy*, 2nd ed., Prentice Hall, Upper Saddle River, New Jersey, pp. 155–172.

Wenk, H. R. and Bulakh, A., 2004, *Minerals: Their Constitution and Origin*, Cambridge University Press, New York, New York, pp. 511–531.

Craig, J. R., Vaughan, D. J. and Skinner, B. J., 2001, *Resources of the Earth, Origin, Use and Environmental Impact*, 3rd ed., Prentice Hall, Upper Saddle River, New Jersey, 520p.

Kessler, S. E., 1994, *Mineral Resources, Economics, and the Environment*, Prentice Hall, Upper Saddle River, New Jersey, 394p.

Chang, L. L. Y., 2002, *Industrial Mineralogy*, Prentice Hall, Upper Saddle River, New Jersey, 472p.

Evans, A. M., 1993, *Ore Geology and Industrial Minerals: An Introduction*, Blackwell, Oxford, 390p.

Guilbert, J. M. and Park, C. F., Jr., 1986, *The Geology of Ore Deposits*, Freeman, New York, 984p.

Background Information: The first part of this exercise is a review of the *metallic* and *nonmetallic ore minerals*, their compositions, and their common uses. With regard to the mineralogy of ores, it is necessary to introduce a few terms and their definitions:

An *orebody* or *ore deposit* is a naturally occurring material from which a mineral or minerals can be extracted at a profit. Because of fluctuations in metal prices and costs associated with mining, many economic deposits of the past are uneconomic today and vice versa. Therefore, the term *mineral deposit* can be used for concentrations of economic minerals that have been ore deposits in the past, are ore deposits in the present, or could become ore deposits in the future. An *ore mineral* is the part of an ore that is economically desirable, as contrasted with *gangue*. *Gangue* is defined as the valueless rock or mineral matter that cannot be avoided in mining and is later separated and discarded. A mineral considered gangue in one deposit may be regarded as ore in another. For example, fluorite in small proportions (as an accessory) in a metallic vein deposit would be considered gangue, whereas fluorite in large concentrations and mined for the fluorite would be ore.

Table 35.1 gives a listing of common metallic ore minerals, their compositions, and common uses of the metals, and Table 35.2 gives a list of nonmetallic ores, that is, *industrial minerals* and *rocks*, their compositions, and common uses.

The second part of this exercise is concerned with the *origin* and *classification of metallic ore minerals*. If you have worked your way through exercises 24 through 30, you will have developed considerable skill in the identification of mineral specimens. Here you will be introduced to the evaluation of groups of minerals, that is, *mineral assemblages* (minerals that occur together) in various ore deposits. A *mineral assemblage* is a group of minerals that are in equilibrium—equilibrium meaning that the various minerals making up the assemblage are in a "happy coexistence" without obvious reaction textures or other indications (as deduced from their chemistry) of chemical disequilibrium. In hand specimen study of mineral assemblages, the observer is limited to the evaluation of textures. Examples of equilibrium textures are (1) the equigranular intergrowth of quartz, feldspars, and biotite in a granite and (2) the intimate intergrowth of pyrite, pyrrhotite, and chalcopyrite (with grains of each of the three sulfides touching grains of the other sulfides) in a high-temperature sulfide occurrence. An example of a reaction texture is a *corona* or corona texture, in which one mineral (or several minerals) has formed at the expense of some earlier mineral (or minerals). An example would be a rim of enstatite completely surrounding an original high-temperature olivine; such a texture implies that the original olivine is not part of (not in coexistence with) the later lower-temperature assemblage which includes enstatite as well as other minerals outside the corona. A mineral assemblage, therefore, records the mineralogy of an ore or rock specimen for a reasonably narrow range of temperature, pressure, and chemical conditions of formation. A series of assemblages, as deduced from careful mineralogical and textural study of ore and

TABLE 35.1 Common Metallic Ore Minerals, Their Compositions, and Their Common Uses

Metal	Ore Mineral	Composition	Examples of Common Uses of the Metal
Aluminum	Bauxite	Mixture of several Al-hydroxides	As Al metal in siding, roofing, in many parts of cars, in packaging and containers, and in electrical applications.
Antimony	Stibnite	Sb_2S_3	As Sb metal in storage batteries, solder, and ammunition.
Chromium	Chromite	$FeCr_2O_4$	As Cr metal in stainless steel, tool steel, and cast iron.
Cobalt	Smaltite*	$(Co,Ni)As_{3-x}$	Co is used in heat- and corrosion-resistant alloys.
	Cobaltite*	$(Co,Fe)AsS$	
Copper	Native copper	Cu	The largest use for Cu metal is in electrical equipment and supplies and as copper and copper alloy wire in electric motors, power transmission lines, and housing and industrial wiring. It is also used in ammunition.
	Bornite	Cu_5FeS_4	
	Chalcocite	Cu_2S	
	Chalcopyrite	$CuFeS_4$	
	Covellite	CuS	
	Cuprite	Cu_2O	
	Enargite	Cu_3AsS_4	
	Malachite	$Cu_2CO_3(OH)_2$	
	Azurite	$Cu_3(CO_3)_2(OH)_2$	
	Chrysocolla	$\sim Cu_4H_4Si_4O_{10}(OH)_8 \cdot nH_2O$	
Gold	Native gold	Au	A large use for gold is in jewelry. It is also used in solid-state devices and in printed circuit boards. Also made into bars and medallions for investment.
	Calaverite*	$AuTe_2$	
Iron	Magnetite	Fe_3O_4	Iron is a major component of steel for the construction of vehicles, mobile equipment, ships, and machinery. Also used in steel frames for large buildings.
	Hematite	Fe_2O_3	
	Limonite	$\sim FeO \cdot OH \cdot nH_2O$	
	Siderite	$FeCO_3$	

TABLE 35.1 (continued)

Metal	Ore Mineral	Composition	Examples of Common Uses of the Metal
Lead	Galena	PbS	Most of the lead is used in the manufacture of lead–acid storage batteries. Also a constituent in special alloys for bearings, and in glass, porcelain enamel, and ceramic glazes.
	Cerussite	$PbCO_3$	
	Anglesite	$PbSO_4$	
Manganese	Pyrolusite	MnO_2	Manganese is essential in the production of virtually all steels, and is important in the production of cast iron.
	Romanechite	$(Ba, H_2O)_2(Mn^{4+}, Mn^{3+})_5O_{10}$	
	Manganite	$MnO(OH)$	
Mercury	Cinnabar	HgS	Mercury is used in mercury batteries and mercury lamps.
Molybdenum	Molybdenite	MoS_2	As an alloying element in steels, cast iron, and other alloys.
Nickel	Pentlandite	$(Fe,Ni)_9S_8$	As nickel metal in many types of alloys.
Silver	Native silver	Ag	The most important use of silver is in photographic materials; also used in electrical products, jewelry, sterling ware and in rear-window defrosters in cars.
	Argentite*	Ag_2S	
Tin	Cassiterite	SnO_2	Tin is used in solder and tinplate, as well as in cans and containers.
	Stannite*	Cu_2FeSnS_4	
Tungsten	Scheelite	$CaWO_4$	Tungsten, as a carbide, is used in cutting and heat-resistant materials. Also a constituent in various specialized alloys.
	Wolframite	$(Fe, Mn)WO_4$	
	Huebnerite*	$MnWO_4$	
Uranium	Uraninite	UO_2	Used in nuclear fuel rods for energy reactors and also in nuclear weapons.
	Pitchblende	massive UO_2	
	Carnotite*	$K_2(UO_2)_2(VO_4)_2 \cdot 3H_2O$	
Zinc	Sphalerite	ZnS	Zinc metal is used in brass making (brass is mainly a copper–zinc alloy) and in galvanizing.
	Smithsonite	$ZnCO_3$	As zinc oxide it is used in paints.
	Hemimorphite	$Zn_4(Si_2O_7)(OH)_2 \cdot H_2O$	
	Zincite*	ZnO	

*The minerals that are starred tend to be less common and have not been described in the preceding exercise; see your text for their descriptions.

TABLE 35.2 Common Nonmetallic Ore Minerals and Some Industrial Rocks, Their Compositions, and Their Common Uses, Arranged by Geologic Mode of Origin

Industrial Mineral or Rock	Composition[a]	Examples of Common Uses
IGNEOUS (INTRUSIVE)		
Olivine	$(Mg,Fe)_2SiO_4$	In high-temperature refractory bricks; also as an abrasive.
Chromite	$FeCr_2O_4$	Cr is an alloying element and the yellow pigment in paint.
Nepheline syenite		Al and alkalies are used in glass manufacture, in glazes and enamel.
Granite		Used as building stone and crushed stone.
PEGMATITIC AND HYDROTHERMAL		
Feldspar	$(K,Na,Ca)Al_{1+x}Si_{3-x}O_8$	Used in glass and ceramics manufacture.
Micas	$KAl_2(AlSi_3O_{10})(OH)_2$, etc.	Used as a heat-insulating material; in electrical circuits (e.g., capacitors). Fine flakes give luster in paint and wallpaper.
Quartz crystal	SiO_2	In radio oscillator circuitry for its piezoelectric property (e.g., in quartz watches). Much of this is synthetic quartz.
Li minerals		
Spodumene	$LiAlSi_2O_6$	In lithium chemicals for lubricating greases, in glass, ceramics, medicine, silicon chips, batteries, and so on.
Lepidolite	$K(Li,Al)_{2-3}(AlSi_3O_{10})(O,OH,F)_2$	
Be minerals		
Beryl	$Be_3Al_2Si_6O_{18}$	In copper–beryllium alloys; as beryllium oxide in ceramics; and as Be metal in aircraft and spacecraft.
Fluorite	CaF_2	Acid-grade CaF_2 for the production of HF. Other grades are used in glass and enamels and as a flux in steelmaking.
IGNEOUS (EXTRUSIVE)		
Basalt and related rocks		In crushed stone, concrete aggregate, and railroad ballast.
Pumice and scoria		In the construction industry for lightweight concrete and as a plaster aggregate.
Perlite (glassy volcanic rock)		Has low thermal conductivity, high sound absorption with many applications.
SEDIMENTARY (CLASTIC)		
Sand and gravel		Construction and road building; outranks all other nonfuel mineral resources (metallic and nonmetallic) in both tonnage and value.
Sandstone		Some is used as facing and trim on buildings; most is used in foundries as disaggregated sand for molds and cores.
Clays		
Kaolinite	$Al_2Si_2O_5(OH)_4$	In ceramics (e.g., china) and in the manufacture of paper, with the filler being kaolinite. High-quality glossy paper contains much kaolinite.
Smectite (also known as "bentonite")	$Na_{0.7}(Al_{3.3}Mg_{0.7})Si_8O_{20}(OH)_4 \cdot nH_2O$	In oil-well-drilling muds; as a bonding and pelletizing agent (as in taconite iron ore pellets); and in products that require absorption properties.
Common clay and shale		Used for building brick, drain tile, and vitrified sewer pipe.

TABLE 35.2 (continued)

Industrial Mineral or Rock	Composition[a]	Examples of Common Uses
Titanium and zirconium minerals		
Ilmenite	$FeTiO_3$	As a white pigment, TiO_2, used in paint, plastics, rubber, and paper.
Rutile	TiO_2	
Zircon	$ZrSiO_4$	In high-temperature refractory brick and sand; in glazes and enamels; and ferroalloys, paint, and pharmaceuticals.
Baddeleyite	ZrO_2	
Rare earth minerals		
Bastnaesite	$CeFCO_3$	Some rare earths are used in specialized alloys; samarium is used in the production of strong permanent magnets; others are used in the glass industry; and as coloring agents.
Monazite	$(Ce,Y)PO_4$	
Xenotime	YPO_4	
Diamonds	C	As gems. Industrial diamonds in the form of grit and powder, are used for grinding wheels and saw blades.
SEDIMENTARY (BIOGENIC)		
Limestone and dolomite	$CaCO_3$ and $CaMg(CO_3)_2$	Limestone is used as crushed stone, and in the manufacture of portland cement. Both limestone and dolomite are used as "agricultural lime".
Diatomaceous earth		In powdered form, a filter aid in clarifying liquids.
Phosphate rock	Mainly $Ca_5(PO_4)_3(F,Cl,OH)$	For the manufacture of fertilizer; also in detergents and animal feed.
Sulfur		
Native sulfur	S	The bulk of sulfur is used in fertilizers; the remainder goes into everything from photography and pharmaceuticals to paint, paper, and petroleum refining.
Recovered sulfur (from crude oil, natural gas, and industrial stack gases)		
In pyrite	FeS_2	
SEDIMENTARY (CHEMICAL)		
Barite	$BaSO_4$	As a weighting agent in the mud used in drilling deep wells.
Salt (halite)	$NaCl$	In the chemical industry, to produce Na and Cl chemicals.
Trona	$Na_3H(CO_3)_2 \cdot 2H_2O$	As a source of Na in glass manufacture.
Gypsum	$CaSO_4 \cdot 2H_2O$	For plaster of Paris, in plasterboard, and as an agent in cement.
Sylvite (and other K minerals)	KCl	For fertilizers, soaps, glass, ceramics, and drugs.
Borate minerals		
Borax	$Na_2B_4O_5(OH)_4 \cdot 8H_2O$	In the manufacture of glass, glass fiber, fire retardants, soaps, detergents, vitreous enamels, and agricultural products.
Kernite	$Na_2B_4O_6(OH)_2 \cdot 3H_2O$	
Ulexite	$NaCaB_5O_6(OH)_6 \cdot 5H_2O$	

TABLE 35.2 (continued)

Industrial Mineral or Rock	Composition[a]	Examples of Common Uses
SURFICIALLY ALTERED		
Vermiculite	$(Mg,Ca)_{0.3}(Mg,Fe,Al)_{3.0}(Al,Si)_4O_{10}(OH)_4 \cdot 8H_2O$	Used in aggregates, in insulation, and in agriculture.
Bauxite	Mixture of 3 Al-hydroxides	For the production of Al; the nonmetallic uses of bauxite are in refractories, chemicals, abrasives, and cement.
METAMORPHIC		
Marble, also serpentine rock known as "verd antique"		As dimension stones (columns, floors, trim) and as exterior facing on large buildings; also as crushed stone.
Asbestos minerals		
Chrysotile	$Mg_3Si_2O_5(OH)_4$	
Amosite	$(Fe, Mg)Si_8O_{22}(OH)_2$	All three have unusual tensile strength and flexibility; they also resist heat and chemical attack. Blended with cotton and rayon fibers, and compressed in molded and cast products; 95% is from chrysotile.[b]
Crocidolite	$Na_2Fe_3{}^{3+}Fe_2{}^{3+}Si_8O_{22}(OH)_2$	
Talc, and soapstone	$Mg_3Si_4O_{10}(OH)_2$	Talc is used as a filler in paint, plastics, paper, and rubber; soapstone, a massive talcose rock, is used for laboratory sinks.
Magnesite	$MgCO_3$	MgO is used as a refractory in steel furnaces and cement kilns; also used in chemicals, among them "milk of magnesia."
Graphite	C	Used in steelmaking, refractories, and foundries.
Corundum and emery	Al_2O_3	Major use is as an abrasive; also used in the production of high-alumina refractory bricks and as a flux in steelmaking.
Garnet	Various compositions	As an abrasive, and as a loose sand for sandblasting.
Wollastonite	$CaSiO_3$	In the production of ceramics such as tiles, insulators, and glazes; also as a filler in plastics because it has a fibrous habit.
Sillimanite minerals (sillimanite, kyanite, and andalusite)	Al_2SiO_5	Used in the manufacture of high-temperature refractories such as heavy-duty refractory bricks.
Pyrophyllite	$Al_2Si_4O_{10}(OH)_2$	In white ceramic products such as floor and wall tile, and in electrical porcelain.

[a]Compositions are given only for minerals, not rock types.

[b]For a discussion of comparative health hazards, see M. Ross, 1982, *A Survey of Asbestos-Related Disease in Trades and Mining Occupations and in Factory and Mining Communities as a Means of Predicting Health Risks of Non-occupational Exposure to Fibrous Minerals*, U.S. Geological Survey, Open-File Report 82-745, 41 p.

SOURCE: Abstracted from P. W. Harben and R. L. Bates, 1984, *Geology of the Nonmetallics*, Metal Bulletin Inc., New York, 392 p.

rock specimens, allows one to infer a possible *depositional*, or *paragenetic*, sequence. An original high-temperature assemblage that has undergone subsequent conditions of lower T, and P, and perhaps more hydrous conditions as well, leaving *relicts* of the original assemblage within a newly formed, later assemblage, is an example.

A *mineral assemblage* as defined earlier is quite different from a *mineral association*. A *mineral association* consists of all minerals that may be found in an ore or rock specimen; they are simply the minerals that occur together irrespective of their relative timing and without implications regarding equilibrium. As such, the mineral association of an ore or rock specimen may well be split into two or more specific mineral assemblages (reflecting quite different conditions of origin as, for example, high-temperature igneous for the original assemblage with a later, low-temperature metamorphic overprint). Therefore, a series of assemblages will together constitute the total mineral association. As a result, the number of minerals present in an association is generally greater than that in an assemblage, and also generally greater than that allowed by the phase rule (as defining equilibrium conditions).

Rock types such as igneous, metamorphic, and sedimentary (see exercises 31 through 33) are formed over various temperature and pressure regimes. Similarly, ore deposits form over a wide range of temperatures, pressures, and geological conditions. The mineral association and the texture and structure of its occurrence can reveal, even in hand specimens, a considerable amount about the conditions of origin of an ore type. The origin of ore deposits and their *classification* is a very large field of knowledge and research, and this exercise is only a brief introduction to the broad field of *economic geology* and *ore petrology*. Figure 35.1 gives a genetic classification of ore deposits and Table 35.3 lists examples of the ore deposits formed by such processes. In Fig. 35.1 and Table 35.3 there are several terms that may be new to you. Most of these are defined in the following list.

Placer deposit – a surficial mineral deposit formed by the mechanical concentration of mineral particles from weathered debris. The mineral concentrated is usually a heavy mineral such as gold, cassiterite, or rutile.

Supergene enrichment – a process by which near-surface sulfides are oxidized and the metals in solution are reprecipitated as either oxides (e.g., cuprite) or carbonates (e.g., malachite) above the water table, or as sulfides (e.g., chalcocite, covellite) below the water table. In the upper layer supergene oxide ores form, in the lower supergene sulfide ores (see Fig. 35.2). The upper oxidized part of such zones is known as the *zone of oxidation and leaching*; there the original minerals have been severely altered, with ubiquitous formation of limonite, giving rise to *gossan*.

Hydrothermal origin – the processes of ore deposition from heated waters of any origin (magmatic, metamorphic, connate, meteoric, etc.).

Epithermal – pertaining to a hydrothermal deposit formed within about 1 km of the earth's surface and in a T range of 50° to 200°C, occurring mainly as veins.

Red beds – sedimentary strata composed largely of sandstone, siltstone, and shale, with locally thin units of conglomerate; they are predominantly red because red hematite coats individual mineral grains.

Fumarole – a volcanic vent from which gases and vapors are emitted.

Hypabyssal – pertaining to an igneous intrusion whose depth is intermediate between that of a plutonic environment and the surface.

Porphyry copper – a large body of rock, typically a granitic porphyry, that contains disseminated chalcopyrite and other sulfide minerals.

Immiscible liquid or melt – said of two components of a liquid or melt that cannot completely dissolve in one another, for example, oil and water, or sulfide melt and silicate melt.

Pegmatite – an exceptionally coarse-grained igneous rock, usually found in irregular dikes, lenses, or veins, especially at the margins of batholiths.

Stratiform ore deposit – a deposit in which the ore minerals constitute one or more sedimentary, metamorphic, or igneous layers. Examples are beds of salt or beds of iron oxide, or layered igneous complexes with layers rich in chromite or platinum metals.

Greisen – a hydrothermally altered granitic rock composed largely of quartz, mica, and topaz. Tourmaline, fluorite, rutile, cassiterite, and wolframite are common associations.

Skarn ore – a silicate gangue or contact-metamorphosed rock composed of Ca-rich silicates such as garnet, pyroxene, and amphibole, and ore minerals such as magnetite and hematite.

With this brief introduction to ore minerals, their chemistry and uses, as well as their classification and occurrence, you are ready to evaluate ore specimens and ore associations in hand specimens in the laboratory. Although large-scale features of ore deposits such as their structure and regional setting cannot be deduced from hand specimen materials, a great deal can be learned by careful observation of *ore* and *host rock* specimens. A host rock is any body of rock that serves as the host for a particular ore mineralization or deposit.

In the hand specimen study of ore minerals, two aspects must be carefully evaluated: (1) the assemblages of the ore minerals and (2) the associations of the ore minerals with the gangue and surrounding host rock. Although textures of ore minerals are best studied in polished sections using a high-magnification microscope, in hand specimen observation a hand lens, a binocular microscope, or both will have to suffice. The observed assemblages are commonly represented graphically. An example of various mineral assemblages in the system Cu–Fe–S is shown in Fig. 35.3. Here the

FIGURE 35.1 Schematic flow sheet for a genetic classification of ore deposits. Dashed boxes denote the movement of material, solid boxes are deposits, and arrows represent genetic relationships and processes. See Table 35.3 for a tabulation of ore deposit types as a function of processes of formation. This flow sheet is a modification by M. T. Einaudi of the original flow sheet published by Skinner and Barton, 1973. (This modification is reproduced, with permission, from the *Annual Review of Earth and Planetary Sciences*, vol. 1, copyright © 1973, by Annual Reviews, Inc., and with permission of Brian J. Skinner, Yale University, New Haven, Connecticut.)

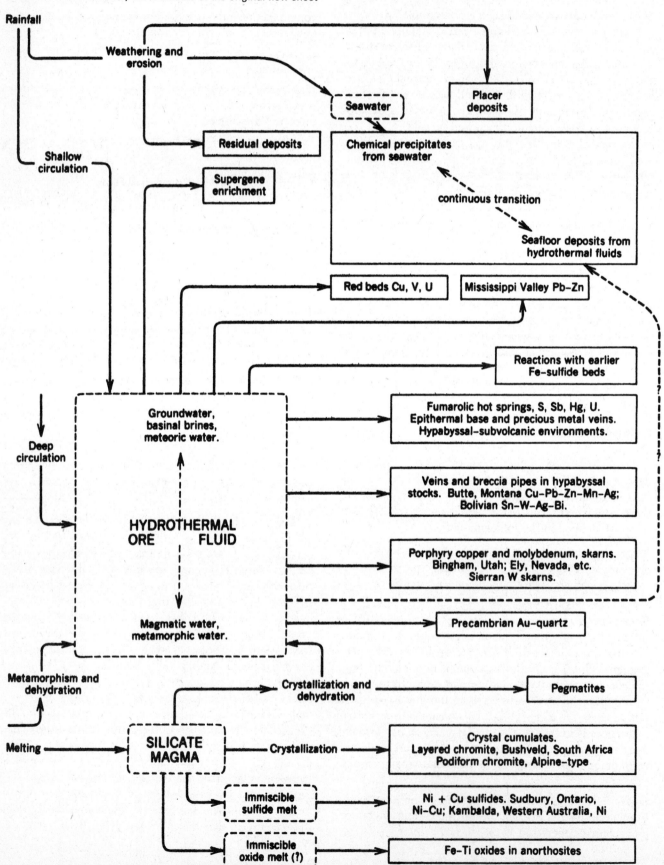

TABLE 35.3 Examples of Ore Deposits Classified According to Processes of Formation

I. Deposits Formed at the Surface by Surface Processes at Ambient Temperatures. The Waters Are Groundwater or Ocean Water.

   A. *Involving chemical or clastic sedimentary processes*

      1. Placers[a] and beach sands: Au, Cr, Ti, Sn, Pt, diamonds (ancient and modern).

      2. Evaporites and brines: K_2O, NaCl, gypsum. Na carbonates and sulfates. Nitrates, Sr, Br, I, $CaCl_2$, MgO.

      3. Precambrian banded iron-formations: Fe and Au (volcanic input?).

      4. Submarine Mn–Cu–Ni–Co nodules; Fe–Mn oozes (modern).

   B. *Weathering*

      1. Laterites, bauxite: Al, Ga; Ni.

      2. Supergene enrichment; oxide and sulfides of Cu.

   C. *Deposited from groundwater in sandstone sequences*

      1. Roll-front (Wyoming-type) sandstone: U.

      2. Tabular (Grants, New Mexico-type) sandstone: U.

II. Deposits Formed at or Near the Surface ($<\frac{1}{2}$ km) by Low to Moderate-Temperature (100°–300°C) Hydrothermal Fluids. The Waters Are Dominantly of Meteoric, Ocean, or Connate Origin.

   A. *Contemporaneous with submarine volcanism*

      1. Ophiolite-type (Cyprus) massive sulfides: *Cu*[b], Ag, Au, Zn. Modern analogues are the "black smokers," massive sulfide mounds in the East Pacific Rise.

      2. Kuroko-type massive sulfides: *Pb, Zu, Cu*, Ag.

      3. Basalt–graywacke-type massive sulfides: *Cu*, Zn.

   B. *Penecontemporaneous with cratonic sedimentary basins, prelithification* (from connate waters or basinal brines)

      1. Sullivan-type (British Columbia) and McArthur River-type (Australia) shale-hosted stratiform massive sulfides: *Pb, Zn*, Ag. Modern analogues are Red Sea metalliferous muds.

      2. Zambian-type stratiform; stratibound copper in siltstone: *Cu*, Co, Ag (at relatively low *T*)

   C. *In cratonic basins but postlithification; open-space filling and replacement* (from connate waters, basinal brines, and metamorphic waters)

      1. Mississippi Valley type; carbonate-hosted: *Pb, Zn*.

      2. Proterozoic unconformity (Athabasca type): *U*.

   D. *Continental hot spring environments* (meteoric water)

      1. Hot-springs type, McLaughlin, California: *Hg, Au*. Modern analogues are geothermal systems

      2. Volcanic sulfur. Modern analogues are active volcanic fumaroles

      3. Epithermal precious metal veins, Comstock, Nevada: *Ag, Au*

stoichiometric compositions of various sulfides and some native elements are plotted on a triangular diagram. The heavy lines between mineral pairs (also known as *tielines*) indicate coexistence. Triangles, therefore, indicate the coexistence of three mineral pairs. For example, the triangle pyrite–pyrrhotite–chalcopyrite means that there are specimens in which pyrite–pyrrhotite, pyrrhotite–chalcopyrite, and chalcopyrite–pyrite occur in contact. An evaluation of the mineral assemblage is the first step in an assessment of the possible origin of the ore and gangue minerals in an ore deposit. With reference to Fig. 35.3, the native copper–chalcocite–digenite–covellite associations are indicative of supergene conditions; the bornite–chalcopyrite–pyrite assemblages are common in hydrothermal veins, or in sulfide ores related to igneous activity; and the native iron–troilite assemblage is found only in meteorites. See exercise 7 for the plotting of such mineral composition diagrams.

Another aspect of the ore and gangue, as well as of host rock specimens, that must be carefully evaluated is their *macroscopic textures*. The texture of a rock or mineral associa-

tion includes the geometric aspects of, and the mutual relations among, its component mineral grains, that is, the size, shape, and arrangement of the constituents, as well as their crystallinity and fabric. The textures of ore bodies vary according to whether the constituent minerals were formed by deposition in an open space from an aqueous solution or from a silicate melt, or by replacement of preexisting rock or ore minerals. Subsequent metamorphism may completely alter any primary textures. The types of textural observations that can be made in hand specimens include the following.

1. In *open space filling:* veins, breccias, and other partly filled openings contain vugs and cavities that can be interpreted as spaces left by incomplete filling of larger open spaces. Other expressions of open-space filling are the *encrustation* (or *crustification*) of earlier formed minerals by later minerals, and *colloform* textures, which are successively deposited, fine onion-skin-like layers that can form only in open spaces. Examples of open space filling are shown in Figs. 35.4a and b.

TABLE 35.3 *(continued)*

III. Deposits Formed at Shallow Levels in the Crust (1–5 km) by Moderate- to High-Temperature (250°–550°C) Hydrothermal Processes in or Near Intermediate to Felsic Igneous Intrusions. The Waters Are of Mixed Meteoric–Magmatic–Metamorphic Origin.

   A. *In or near cupolas of small porphyry stocks*

      1. Cordilleran base–metal lodes, veins, Butte, Montana: *Cu, Pb, Zn,* Ag, Mn

      2. Porphyry copper–molybdenum, Bingham, Utah: *Cu, Mo.*

      3. Climax-type (Colorado) high-silica rhyolite porphyry: *Mo,* Sn, W, F, Zn, Pb.

      4. Greisen: *Sn,* W, Mo, F, Zn.

   B. *Replacement deposits in carbonate rocks, near felsic stocks*

      1. Massive sulfide–silica–carbonate replacement bodies: *Pb, Zn,* Ag, Mn, Cu.

      2. Base–metal sulfide, magnetite, and tin skarns: *Fe, Cu, Zn, Pb, Sn,* and the like.

   C. *In a mesabyssal plutonic and metamorphic environment*

      1. Metamorphic gold–quartz veins: *Au.*

      2. Tungsten-rich skarns.

IV. Deposits Formed by Magmatic Processes

   A. *Involving magmas with highly volatile substances*

      1. Pegmatites: *Li, Sn, Ta, Nb, Bi, Be, REE.*

      2. Carbonatite–alkalic complexes: *Nb, P, REE, U, Cu.*

      3. Kimberlites: diamond.

   B. *Involving fractional crystallization*

      1. Stratiform Bushveld-type (South Africa) chromite: *Cr.*

      2. Podiform Alpine-type chromite in peridotite and gabbro: *Cr.*

      3. Stratiform vanadiferous magnetite: *V.*

   C. *Involving immiscible oxide liquid*

      1. Anorthosite massifs: *Ti.*

   D. *Involving immiscible sulfide liquid*

      1. Mafic Sudbury type (Ontario) and Duluth type: *Ni, Cu* sulfide.

      2. Ultramafic type (komatiitic): *Ni, Cu* sulfide.

      3. Zoned ultramafic-felsic, Alaskan type: platinum group metals

      4. Gabbroic zone, Merensky Reef type (Zimbabwe): platinum group metals, Cu, Ni, Ag, Au

[a]Many of the terms are defined in the text.

[b]Major metal sources are in italics.

Source: By permission of M. T. Einaudi, Stanford University, Palo Alto, California.

2. *Magmatic textures* are the result of precipitation from a silicate melt. Oxide minerals such as chromite commonly crystallize early from a melt and may therefore be in euhedral crystals. These crystals tend to occur as well-defined layers in the host rock and are the result of settling and accumulation of the early crystals in the melt. When oxide and silicate minerals crystallize simultaneously, textures typical of granites develop (in which all mineral grains are anhedral) as a result of the mutual interference of the grains during growth. Sulfides, because of their lower temperatures of crystallization, generally form later than associated silicates and will commonly occur as anhedral grains interstitial to silicate grains.

3. *Replacement textures* are formed when one mineral dissolves and another is simultaneously deposited in its place, without the intervening development of appreciable open spaces, and commonly without a change in volume. Replacement processes can affect ore and gangue minerals. Pseudomorphism in which a preexisting mineral has been replaced by another is the most obvious example of replacement. A well-known example is that of quartz pseudomorphous after fluorite; the crystal form of fluorite is completely preserved by the later quartz. An original texture of a sedimentary rock may also be pseudomorphic, as shown, for example, by hematite oolites in banded iron-formations. Other evidence for replacement, although more difficult to assess in hand specimens, is given by islands of unreplaced host mineral or wall rock, by the nonmatching nature of walls or borders of a fracture, and by replacement envelopes (see Fig. 35.4c).

Yet another aspect of ore deposits and their mineral associations that can be observed in hand specimens is

FIGURE 35.2 Diagramatic representation of weathering and enrichment of a vein rich in primary Cu sulfides. (From *Manual of Mineral Science*, 23rd ed., p. 352.)

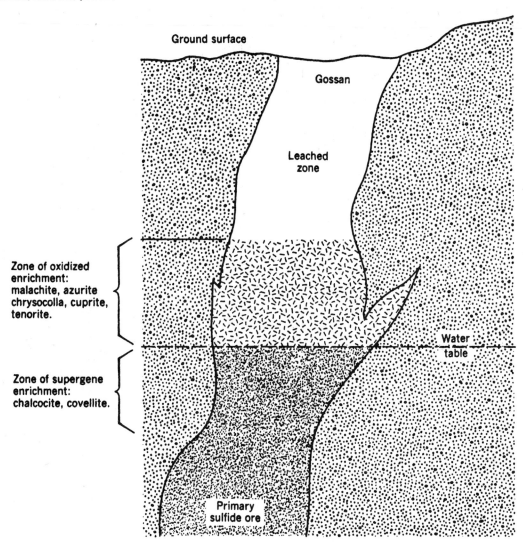

FIGURE 35.3 Some of the most common sulfides represented in the Cu–Fe–S system. Many of these sulfides (e.g., bornite and chalcopyrite) show some solid-solution of Cu and Fe in particular; this is not shown in the diagram. Tielines connect commonly occurring pairs of minerals. Triangles indicate coexistences of the three sulfides. The Fe–S coexistence is common in iron meteorites. (From *Manual of Mineral Science*, 23rd ed., p. 355.)

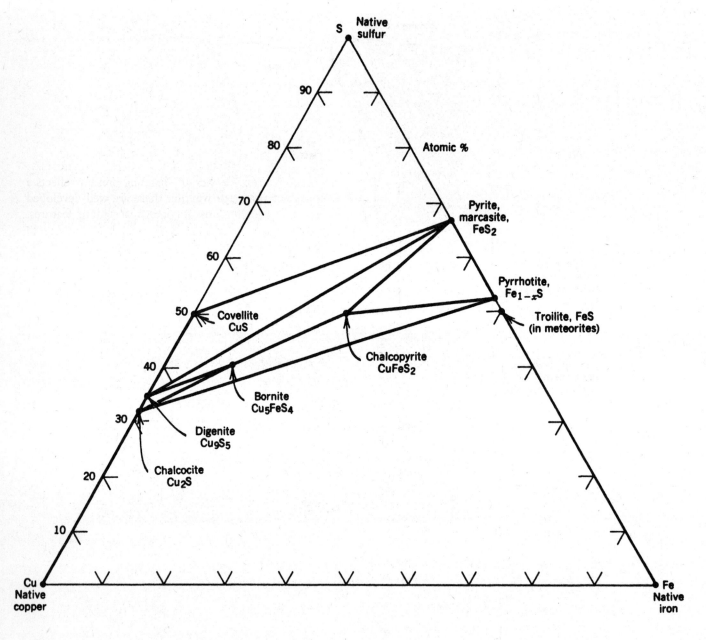

wall rock alteration. This alteration is generally best seen in the *country rocks* (or *host rocks*) that enclose hydrothermal ore deposits. The heated circulating solutions may have altered the original mineralogy of the host rock over distances several inches to many feet away from a vein. Such alteration may be expressed in well-defined zones or envelopes. The most obvious types of alteration that can be observed without the aid of a microscope are (1) kaolinite formation at the expense of plagioclase, (2) montmorillonite formation after amphiboles and plagioclase, and (3) sericite forming as an alteration product of all primary rock-forming silicates such as feldspars, micas, and mafic minerals. Hand specimens of primary igneous rocks, such as from porphyry copper deposits (see

Fig. 35.4*d*), may contain sulfide veinlets cutting across them, in association with wall rock alteration.

MATERIALS

This exercise is most successful if a range of hand specimens for several ore deposits of diverse origin are available for study. Suites of ore specimens as well as host rock or wall rock specimens for various deposits (classified in Table 35.3) of diverse origin would be (1) magmatic (intrusive) ores, (2) porphyry-type ores, (3) pegmatitic ores, (4) ores with volcanic associations, (5) vein ores of various types, (6) skarn ores, (7) ores that are the result of chemical sedimentation (e.g., evaporites and banded iron-formations), (8) stratiform ores formed in sedimentary environments, and (9) Witwatersrand-type placer ores. If at all possible the suite should include average ore types as well as typical host rocks, in addition to the more unusual mineralogical specimens that are commonly part of university and museum collections. Use of a hand lens, a binocular microscope, or both is essential in the recognition of fine-grained intergrowths and alteration of minerals. A photocopy of a triangular diagram (such as Fig. 7.2) is needed if you are asked to represent associations graphically.

If no specimen material is available, this exercise will provide you with an introduction to and review of metallic and nonmetallic ore minerals, their genesis, and classifications.

ASSIGNMENT

This assignment is based on the availability of representative suites of ore and wall rock (or host rock) specimens, for ore deposits of various origins.

1. For each suite of minerals and rocks (for a specific ore deposit type) handle and observe as many of the specimens as are available. In exercises 24 through 30 you have in all likelihood learned mineral identification by the study of unusual mineral specimens, that is, mineral specimens that consist commonly of only one mineral in a generally coarse-grained form. Such mineral specimens are rare and are generally acquired specifically for courses in mineralogy. In this exercise, you must extend your mineral identification skills to more usual specimens, to specimens in which several minerals may be closely intergrown with grain sizes much smaller than you have become accustomed to. After careful study of all of the specimens for a deposit, list in Table 35.4 all the minerals that are found in the ore specimens, and all the minerals that make up the wall rock. In the ore specimens you should distinguish between ore and gangue minerals and the assemblages among them. For complex Cu–Fe–S ores you may wish to plot the sulfide assemblages on a triangular diagram, as outlined in exercise 7.

2. Study and describe the textures exhibited by the ore minerals as well as the wall rock or host rock samples. For example, decide whether the specimens have magmatic or volcanic textures; whether any banding present reflects a sedimentary origin; or whether there are well-developed crosscutting relationships and open-space-filling features, as in vein deposits, and so on. Report your findings in Table 35.4.

3. Note the presence of wall rock alteration and describe its features in Table 35.4.

4. On the basis of your observations of the mineralogy and textures of each suite of specimens, decide on a possible origin for each ore and wall rock suite. Consult Table 35.3 and then enter your conclusions about the origin of each in Table 35.4.

5. In the last column of Table 35.4 enter the metal (or metals) for which specific ore specimens (or ore suites) would be mined.

FIGURE 35.4 *Open-space fillings.* (*a*) A cross section of a vein assemblage showing encrustation, which is crustiform banding. (*b*) Colloform banding of malachite in what was originally an open space in the wall rock. *Replacement.* (*c*) Replacement envelope localized by fractures. Notice the irregular margins and the widening of the envelope at intersections. *Wall rock alteration.*

(*d*) A polished surface of two veinlets cutting across each other in a porphyry copper deposit. The large phenocrysts in the host granite are plagioclase partially altered to montmorillonite. The veinlets are mostly quartz and sericite with the horizontal veinlet having pyrite along the central seam.

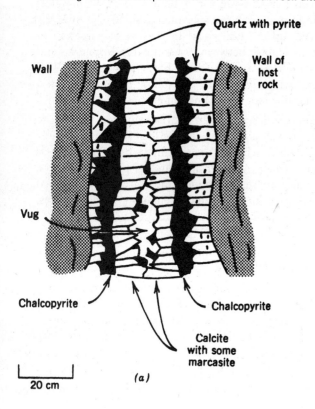

Quartz with pyrite

Wall

Wall of host rock

Vug

Chalcopyrite

Chalcopyrite

Calcite with some marcasite

20 cm

(*a*)

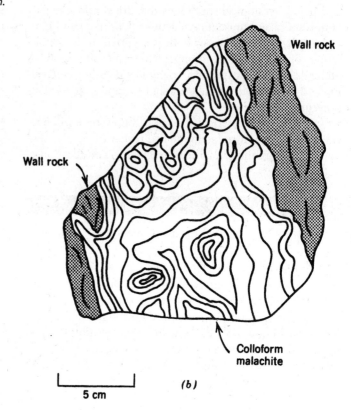

Wall rock

Wall rock

Colloform malachite

5 cm

(*b*)

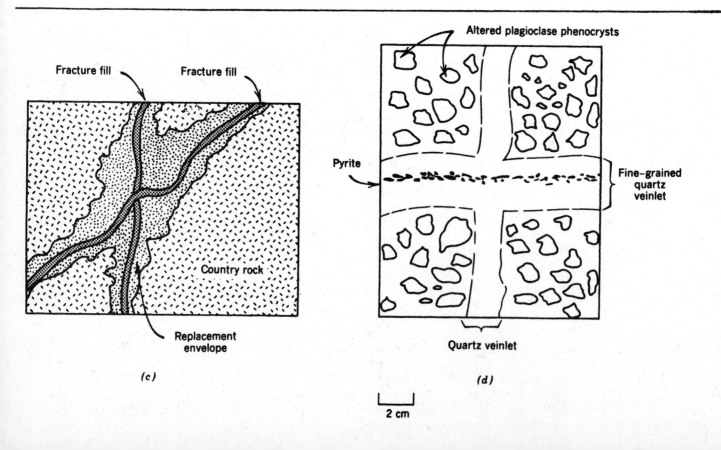

Fracture fill

Fracture fill

Country rock

Replacement envelope

(*c*)

Altered plagioclase phenocrysts

Pyrite

Fine-grained quartz veinlet

Quartz veinlet

(*d*)

2 cm

TABLE 35.4 Record of Mineralogical and Textural Findings on Suites of Hand Specimens from Ore Deposits, and Your Brief Conclusions about Their Origin

Consult Table 35.3 and Fig. 35.1.

Rock Specimen Number	Mineral Assemblages: Distinguish assemblages of ore minerals from gangue; also record the mineralogy of the host rock	Types of Textures: Note the type of possible wall rock alteration	Brief Statement about the Origin of a Suite of Specimens	The Metal or Metals for Which These Specimens are an Ore

Please make photocopies of this form if more copies are needed.

NOTES

NOTES

NOTES

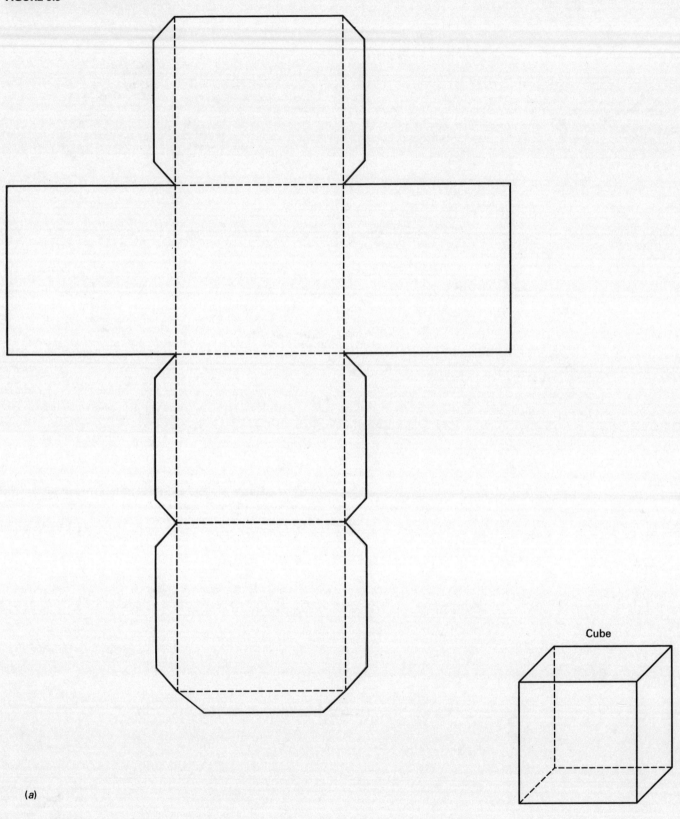

Cube

(a)

Insert 1

Student Name

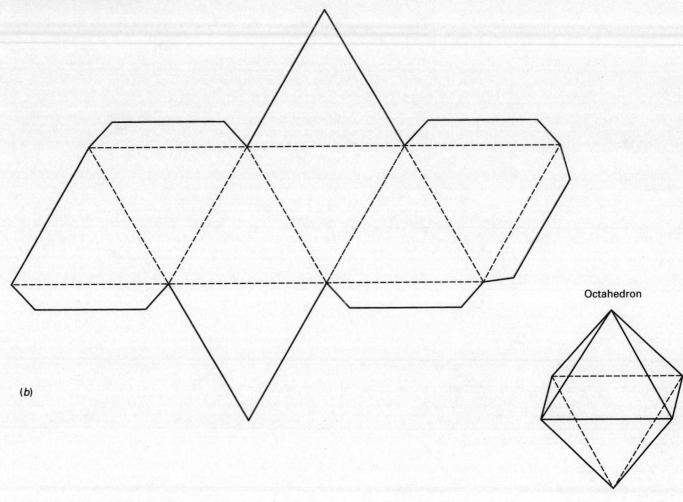

Octahedron

(b)

(c)

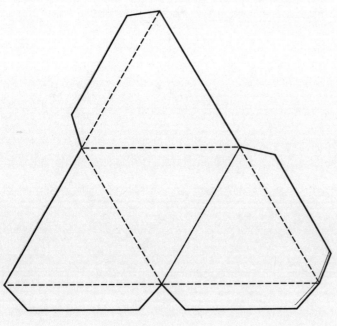

Tetrahedron

Insert 2

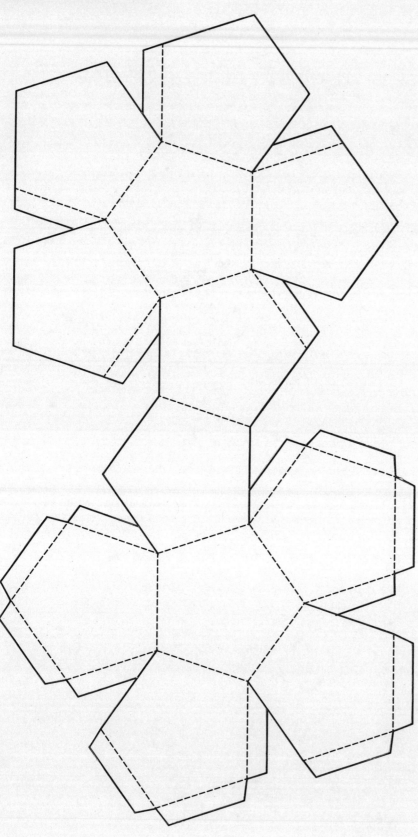

Pyritohedron

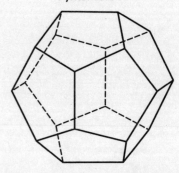

(d)

Insert 3

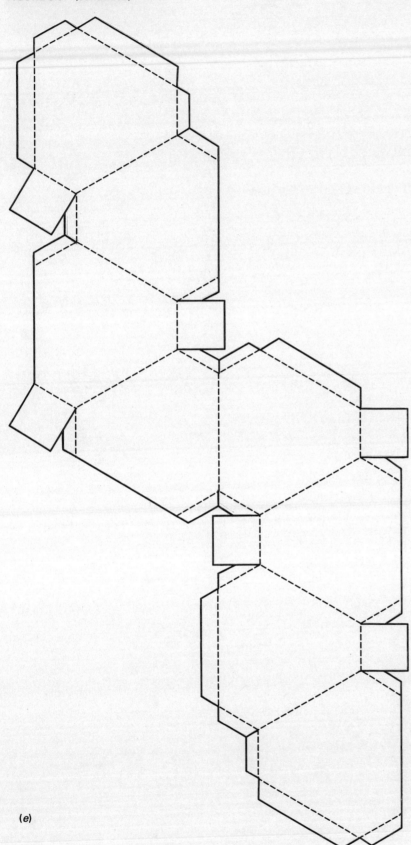

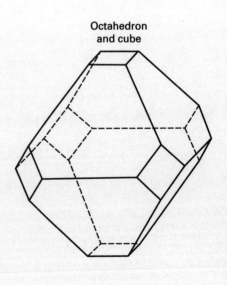

Octahedron
and cube

(e)

Student Name

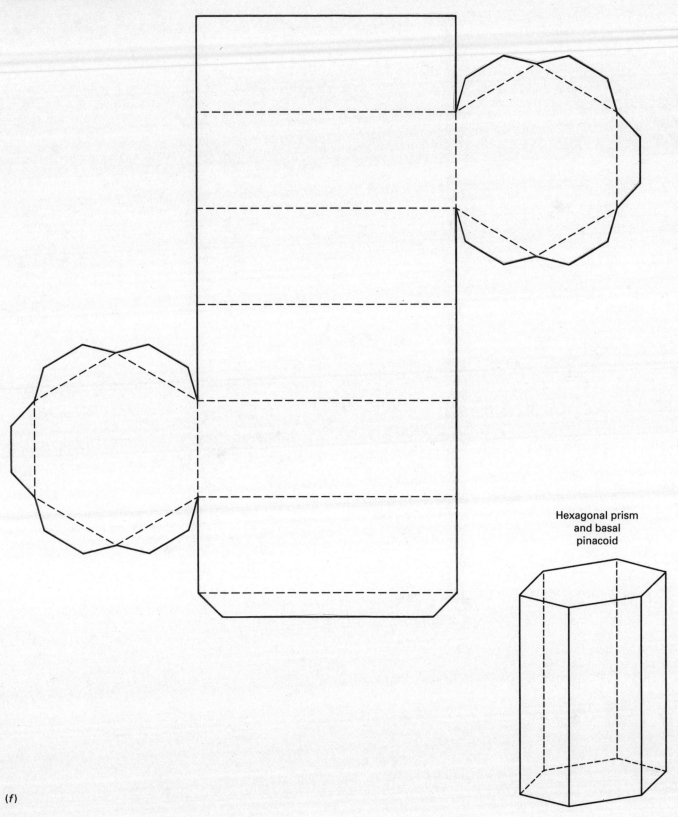

Hexagonal prism
and basal
pinacoid

(f)

Student Name

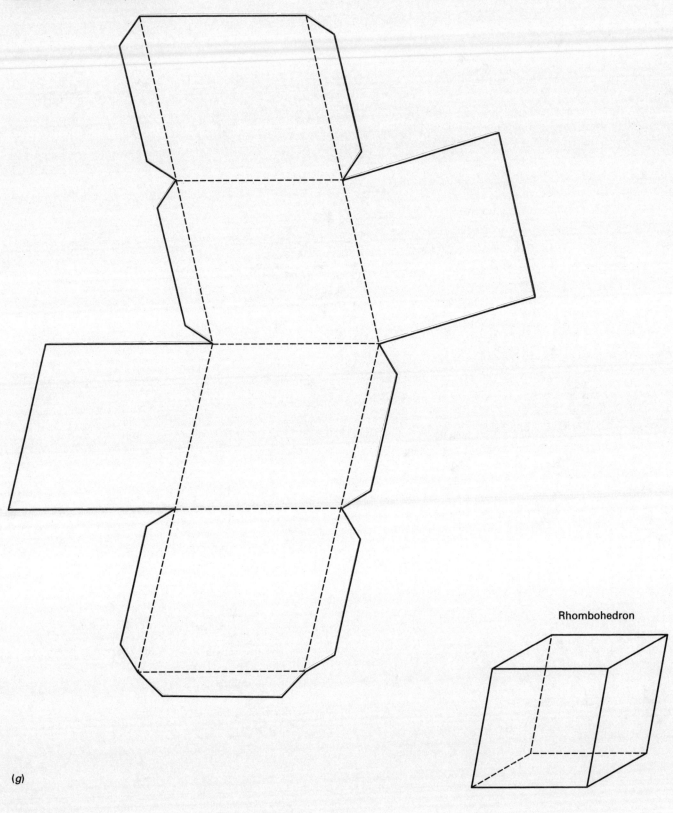

Rhombohedron

(g)

Student Name

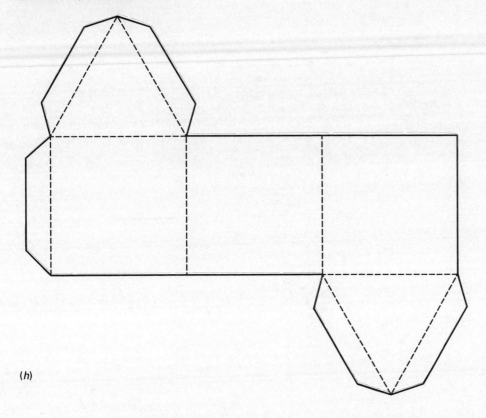

(h)

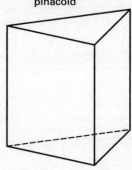

Trigonal prism and basal pinacoid

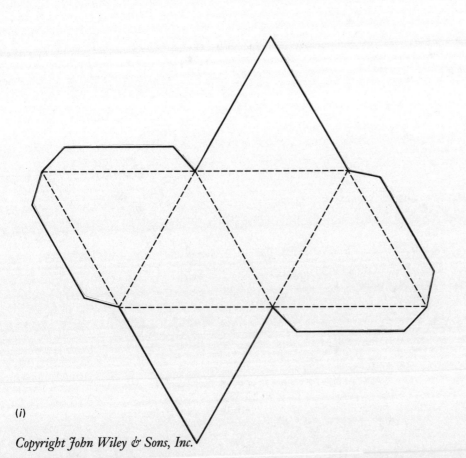

(i)

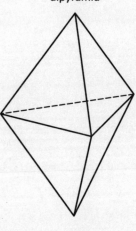

Trigonal dipyramid

Insert 7

Student Name

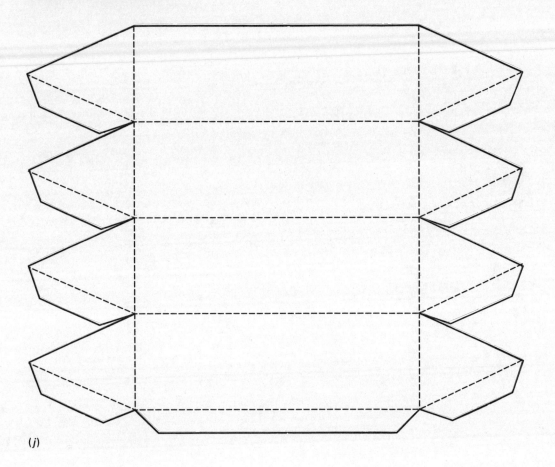

(*j*)

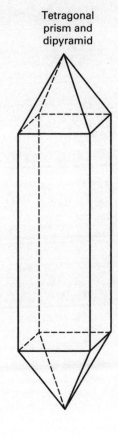

Tetragonal
prism and
dipyramid

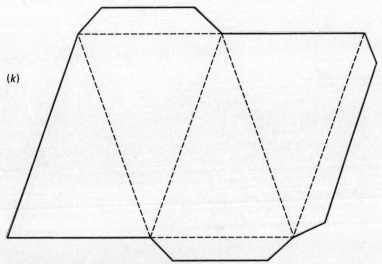

(*k*)

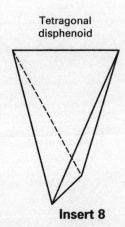

Tetragonal
disphenoid

Insert 8

Student Name

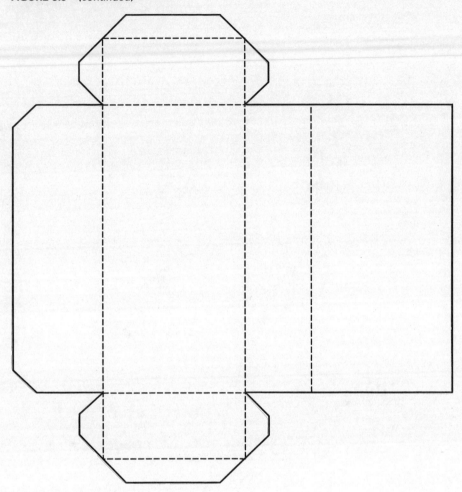

(*l*)

Orthorhombic
pinacoids

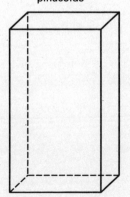

Student Name

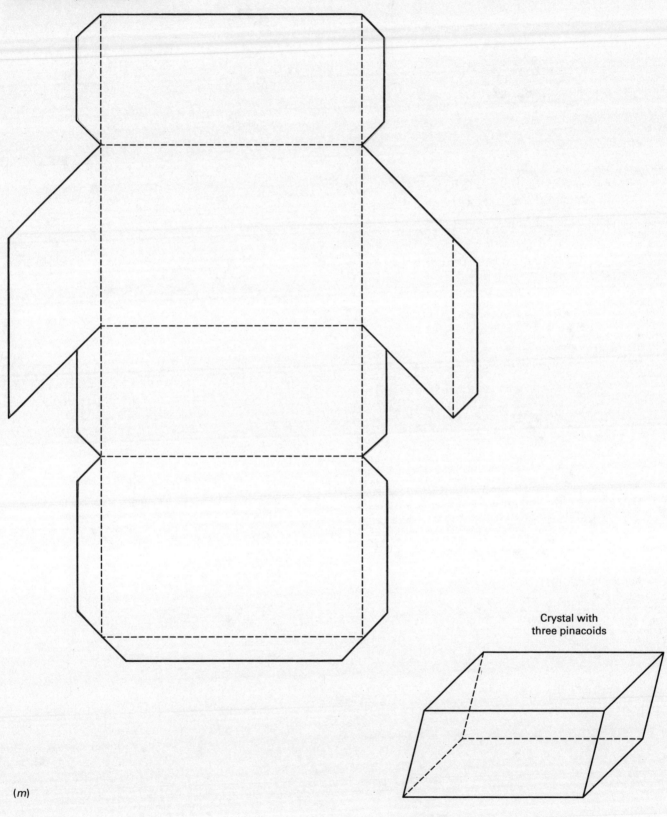

Crystal with
three pinacoids

(*m*)